Algorithms, Fractals, and Dynamics

Algorithms, Fractals, and Dynamics

Edited by

Y. Takahashi

University of Tokyo
Tokyo, Japan

Plenum Press • New York and London

Library of Congress Cataloging-in-Publication Data

Algorithms, fractals, and dynamics / edited by Y. Takahashi.
    p.   cm.
    "Proceedings of Hayashibara Forum '92: International Symposium on
New Bases for Engineering Science, Algorithms, Dynamics, and
Fractals, held November 23-28, 1992, in Okayama, Japan; and of a
symposium on algorithms, fractals, and dynamics, held November
30-December 2, 1992, in Kyoto, Japan"--T.p. verso.
    Includes bibliographical references and index.
    ISBN 978-1-4613-7996-6
    1. Differentiable dynamical systems--Congresses. 2. Fractals-
-Congresses. 3. Algorithms--Congresses.   I. Takahashi, Y.
(Yōichirō), 1946-   . II. Hayashibara Forum '92: International
Symposium on New Bases for Engineering Science, Algorithms,
Dynamics, and Fractals (1992 : Okayama-shi, Japan)
QA614.8.A43 1995
514'.74--dc20   95-35773
                                                              CIP

Proceedings of Hayashibara Forum '92: International Symposium on New Bases for Engineering
Science, Algorithms, Dynamics and Fractals, held November 23–28, 1992, in Okayama, Japan; and of a
symposium on Algorithms, Fractals and Dynamics, held November 30–December 2, 1992,
in Kyoto, Japan

ISBN 978-1-4613-7996-6   ISBN 978-1-4613-0321-3   (eBook)
DOI 10.1007/978-1-4613-0321-3

© 1995 Plenum Press, New York
Softcover reprint of the hardcover 1st edition 1995

A Division of Plenum Publishing Corporation
233 Spring Street, New York, N. Y. 10013

10 9 8 7 6 5 4 3 2 1

Preface

In 1992 two successive symposia were held in Japan on algorithms, fractals and dynamical systems.

The first one was Hayashibara Forum '92: International Symposium on New Bases for Engineering Science, Algorithms, Dynamics and Fractals held at Fujisaki Institute of Hayashibara Biochemical Laboratories, Inc. in Okayama during November 23-28 in which 49 mathematicians including 19 from abroad participated. They include both pure and applied mathematicians of diversified backgrounds and represented 11 countries. The organizing committee consisted of the following domestic members and Mike KEANE from Delft:

Masayosi HATA, Shunji ITO, Yuji ITO,
Teturo KAMAE (chairman), Hitoshi NAKADA,
Satoshi TAKAHASHI, Yoichiro TAKAHASHI,
Masaya YAMAGUTI

The second one was held at the Research Institute for Mathematical Science at Kyoto University from November 30 to December 2 with emphasis on pure mathematical side in which more than 80 mathematicians participated.

This volume is a partial record of the stimulating exchange of ideas and discussions which took place in these two symposia.

October 23, 1994   Yoichiro Takahashi

Contents

ASSOCIATED ACTIONS AND UNIQUENESS OF COCYCLES

Jon Aaronson[1], Toshihiro Hamachi[2], and Klaus Schmidt[3]

[1] School of Mathematical Sciences, Tel Aviv University
69978 Tel Aviv, Israel
[2] Department of Mathematics, College of General Education
Kyushu University, Roppon-matsu, Chuou-ku, Fukuoka, 810, Japan
[3] Mathematics Institute, Warwick University, Coventry CV4 7AL, U.K.

ABSTRACT We use the methods of orbital ergodic theory to show the existence of many strange cocycles. Any conservative ergodic flow is the associated action for some recurrent cocycle of an ergodic probability preserving transformation, and this cocycle is determined uniquely up to cohomology via orbit equivalence.

Introduction

Let $(X, \mathcal{B}, m)$ be a non-atomic Lebesgue probability space, and let Z be a countable group of invertible, nonsingular transformations of X. Let G be a locally compact topological group G. A (G-valued) *cocycle* of Z is a measurable function $\varphi : Z \times X \to G$ which satisfies the *cocycle equation*:

$$\varphi(zz', x) = \varphi(z, z'x)\varphi(z', x) \quad (z, z' \in Z,\ x \in X).$$

If $Z = \mathbb{Z} = \{S^n : n \in \mathbb{Z}\}$ where $S : X \to X$ is an invertible nonsingular transformation of X, then given any measurable function $\varphi : X \to G$ we may define $\varphi : \mathbb{Z} \times X \to G$ (the *cocycle* of S) by

$$\varphi(n, x) = \begin{cases} \varphi(S^{n-1}x)\varphi(S^{n-2}x)\dots\varphi(x) & n \geq 1, \\ e & n = 0, \\ \varphi(S^n x)\varphi(S^{n+1}x)\dots\varphi(S^{-1}x) & n \leq -1. \end{cases}$$

The *skew product* action of Z on $X \times G$ is defined by $z_\varphi(x, y) = (zx, \varphi(x)y)$, $(z \in Z)$. See [Sch].

Aaronson would like to thank the mathematics department of Keio university, and Hamachi would like to thank the mathematics institute of Warwick university for hospitality provided while this paper was written.

The transformations $\{z_\varphi : z \in Z\}$ are nonsingular with respect to the measure $m \times m_G$ where m_G is left Haar measure on G, and the cocycle φ is called *recurrent* if the skew product action Z_φ is conservative.

Recall that the *invariant factor* of a countable group of non-singular transformations $\mathcal{T}$ of the Lebesgue probability space $(Y, \mathcal{A}, p)$ is a Lebesgue probability space $\Omega(\mathcal{T}) = (\Omega, \mathcal{A}', p')$ equipped with a factor map $\pi : Y \to \Omega$ such that $p \circ \pi^{-1} \sim p'$, and

$$\pi^{-1}\mathcal{A}' = \mathcal{I}(\mathcal{T}) := \{A \in \mathcal{A} : \tau^{-1}A = A \mod 0 \; \forall \, \tau \in \mathcal{T}\}.$$

The invariant factor is isomorphic to the measure space of ergodic components of p (or *ergodic decomposition* of $\mathcal{T}$).

Define a G-action on $X \times G$ by $Q_g(x, y) = (x, yg)$. Clearly,

$$z_\varphi \circ Q_g = Q_g \circ z_\varphi, \; \forall \, g \in G, \; z \in Z,$$

and so $Q_g \mathcal{I}(Z_\varphi) = \mathcal{I}(Z_\varphi)$, whence Q acts on the invariant factor of Z_φ. It follows from the ergodicity of Z on X, that $\{z_\varphi \circ Q_g : g \in G, \; z \in Z\}$ acts ergodically on $X \times G$, whence Q is ergodic on $\Omega(Z_\varphi)$.

We call this action of Q on Ω, the *associated group action of the cocycle*, (the G-action of φ). It seems to have been considered first in [Ma], and is called the *Mackey range* of the cocycle in [B-G],[G-S1], and [G-S2].

Remarks

(1) The associated $\mathbb{R}$-action of a non-negative cocycle of an automorphism is precisely the so called *special flow* (see §3) built under the cocycle with that automorphism as base. In this case, the cocycle is non-recurrent. The concept of associated group action has been used in [Ka] to generalise the notion of special flow for multidimensional group actions.

(2) In the setting where S is an invertible, non-singular transformation, and

$$\varphi = \log S', \qquad (S' := \frac{dm \circ S}{dm}),$$

the $\mathbb{R}$-action of φ has been considered and called the *associated*, or *Krieger flow*. The cocycle φ is recurrent if S is conservative.

Theorem 1. [B-G]. *Let S be an ergodic probability preserving automorphism, and let G be a locally compact, second countable amenable group.*

For any non-singular, conservative ergodic free action $T = \{T_g : g \in G\}$, there is a recurrent G-valued cocycle of $\{S^n : n \in \mathbb{Z}\}$ whose action is isomorphic to T.

For the convenience of the reader, we sketch a proof of theorem 1 (different from that of [B-G]) for the case $G = \mathbb{R}$. Our main result is that when the $G = \mathbb{R}$, the cocycle appearing in theorem 1 is unique up to cohomology via orbit equivalence. This strengthens theorem 5.12 in [B-G].

Thorem 2. (Uniqueness of cocycle) *Let S and S' be ergodic probability preserving automorphisms of X and X' respectively, and suppose that φ and φ' are two cocycles* *having associated actions which are conjugate, conservative, ergodic, nonsingular free $\mathbb{R}$ actions.*

Then $\exists$ an orbit equivalence from S to S', i.e. a measure preserving map $\pi : X \to X'$ such that

$$\{S'^n \pi(x) : n \in \mathbb{Z}\} = \{\pi S^n(x) : n \in \mathbb{Z}\} \quad a.e.x,$$

and a measurable function $g : X \to \mathbb{R}$ such that

$$\varphi(x) - \varphi'(\pi(x)) = g(x) - g(Sx). \quad a.e.x$$

After completing this article, we were informed by Golodets and Sinel'shchikov that they have also obtained theorem 2 by a different method in [G-S2].

Remarks

(1) It is not hard to show that if two cocycles are cohomologous via orbit equivalence (as in theorem 2), then their actions are isomorphic.

(2) Theorems 1, and 2 may be considered as a "measure preserving analogue" of Krieger's theorem ([Kri], see also [K-W], [H-O1]).

(3) Cocycles having trivial, transitive, and periodic associated actions are also unique up to cohomology via orbit equivalence ([G-S1]). Indeed, in case the associated actions in the theorem are transitive, it is well known (see [Sch]) that both cocycles are coboundaries, and we may assume that $\varphi', \varphi \equiv 0$; the result now following from Dye's theorem [Dye].

Let $\hat{G}$ denote the group of *characters* of G (continuous homomorphisms $G \to \mathbb{T}$). Clearly if φ is a G-valued cocycle, and $a \in \hat{G}$, then $a \circ \varphi$ is a $\mathbb{T}$-valued cocycle. It is natural to ask whether, for a recurrent φ of an ergodic probability preserving transformation S, the ergodicity of $S_{a \circ \varphi} \ \forall a \in \hat{G}$ implies the ergodicity of S_φ.

This is true for compact, Abelian G, (see [F] and [P]). The (apparently well known fact) that it is false for $G = \mathbb{R}$ follows from theorem 1.

Example. For any ergodic probability preserving automorphism S, there is a recurrent cocycle $\varphi : X \to \mathbb{R}$ such that $S_{a \circ \varphi}$ is ergodic for every $a \in \mathbb{R}$, but S_φ is not ergodic.

The example is obtained by choosing a recurrent $\varphi : X \to \mathbb{R}$ with a non-trivial, weakly mixing associated $\mathbb{R}$-action. This means that S_φ is not ergodic (as φ's $\mathbb{R}$-action is non-trivial), and also that there are no measurable solutions $a \in \mathbb{R}$, $\psi : X \to \mathbb{T}$ of the functional equation

$$e^{ia\varphi} = \frac{\psi \circ S}{\psi},$$

as such a solution would give rise to an eigenvalue of φ's associated $\mathbb{R}$-action, which is assumed weakly mixing. It follows from [F] that $S_{a \circ \varphi}$ is ergodic for every $a \in \mathbb{R}$. An analogous example can be constructed for the case $G = \mathbb{Z}$.

The organisation of the paper is as follows. After reviewing some definitions from orbital ergodic theory in §1, we sketch a proof of theorem 1 in §2. In §3, we study special flow representations of the associated action, laying the foundations for our proof of theorem 2 in §4. In §4, as an introduction to the ideas involved in the proof (copying towers in an appropriate manner), we first sketch a proof of the uniqueness of ergodic cocycles (first established in [G-S1]). Theorem 2 is then reduced to a "relative Dye theorem" which is more easily established.

§1 Nonsingular equivalence relations

Let $(X, \mathcal{B}, m)$ be a non-atomic standard probability space. A measurable equivalence relation $\mathcal{R} \in \mathcal{B} \otimes \mathcal{B}$ on X is said to be *countable* if $\mathcal{R}_x := \{y \in X : (x, y) \in \mathcal{R}\}$ is countable $\forall x \in X$. For example, if Z is a countable group of nonsingular transformations of X, then the equivalence relation *generated* by Z,

$$\mathcal{R}(Z) = \{(x, zx) : x \in X, \ z \in Z\}$$

is countable. It is also "nonsingular" in a reasonable sense which we proceed to explain.

A *partial* non-singular transformation of $(X, \mathcal{B}, m)$ is a triple (ϕ, A, B) where $A, B \in \mathcal{B}$ and $\phi : A \to B$ is an invertible, m-non-singular transformation. It will be natural to sometimes write $\phi = (\phi, A, B)$ and $A = \mathrm{Dom}\,\phi$, $B = \mathrm{Im}\,\phi$. A *partial transformation of* $\mathcal{R}$ is a partial transformation ϕ satisfying $(x, \phi x) \in \mathcal{R} \ \forall \ x \in \mathrm{Dom}\,\phi$. The collection of partial transformations of $\mathcal{R}$ is denoted by $[\mathcal{R}]_*$, and is known as the *groupoid* of $\mathcal{R}$. The *full group* of $\mathcal{R}$ is that subset $[\mathcal{R}] = \{\tau \in [\mathcal{R}]_* : \mathrm{Dom}\,\tau = \mathrm{Im}\,\tau = X\}$.

The measurable, countable equivalence relation $\mathcal{R}$ is called *nonsingular* if $\mathcal{R} = \{(x, \phi x) : \phi \in [\mathcal{R}]_*, \ x \in \mathrm{Dom}\,\phi\}$. It is known (see [Fe-Mo]) that every countable, nonsingular equivalence relation is generated by a countable group of nonsingular transformations. The notions of *conservativity*, *ergodicity*, and *invariant factor* are defined with reference to the generating group of nonsingular transformations.

A measurable equivalence relation is said to be *of type* II_1 if all its partial transformations are probability preserving, and is said to be *hyperfinite* if it is generated by a single nonsingular automorphism. In particular, if $\mathcal{R}$ is a measurable, countable hyperfinite equivalence relation of type II_1 on X, then

$\exists$ an invertible, probability preserving transformation $S : X \to X$ such that $\mathcal{R} = \{(x, S^n x) : x \in X, \ n \in \mathbb{Z}\}$.

Let G be a locally compact topological group G. If $\mathcal{R} = \mathcal{R}(Z)$ where Z is a freely acting group of automorphisms of X, and $\varphi : Z \times X \to G$, is a cocycle of Z, we may define an *orbit cocycle* of $\mathcal{R}$, $\varphi : \mathcal{R} \to G$ by

$$\varphi(x, zx) := \varphi(z, x),$$

which satisfies

$$\varphi(x, z) = \varphi(x, y)\varphi(y, z) \text{ whenever } (x, y), \ (y, z) \in \mathcal{R}.$$

Orbit cocycles of $\mathcal{R}(Z)$ also give rise to cocycles of Z in this way.

The cocycle φ generates the equivalence relation

$$\mathcal{R}_\varphi := \{((x, u), (y, v)) \in (X \times G) \times (X \times G) : (x, y) \in \mathcal{R}, \ vu^{-1} = \varphi(x, y)\}$$

on $X \times G$, and is called *recurrent* if $\mathcal{R}_\varphi$ is conservative. Note that in case $\mathcal{R} = \mathcal{R}(Z)$, we have $\mathcal{R}_\varphi = \mathcal{R}(Z_\varphi)$. As before, the *associated action* of φ is the G-action $(x,y) \mapsto (x,yg)$ restricted to the invariant factor of $\mathcal{R}_\varphi$.

§2 Sketch of a proof of theorem 1

Let T be the given non-singular, conservative, ergodic $\mathbb{R}$-action considered (without loss of generality) acting on X. Let R be a conservative ergodic automorphism of a Lebesgue space $(Y, \mathcal{F}, \rho)$ such that $\tilde{R} : Y \times \mathbb{R} \to Y \times \mathbb{R}$ defined by

$$\tilde{R}(y,u) = (Ry, u - \log R'(y))$$

is ergodic. Here, $R' := \frac{d\rho \circ R}{d\rho}$, and the measure on $Y \times \mathbb{R}$ is $d\mu(x,y) = d\rho(x)e^u du$. Note that $\mu \circ \tilde{R} = \mu$. Such an automorphism R (of *type* III_1) is constructed e.g. in [H-O1].

Define an infinite, σ-finite measure space $W = Y \times \mathbb{R} \times X$ equipped with the measure $d\nu(y,u,x) := d\rho(y)e^u du dm(x)$.

Define measure preserving automorphisms of W by

$$\hat{R}(y,u,x) := (Ry, u - \log R'(y), x),$$

$$\hat{T}_q(y,u,x) := (y, u - \log T_q'(y), T_q x) \quad (q \in \mathbb{Q}).$$

The measure preserving automorphisms $\hat{R}$ and $\hat{T}_q$ evidently commute. The group $Z := \{\hat{R}^m \hat{T}_q : m \in \mathbb{Z}, q \in \mathbb{Q}\}$ is countable, amenable ($\because$ Abelian), and acts ergodically on W (by the ergodicity of $\tilde{R}$ on $X \times \mathbb{R}$, and $\{T_q : q \in \mathbb{Q}\}$ on X). We have taken the idea for the construction of $\hat{R}$ and $\hat{T}_q, q \in \mathbb{Q}$ from [Ham1].

We produce first a recurrent $\mathbb{R}$-valued cocycle of Z whose action is isomorphic to T. The relevant cocycle is defined by

$$\varphi(\hat{R}^m \hat{T}_t, (y,u,x)) = -t.$$

We show that φ is recurrent, and that the φ's action is isomorphic to T.

The transformation $\hat{R}_\varphi = \tilde{R} \times \mathrm{Id}$ on $W \times \mathbb{R} = (Y \times \mathbb{R}) \times (X \times \mathbb{R})$ is evidently conservative, and as $\{\hat{R}^m : m \in \mathbb{Z}\} \subset Z$, the skew product action of Z is also conservative, whence the recurrence of φ.

Next, we identify the invariant factor of the $\dot{Z}$-action on $W \times \mathbb{R}$. It is not hard to see that any invariant measurable set is of form

$$B = \{(T_t x, -t) : t \in \mathbb{R}, x \in B_0\} = B(B_0)$$

where $B_0 \subset X$ is measurable. Evidently,

$$Q_s B(B_0) = B(T_s B_0)$$

and the action of φ is now clearly isomorphic to T.

To establish the theorem, it is sufficient to find a recurrent cocycle of S whose action is the same as that of φ.

By [C-F-W], $\mathcal{R}$ is hyperfinite, and there is an ergodic measure preserving automorphism U of W such that $[U] = [Z]$, and a recurrent cocycle ψ of U whose action is the same as that of φ.

Fix a measurable set $A \subset W$ such that $\nu(A) = 1$, and consider the induced transformation $(U_\psi)_{A \times \mathbf{R}}$. It follows that

$$(U_\psi)_{A \times \mathbf{R}} = (U_A)_\phi$$

where

$$\phi(x) := \sum_{k=0}^{r_A(x)-1} \psi(U^k x), \quad r_A(x) = \min\{n \geq 1 : U^n x \in A\}.$$

This means that ϕ is a recurrent cocycle of U_A, since $(U_A)_\phi = (U_\psi)_{A \times \mathbf{R}}$ is conservative.

Moreover, the action of ϕ is the isomorphic to that of ψ. The isomorphism is constructed as follows,

for each $(U_\psi)_{A \times \mathbf{R}}$-invariant measurable set B, set

$$B^* = \bigcup_{n=-\infty}^{\infty} (U_\psi)^n B.$$

Obviously, $B^* \in \mathcal{I}(U_\psi)$ and $B^* \cap (A \times \mathbf{R}) = B \mod \nu$. This and the fact that

$$\bigcup_{n=-\infty}^{\infty} (U_\psi)^n (A \times \mathbf{R}) = W \times \mathbf{R} \mod \nu$$

ensure that $B \mapsto B^*$ is a bijection from $\mathcal{I}((U_A)_\phi)$ to $\mathcal{I}(U_\psi)$, and hence that the induced point mapping $\Omega((U_A)_\phi) \to \Omega(U_\psi)$ is a measure space isomorphism.

The fact that

$$(B^*) \circ Q_t \equiv (B \circ Q_t)^* \quad (B \in \mathcal{I}((U_A)_\phi), \ t \in \mathbf{R})$$

ensures that the actions of ψ and ϕ are isomorphic.

Lastly, by Dye's theorem ([Dye]) we may suppose without loss of generality that $A = X$, $[U_A] = [S]$, and that ϕ is a cocycle of S.

§3 Lacunarity, and special flow representations of the associated action

Definition. Let $(Z, \mathcal{C}, p)$ be a standard σ-finite measure space, let $U : Z \to Z$ be an invertible, nonsingular transformation, and let $f : Z \to (0, \infty)$ be measurable. Define

$$W = \{(z, u) : z \in Z, \ 0 \leq u < f(z)\},$$

and for $t \in \mathbf{R}$, $(z, u) \in W$:

$$T_t(z, u) = (U^n z, u + t - f(n, z)) \quad \text{for } f(n, z) \leq t + u < f(n + 1, z)$$

where $f : \mathbb{Z} \times Z \to \mathbf{R}$ is the cocycle of U determined by f. The flow T on W is nonsingular with respect to the product measure of p with Lebesgue measure.

The triple (Z, U, f) is a *special flow representation* of T with *base transformation* U built under the *height function* f. The flow T is said to be *represented* by (Z, U, f).

An isomorphism of special flow representations (Z, U, f) and (Z', U', f') is a measure space isomorphism $\pi : Z \to Z'$ satisfying $\pi U = U' \pi$ and $f' \circ \pi = f$. Clearly, isomorphism of special flow representations entails isomorphism of the represented flows (but not vice versa).

By the Krengel-Kubo theorem ([Kre] , [Ku]), any conservative, nonsingular free $I\!R$-action T of a Lebesgue space has a special flow representation (Z, U, f) with a nonsingular base automorphism U of a Lebesgue space Z built under a ceiling function f which is bounded away from 0.

A periodic flow has a "trivial" special flow representation (Z, U, f) where U is the identity on the one-point space Z, and f is constant.

In this section, to any recurrent cocycle, we associate a special flow representation of its associated action (the K-representation, see below). Although the associated actions of cohomologous cocycles are isomorphic, their K-representations need not be.

Let $(X, \mathcal{B}, m)$ be a standard probability space, and let $\mathcal{R} \in \mathcal{B} \otimes \mathcal{B}$ be a countable, measurable equivalence relation. A cocycle $\varphi : \mathcal{R} \to I\!R$ is called *lacunary* if

$$\tilde{f}(x) := \inf\{\varphi(x, y) : \varphi(x, y) > 0, \ (x, y) \in \mathcal{R}\}$$

is bounded below (i.e. $\exists \ \epsilon > 0$ such that $\tilde{f} > \epsilon$ a.e.). Suppose that $\varphi : \mathcal{R} \to I\!R$ is a recurrent, lacunary cocycle. Define the subrelation $\mathcal{S} = \mathcal{S}_\varphi$ of $\mathcal{R}$ by setting

$$\mathcal{S} = \{(x, y) | \varphi(x, y) = 0\}.$$

As before, let $\Omega(\mathcal{S})$ be the invariant factor of $\mathcal{S}$, and $\mathcal{I}(\mathcal{S})$ be the $\mathcal{S}$-invariant sets in X. The invariant factor map is $\pi : X \to \Omega(\mathcal{S})$ such that $x \in \pi(x)$, and $\pi^{-1}\mathcal{B}(\Omega(\mathcal{S})) = \mathcal{I}(\mathcal{S})$. Let $\overline{m} = m \circ \pi^{-1}$, and let $\{m_z : z \in \Omega\}$ denote the induced conditional probabilities:

$$\int_A m_z(B) d\overline{m} = m(B \cap \pi^{-1}A) \quad (A \in \mathcal{B}(\Omega), \ B \in \mathcal{B}(X)).$$

Proposition 3.1. *If φ is recurrent, then the conditional probabilities $\{m_z : z \in \Omega\}$ are $\overline{m}$-almost all non-atomic.*

Proof. Suppose otherwise, that for some $\delta > 0$, m_z has an atom of mass at least δ for each $z \in A \in \mathcal{B}(\Omega)$, where $\overline{m}(A) > 0$. Then, $\exists \ \alpha : A \to X$ measurable such that $m_z(\{\alpha(z)\}) \geq \delta \ \forall z \in A$. The set $E = \alpha(A) \in \mathcal{B}(X)$ mod 0, and $m(E) > 0$. Moreover for a.e. $z \in \Omega$, $\#(E \cap \pi^{-1}\{z\}) \leq \frac{1}{\delta}$, whence $\#(E \cap \mathcal{S}_x) \leq \frac{1}{\delta}$ for a.e. $x \in X$.

Let $\epsilon > 0$ be such that

$$\min\{|\varphi(x, y)| : \ (x, y) \in \mathcal{R}, \ \varphi(x, y) \neq 0\} > 2\epsilon.$$

Let $(x, u) \in \tilde{E} := E \times (-\epsilon, \epsilon)$ and $(x, y) \in \mathcal{R}$. If $(y, u + \varphi(x, y)) \in \tilde{E}$, then, $\varphi(x, y) = 0$ and hence, $y \in \mathcal{S}_x \cap E$. This contradicts the recurrence of φ, as $\#(\tilde{E} \cap (\mathcal{R}_\varphi)_{(x,u)}) \leq \frac{1}{\delta}$ for $(x, u) \in \tilde{E}$. $\square$

As a corollary of proposition 3.1, we have

Hopf equivalence on ergodic components
If A, $B \in \mathcal{B}(X)_+$, and $m_z(A) = m_z(B)$ for $\overline{m}$-a.e. $z \in \Omega(\mathcal{S})$, then $\exists \ g \in [\mathcal{S}]_*$ such that $gA = B$ mod m,

the proof being Hopf's classical exhaustion argument which works in this situation because of proposition 3.1.

Note that the function $\tilde{f}$ (involved in the definition of lacunarity above) is $\mathcal{S}$-invariant. Choose a suitable measurable $f : \Omega(\mathcal{S}) \to \mathbb{R}$ so that $\tilde{f} = f \circ \pi$.

Theorem 3.2 (c.f. [Katz],[Kri]). *Let $\mathcal{R}$ be an ergodic hyperfinite equivalence relation of type II_1, and let $\varphi : \mathcal{R} \to \mathbb{R}$ be a lacunary, recurrent cocycle of $\mathcal{R}$ whose action T is free, then there is a nonsingular automorphism U of $(\Omega(\mathcal{S}), \overline{m})$ such that*

$$U(\pi x) = \pi y \text{ where } \varphi(x,y) = -f(\pi x),$$

and that $(\Omega(\mathcal{S}), U, f)$ is a special flow representation for T.

Proof. Let $\epsilon > 0$ be such that

$$\min\{|\varphi(x,y)| \ : \ (x,y) \in \mathcal{R}, \varphi(x,y) \neq 0\} > 2\epsilon.$$

We note that for a.e. x,

$$|\varphi(y,x) - \varphi(z,x)| = |\varphi(y,z)| > 2\epsilon \text{ or } = 0 \quad ((y,x),(z,x) \in \mathcal{R})$$

whence, there exists a sequence of measurable functions $\xi(k,x), -\infty < k < \infty$, such that

$$\begin{aligned}
\xi(0,x) &= 0, \\
\xi(k,x) + 2\epsilon &\leq \xi(k+1,x), \quad -\infty < k < \infty \\
\{\varphi(y,x)|(y,x) \in \mathcal{R}\} &= \{\xi(k,x)| - \infty < k < \infty\} \quad \text{a.e.} x
\end{aligned}$$

The cocycle property implies that for a.e. x,

$$\xi(l+k,x) = \xi(k,x) + \xi(l,y), \quad \forall l \in \mathbb{Z}, \forall y : (y,x) \in \mathcal{R}$$

where k is determined by

$$\xi(k,x) = \varphi(y,x).$$

Obviously each $\xi(l,x)$ is an $\mathcal{S}$-invariant function. By definition of the function f,

$$f(\pi(x)) = \xi(1,x).$$

Firstly we will show the existence of an automorphism U on $(\Omega, \overline{m})$ satisfying for a.e.x,

$$U\pi(x) = \pi(y)$$

where

$$\varphi(y,x) = f(\pi(x)).$$

Let

$$(\widehat{X}, \widehat{m}) = (X \times \mathbb{Z}, m \times n),$$

where n is the counting measure on $\mathbb{Z}$. Define an equivalence relation $\widehat{\mathcal{R}}$ on $\widehat{X}$ by

$$((x,i),(y,j)) \in \widehat{\mathcal{R}} \text{ if } \varphi(y,x) = \xi(i-j,x).$$

Define a map $\hat{\pi} : \widehat{X} \to \Omega$ by setting

$$\hat{\pi}(x, i) = \pi(y)$$

where

$$((x, i), (y, 0)) \in \widehat{\mathcal{R}}.$$

We note that a measurable subset $\widehat{A} \subset \widehat{X}$ of positive measure is $\widehat{\mathcal{R}}$-invariant if and only if $\widehat{A}$ is of the form :

$$\widehat{A} = \cup_{i=-\infty}^{\infty} A_i \times \{i\}$$

where each A_i is an $\mathcal{S}$-invariant set of positive measure, and for any i and j and for a.e.$x \in A_i$, $\exists y \in A_j$ such that

$$((x, i), (y, j)) \in \widehat{\mathcal{R}}.$$

In this case

$$\hat{\pi}(\widehat{A}) = \pi(A_0) \quad \text{and} \quad \widehat{A} = \hat{\pi}^{-1}(\pi(A_0)),$$

whence, $\hat{\pi}$ induces a measure space isomorphism from the invariant factor space of $\widehat{\mathcal{R}}$ onto the invariant factor space of $\mathcal{S}$. So, the map $\hat{\pi}$ is considered to be an invariant factor map of $\widehat{\mathcal{R}}$.

We note $\xi(n, x) = f(n, \pi(x))$. Since the automorphism $(x, i) \to (x, i - 1)$ of $\widehat{X}$ commutes with $\widehat{\mathcal{R}}$-equivalence relation, it also acts on the invariant factor of $\widehat{\mathcal{R}}$. We denote this factor automorphism by U,

$$U\hat{\pi}(x, i) = \hat{\pi}(x, i - 1).$$

In other words, U satisfies

$$U\pi(x) = \pi(y)$$

where

$$\varphi(y, x) = \xi(1, x) = f(\pi(x)),$$

(see [H-O2]).

Next we show that (Ω, U, f) is a special flow representation of T. Let

$$W = \{(z, u) | z \in \Omega, 0 \le u < f(z)\}.$$

Let $(x, u) \in X \times \mathbb{R}$ and $\xi(n - 1, x) \le u < \xi(n, x)$. Choose $y \in X$ so that

$$\varphi(y, x) = \xi(n - 1, x)$$

and define a map $\pi : X \times \mathbb{R} \to W$ by setting

$$\pi(x, u) = (\pi(y), u - \xi(n - 1, x)) = (U^{n-1}\pi(x), u - f(n - 1, \pi(x))).$$

This map π is well-defined and is an invariant factor map of $\mathcal{R}_\varphi$ onto W. Obviously, $\pi(x, u + t) = T_t\pi(x, u)$. $\qquad\square$

In the sequel, we'll denote $U = U_\varphi$, and $f = f_\varphi$ and call $(\Omega(\mathcal{S}_\varphi), U_\varphi, f_\varphi)$ the K-*representation* of the action corresponding to φ.

The next proposition says that any special flow representation of the action is isomorphic to the K-representation of some cohomologous, lacunary cocycle.

Proposition 3.3. *Suppose that (Z, U, f) is a special flow representation of the free, conservative action of the recurrent cocycle φ, and $\inf f > 0$, then there is a lacunary cocycle ψ, cohomologous to φ, and a measure space isomorphism $\pi : \Omega(S_\psi) \to Z$ such that $\pi \circ U_\psi = U \circ \pi$, and $f \circ \pi = f_\psi$.*

Remark. If the flow is periodic, then the result is proven in [Sch], and the lacunary cocycle ψ satisfies the following properties:

(1) For a.e.x, $\{\psi(y, x) : (y, x) \in \mathcal{R}\} = \{n\lambda : n \in \mathbb{Z}\}$ where $\lambda > 0$ is the period of the flow.

(2) $S_\psi = \{(x, y) \in \mathcal{R} : \psi(x, y) = 0\}$ is ergodic.

Proof. Set
$$W := \{(z, t) : z \in Z, 0 \le t < f(z)\}.$$
Let $\phi : X \times \mathbb{R} \to W$ be the invariant factor map of $\mathcal{R}_\varphi$, i.e. such that
$$\phi(x, y + t) = T_t \phi(x, y), \ \& \ \phi^{-1} B(W) = \mathcal{I}(\mathcal{R}_\varphi). \cdot$$
There are measurable maps $\zeta : X \to Z$, and $\eta : X \to \mathbb{R}$ such that $0 \le \eta(x) < f(\zeta x)$, and
$$\phi(x, 0) = (\zeta x, \eta(x)) \quad \text{for } x \in X.$$

Now, if $(x, y) \in \mathcal{R}$ then
$$
\begin{aligned}
(\zeta x, \eta(x)) \ &= \ \phi(x, 0) \\
&= \ \phi(y, \varphi(x, y)) \\
&= \ T_{\varphi(x,y)} \phi(y, 0) \\
&= \ (U^n \zeta y, \eta(y) + \varphi(x, y) - f(n, \zeta y))
\end{aligned}
$$

where $n = n(x, y) \in \mathbb{Z}$ is defined by
$$f(n, \zeta y) \le \eta(y) + \varphi(x, y) < f(n + 1, \zeta y).$$

Since $n(x, y)$ is also the unique $n \in \mathbb{Z}$ such that $\zeta x = U^n \zeta y$, it's clear that $n : \mathcal{R} \to \mathbb{Z}$ is an orbit cocycle.

Now define $\psi : \mathcal{R} \to \mathbb{R}$ by
$$\psi(x, y) = f(n(x, y), \zeta y).$$

Claim 1. *The function ψ is a lacunary orbit cocycle, and the cocycles φ and ψ are cohomologous.*

Proof. Suppose $(x, y), (y, z) \in \mathcal{R}$, and $n(x, y) = k$, $n(y, z) = \ell$.
Then $\zeta y = U^\ell \zeta z$, and
$$
\begin{aligned}
\psi(x, y) + \psi(y, z) \ &= \ f(k, \zeta y) + f(\ell, \zeta z) \\
&= \ f(k, U^\ell \zeta z) + f(\ell, \zeta z) \\
&= \ f(k + \ell, \zeta z) \\
&= \ \psi(x, z)
\end{aligned}
$$

since $k + \ell = n(x, z)$. Thus, ψ is cocycle.

Lacunarity follows from $\inf f > 0$.

Lastly, equating second coordinates in the equation derived above from $\phi(x, 0) = \phi(y, \varphi(x, y))$, we have

$$\eta(x) = \eta(y) + \varphi(x, y) - f(n(x, y), \zeta y) \Rightarrow \varphi(x, y) - \psi(x, y) = \eta(x) - \eta(y).$$

$\square$

Claim 2

$$\min\{\psi(x, y) : y \in X, \psi(x, y) > 0\} = f(U^{-1}\zeta x), \quad a.e.x$$

Proof. We prove that for a.e.$x, \exists y$ such that $(x, y) \in \mathcal{R}$, and $n(x, y) = 1$ which suffices, since

$$\psi(x, y) = f(n(x, y), \zeta y) = f(n(x, y), U^{-n(x,y)}\zeta x).$$

To do this, we must show that for a.e. $x, \exists y$ such that $(x, y) \in \mathcal{R}$, and $\zeta x = U\zeta y$. To show this, note that

$$\phi(x, -f(U^{-1}\zeta x)) = T_{-f(U^{-1}\zeta x)}(\zeta x, \eta(x)) = T_{\eta(x)}(U^{-1}\zeta x, 0)$$

whence $\exists y \in X$, for which $(x, y) \in \mathcal{R}$ and $\zeta y = U^{-1}\zeta x$. This establishes the claim.

$\square$

The map $\zeta : X \to Z$ is apparently $\mathcal{S}_\psi$-invariant, and hence induces a measure space isomorphism $\pi : \Omega(\mathcal{S}_\psi) \to Z$. It follows from claims 1 and 2 that

$$\pi U_\psi = U\pi, \ \& \ f \circ \pi = f_\psi.$$

$\square$

Corollary 3.4. *For any measurable subset $A \subset X$ of positive measure and for each integer n, there exists a partial transformation $g \in [\mathcal{R}]_*$ such that*

$$Dom\,(g) \subset A$$
$$n(gx, x) = n, \quad and \quad \psi(gx, x) = f(n, \pi(x)), \quad (x \in Dom\,(g)),$$
$$\pi(gx) = U^n\pi(x), \quad (x \in Dom\,(g)).$$

Remarks

(1) Suppose that $\mathcal{R}$ is measure preserving (indeed that $m \circ g = m \forall g \in [\mathcal{R}]$). In general, it is not possible to find $g \in [\mathcal{R}]$ such that $\pi(gx) = U\pi x$. If this were the case, then $\overline{m}$ would be U-invariant. By theorem 1, the action may be of type III, whence the absence of U-invariant, absolutely continuous probabilities.

(2) It is not hard to show, using Hopf equivalence, that if $\overline{m}$ is U-invariant, then $\exists g \in [\mathcal{R}]$ such that $\pi(gx) = U\pi x$.

(3) As the examples below show, it may be that there is a U-invariant, probability $\mu \sim \overline{m}$, but $\mu \neq \overline{m}$, whence again there is no $g \in [\mathcal{R}]$ such that $\pi(gx) = U\pi x$.

Example. For $a_n \geq 4$ $(n \in I\!N)$ let

$$X = \prod_{n=1}^{\infty} \{0, 1, \ldots, a_n - 1\}, \quad m = \prod_{n=1}^{\infty} \{\frac{1}{a_n}, \ldots, \frac{1}{a_n}\},$$

and

$$\mathcal{R} = \{(x, y) \in X \times X : \#\{k \in I\!N : x_k \neq y_k\} < \infty\}.$$

For $d_n > 0$ $(n \in I\!N)$ such that $d_{n+1} > \sum_{k=1}^{n} d_k$, and $2 \leq b_n < a_n$, define, for $n \in I\!N$, $\beta_n : \{0, 1, \ldots, a_n - 1\} \to I\!R$ by

$$\beta_n(k) = d_n 1_{[0, b_n - 1]}(k).$$

Now define $\varphi : \mathcal{R} \to I\!R$ by

$$\varphi(x, y) = \sum_{n=1}^{\infty} (\beta_k(x_k) - \beta_k(y_k)).$$

We have that φ is recurrent since $\#(S_\varphi)_x = \infty \ \forall \ x \in X$, and lacunary with

$$f(\pi x) = \inf\{\varphi(x, y) : (x, y) \in \mathcal{R}, \ \varphi(x, y) > 0\} = d_{\ell(x)} - \sum_{k=1}^{\ell(x)-1} d_k$$

where

$$\ell(x) = \min\{k \geq 1 : \varphi_k(x_k) = d_k\}.$$

Also

$$\Omega(S_\varphi) = \{0, 1\}^{I\!N}, \quad \overline{m} = \prod_{n=1}^{\infty} \{\frac{b_n}{a_n}, \frac{a_n - b_n}{a_n}\},$$

and

$$U(1, \ldots, 1, 0, \ldots) = (0, \ldots, 0, 1, \ldots), \text{ the adding machine.}$$

Set

$$\mu = \prod_{n=1}^{\infty} \{\frac{1}{2}, \frac{1}{2}\}.$$

There is a $\overline{m}$-a.c. U-invariant probability iff $\mu \sim \overline{m}$ iff

$$\sum_{n=1}^{\infty} |\frac{2b_n}{a_n} - 1| < \infty$$

and in this case, setting

$$p_0(n) = \frac{b_n}{a_n}, \quad p_1(n) = 1 - p_0(n)$$

we have that

$$h(x) = \frac{d\overline{m}}{d\mu}(x) = \prod_{n=1}^{\infty} 2p_{x_n}(n).$$

Fixing, for example $a_n = 2^{n+1} + 1$, $b_n = 2^n$, & $d_n = 2^n$, we obtain $\overline{m} \sim \mu$ but $h \neq 1$.

A *normalizer* of $\mathcal{R}$ is an automorphism of X such that

$$(Ry, Rx) \in \mathcal{R} \text{ for a.e.}(y, x) \in \mathcal{R}.$$

By $N[\mathcal{R}]$ we denote the set of all normalizers of $\mathcal{R}$. By $N[\mathcal{R}]_*$, we denote the set consisting of all partial transformations R satisfying

$$(Rx, Ry) \in \mathcal{R} \text{ for a.e.} (x,y) \in \mathcal{R}_{\text{Dom}\,R}.$$

Lemma 3.5. *Let $R \in N[\mathcal{S}]_* \cap [\mathcal{R}]_*$, then $\exists$ a measurable function $n : \pi(Im(R)) \to \mathbb{N}$ such that for a.e. $z \in \pi(Im(R))$ and for m_z-a.e.$x \in Im(R)$*

$$\pi(R^{-1}x) = U^{-n(z)}z.$$

This is uniquely determined if U is aperiodic.

Proof. Since $R \in [\mathcal{R}]_*$, we see that for a.e. $x \in Im(R)$, $\exists$ an integer $n = n(x)$ such that

$$\pi(R^{-1}x) = U^{-n}z.$$

We show that n depends only on $\pi(x)$. If not, there are integers k , l and measurable subsets $A \subset X, B \subset X$, and $W \subset \Omega$ of positive measure with the following properties:

(1) For $z \in W$, $m_z(A) > 0$, and $m_z(B) > 0$.

(2) For any $z \in W$,

$$\pi(R^{-1}x) = \begin{cases} U^{-k}z, & \text{if } \pi(x) = z, \quad x \in A, \\ U^{-l}z, & \text{if } \pi(x) = z, \quad x \in B. \end{cases}$$

Since $\mathcal{R}$ is ergodic, there exists $\phi \in [\mathcal{R}]_*$ such that

$$\text{Dom}\,(\phi) \subset A \cap Im\,(R), \quad Im\,(\phi) \subset B \cap Im\,(R).$$

If $x \in \text{Dom}(\phi)$ and $z = \pi(x)$ then

$$U^{-k}z = \pi(R^{-1}x) = \pi(R^{-1} \circ \phi(x)) = U^{-l}z.$$

Contradiction. □

Lemma 3.6. *Assume that $\mathcal{R}$ is an ergodic equivalence relation of type II_1. Let $R \in N[\mathcal{S}]_* \cap [\mathcal{R}]_*$ and E be a measurable subset of $Dom(R)$. Then, for a.e. $z \in \pi(Im(R))$,*

$$m_z(RE) = \frac{d\overline{m}U^{-n(z)}}{d\overline{m}}(z)m_{U^{-n(z)}z}(E)$$

where $n(z)$ is as in Lemma 3.5.

Proof. Let $g(z) \in L^\infty(\Omega, \overline{m})$. We may suppose that E satisfies for some integer n ,

$$\pi(Rx) = U^n\pi(x) \quad (x \in E),$$

an arbitrary measurable subset $E \subset \mathrm{Dom}(R)$ being a countable disjoint union of such sets. Then, for $f \in L^\infty(Z)$, $[g \neq 0] \subset \mathrm{Im}\, R$,

$$
\begin{aligned}
\int g(z) m_z(R(E)) d\overline{m}(z) &= \int g(\pi(x)) 1_{R(E)}(x) m(dx) \\
&= \int g(\pi(Ry)) 1_E(y) m(dy) \quad \text{(use that R is m-preserving)} \\
&= \int g(U^n \pi(y)) 1_E(y) m(dy) \\
&= \int g(U^n z) \overline{m}(dz) \int_{\pi(y)=z} 1_E(y) m_z(dy) \\
&= \int g(z) \frac{d\overline{m} U^{-n}}{d\overline{m}}(z) \overline{m}(dz) \int_{\pi(y)=U^{-n}z} 1_E(y) m_{U^{-n}z}(dy) \\
&= \int g(z) \frac{d\overline{m} U^{-n}}{d\overline{m}}(z) m_{U^{-n}z}(E) \overline{m}(dz).
\end{aligned}
$$

$\square$

Remark. A relevant idea is seen in [Ham2].

§4 Orbit equivalences, and the proof of Theorem 2

In order to establish theorem 2, we need to construct an orbit equivalence. This will be done by appropriately copying generating sequences of *towers*.

Recall (from [Ham2]), that a *tower* ξ of an equivalence relation $\mathcal{R}$ on X consists of a finite partition $\mathcal{P}_\xi = \{E_\alpha : \alpha \in \Lambda\}$ of X, and a finite family of partial transformations $\mathcal{T}_\xi = \{e_{\alpha,\beta} \in [\mathcal{R}]_* : \alpha, \beta \in \Lambda\}$ satisfying

$$
\mathrm{Dom}\,(e_{\alpha,\beta}) = E_\beta, \quad \mathrm{Im}\,(e_{\alpha,\beta}) = E_\alpha,
$$

and

$$
e_{\alpha,\beta} e_{\beta,\gamma} = e_{\alpha,\gamma}, \quad e_{\alpha,\alpha} = \mathrm{Id}|_{E_\alpha}.
$$

In order to introduce the method, and use of such copyings (see [K-W] and [H-O1]) we first show that any two $\mathbb{R}$-valued ergodic cocycles of countable, hyperfinite, equivalence relations of type II_1 are cohomologous via orbit equivalence. This was first established in [G-S1]. One needs the following

Lemma. *Suppose that φ is an ergodic $\mathbb{R}$-valued cocycle of the ergodic, hyperfinite equivalence relation $\mathcal{R}$ of type II_1. If $\mathcal{P} = \{E_\alpha : \alpha \in \Lambda\}$ is a measurable partition of $Y \in \mathcal{B}$ into sets of equal measure, $r_{\beta,\alpha} \in \mathbb{R}$, $(\alpha, \beta \in \Lambda)$, and $\epsilon > 0$, then there is a tower ξ of $\mathcal{R}_Y$ such that*

$$
\mathcal{P}_\xi = \mathcal{P},
$$

and

$$
|\varphi(e_{\beta,\alpha} x, x) - r_{\beta,\alpha}| < \epsilon \ \text{a.e. on } E_\alpha, \ \forall\, \alpha, \beta \in \Lambda.
$$

Sketch of proof. First note that the tower ξ can be split into a disjoint union of towers $\xi_i = \xi \cap Y_i$ $(i \in \mathbb{N})$ so that

$$
|\varphi(e_{\beta,\alpha} x, x) - \varphi(e_{\beta,\alpha} y, y)| < \epsilon \ \forall\, \alpha, \beta \in \Lambda, \ x, y \in E_\alpha \cap Y_i, \ i \in \mathbb{N}.
$$

14

Because of the ergodicity of $\varphi : \mathcal{R} \to \mathbb{R}$, we have that (see [Sch]) for any $r \in \mathbb{R}$, $\epsilon > 0$, and $A, B \in \mathcal{B}_+$

$$\exists R \in [\mathcal{R}]_* \ni \mathrm{Dom}\, R \subset A, \ \& \ m(\{x \in A : \ Rx \in B, \ |\varphi(Rx, x) - r| < \epsilon\}) > 0,$$

whence, by Hopf equivalence, if $r \in \mathbb{R}$, $\epsilon > 0$, and $A, B \in \mathcal{B}_+$, $m(A) = m(B)$ then

$$\exists R \in [\mathcal{R}]_* \ni \mathrm{Dom}\, R = A, \ \mathrm{Im}\, R = B, \ \& \ |\varphi(Rx, x) - r| < \epsilon \text{ a.e. on } A.$$

The proof is completed by Hopf's exhaustion method. $\qquad\square$

Now suppose that $\mathcal{R}$ and $\mathcal{R}'$ are countable, hyperfinite, equivalence relations of type II_1 and that $\varphi : \mathcal{R} \to \mathbb{R}$, $\varphi' : \mathcal{R}' \to \mathbb{R}$ are ergodic cocycles. The lemma is used (as in [K-W],[H-O1]) to obtain isomorphic sequences of towers $\xi_n = (\mathcal{P}_n, \mathcal{T}_n)$ for $\mathcal{R}$, and $\xi'_n = (\mathcal{P}'_n, \mathcal{T}'_n)$ for $\mathcal{R}'$, where

$$\mathcal{P}_n = \{E_\alpha : \alpha \in \Lambda_n\}, \ \mathcal{T}_n = \{e_{\alpha,\beta} : \alpha, \beta \in \Lambda_n\},$$

which are *generating* in the sense that the σ-algebra $\mathcal{B}$ is generated by the sets $E_\alpha \in \mathcal{P}_n$, $(n \geq 1)$, and that

$$\mathcal{R} = \bigcup_{n \geq 1} \{(y, x) | \text{for some } \alpha \text{ and } \beta \text{ in } \Lambda_n, x \in E_\alpha, \ y = e_{\beta,\alpha} x\}.$$

These towers are obtained together with a sequence of parameters $\alpha_n \in \Lambda_n$ and satisfy the following.

(1) The tower ξ_0 is trivial in the sense that $|\Lambda_0| = 1$.

(2) ξ_{n+1} refines ξ_n in the sense that

$$\begin{aligned}
\Lambda_{n+1} &= \Lambda_n \times \Gamma_{n+1} \quad (\Gamma_{n+1} \text{ a finite set and } \Lambda_1 = \Gamma_1) \\
E_\alpha &= \bigcup_{\gamma \in \Gamma_{n+1}} E_{(\alpha,\gamma)} \quad (\alpha \in \Lambda_n) \\
e_{(\alpha,\gamma),(\beta,\gamma)} &= e_{\alpha,\beta} \text{ on } E_{(\beta,\gamma)}, \quad (\alpha, \beta \in \Lambda_n, \text{ and } \gamma \in \Gamma_{n+1})
\end{aligned}$$

(3)
$$E_{\alpha_{n+1}} \subset E_{\alpha_n}$$

(4) For each $n \geq 1$ and for each $\gamma \in \Gamma_{n+1}$, $\exists r_{\alpha_{n+1},(\alpha_n,\gamma)} \in \mathbb{R}$ such that

$$|\varphi(e_{\alpha_{n+1},(\alpha_n,\gamma)}x, x) - r_{\alpha_{n+1},(\alpha_n,\gamma)}| < \frac{1}{2^n} \quad (\text{ for a.e. } x \in E_{(\alpha_n,\gamma)})$$

(5) The towers ξ'_n satisfy the analogous properties (1')-(4') with the parameter sets Λ_n.

The towers are obtained by means of the following refinement process ([K-W], [H-O1]). A product refinement ξ_{n+1} of ξ_n is obtained by choosing a "base element" E_{α_n} of ξ_n, constructing a tower $\{E_{(\alpha_n,\gamma)} : \gamma \in \Gamma_n\}$ which generates $\mathcal{R}_{E_{\alpha_n}}$ up to some fixed precision, and such that

$$|\varphi(e_{(\alpha_n,\gamma'),(\alpha_n,\gamma)}x, x) - \varphi(e_{(\alpha_n,\gamma'),(\alpha_n,\gamma)}y, y)| < \frac{1}{2^{n+1}} \text{ a.e. on } E_{(\alpha_n,\gamma)} \ \forall \ \gamma, \gamma' \in \Gamma_n.$$

This refinement is copied in a measure preserving way to obtain a refinement ξ'_{n+1} of ξ'_n, which refinement is then refined to ξ'_{n+2} by the same process, and then copied back.

Note that it follows from property (4) that for each $n \geq 1$ and for each $\beta, \beta' \in \Lambda_n$, $\exists r_{\beta',\beta} \in I\!\!R$ such that

$$|\varphi(e_{\beta',\beta}x, x) - r_{\beta',\beta}| < 3 \quad \text{a.e. on } E_\beta.$$

The natural correspondences between the towers ξ_n and ξ'_n generate an orbit equivalence Φ of $\mathcal{R}$ with $\mathcal{R}'$ such that

$$\Phi E_\beta = E'_\beta, \quad \Phi \circ e_{\alpha,\beta} = e'_{\alpha,\beta} \circ \Phi, \quad (\alpha, \beta \in \Lambda_n, n \geq 1).$$

It follows that for a.e. $(x, y) \in \mathcal{R}$,

$$|\varphi'(\Phi y, \Phi x) - \varphi(y, x)| < 6,$$

whence $\exists \, \eta : X \to I\!\!R$ bounded and measurable such that

$$\varphi'(\Phi y, \Phi x) - \varphi(y, x) = \eta(y) - \eta(x).$$

In case φ and φ' are ergodic $Z\!\!\!Z$ valued cocycles, an adjustment of the above shows that there is an orbit equivalence $\Phi : X \to X'$ such that

$$\varphi'(\Phi x, \Phi y) = \varphi(x, y) \quad \text{a.e. on } \mathcal{R}.$$

In case φ and φ' have isomorphic periodic actions, they are also cohomologous via orbit equivalence. To see this, we may suppose that φ and φ' satisfy the conditions in the remark after proposition 3.3. The result now reduces to the uniqueness of ergodic $Z\!\!\!Z$-valued cocycles.

We now turn to the

Proof of theorem 2.
Let $\mathcal{R}$ and $\mathcal{R}'$ be hyperfinite of type II_1. Let φ and φ' be recurrent orbit cocycles of $\mathcal{R}$ and $\mathcal{R}'$ respectively, having isomorphic associated actions. By proposition 3.3, we may assume without loss of generality, that φ and φ' are lacunary, and have isomorphic K-representations.
Let φ and φ' have (respectively):
kernels $S = S_\varphi$, and $S' = S'_\varphi$;
K-representations $(\Omega, U, f) := (\Omega(S), U_\varphi, f_\varphi)$ and $(\Omega', U', f') := (\Omega(S'), U_{\varphi'}, f_{\varphi'})$.
Suppose that $\mu \sim \overline{m}$ and $\mu' \sim \overline{m}'$ are probabilities, and that $\theta : (\Omega, \mathcal{B}(\Omega), \mu) \to (\Omega', \mathcal{B}(\Omega'), \mu')$ is a measure space isomorphism satisfying

$$\theta \circ U = U' \circ \theta, \quad f' \circ \theta = f, \; \& \; \mu \circ \theta^{-1} = \mu'.$$

Let:

(1) $\pi : X \to \Omega$ and $\pi' : X' \to \Omega'$ be the invariant factor maps,

(2) $h = \frac{d\overline{m}}{d\mu}$, $h' = \frac{d\overline{m}'}{d\mu'}$ where $\overline{m} = m \circ \pi^{-1}$ and $\overline{m}' = m' \circ \pi'^{-1}$.

One way to establish the theorem would be to obtain an orbit equivalence $X \to X'$ extending $\theta : \Omega \to \Omega'$. For this to be possible, we would need

$$\overline{m}' \circ \theta^{-1} = \overline{m} \quad (\Leftrightarrow h' \circ \theta = h).$$

Indeed, our first task will be to reduce to this situation, which will yield an orbit equivalence as above, and establish the theorem without coboundary (see lemma 4.1 below).

The reduction will be done by restricting to subsets $Y \in \mathcal{B}(X)$, $Y' \in \mathcal{B}(X')$ in such a way as to deform the measures appropriately.

We now describe this process of restriction. For $Y \in \mathcal{B}(X)$ such that $\pi(Y) = \Omega$,

(1) let $\mathcal{R}_Y = \mathcal{R} \cap (Y \times Y)$, $\varphi_Y := \varphi|_{\mathcal{R}_Y}$;

(2) note that the kernel of φ_Y is given by $\mathcal{S}_{\varphi_Y} = \mathcal{S}_\varphi \cap (Y \times Y) := \mathcal{S}_Y$,

(3) and the invariant factor for $\mathcal{S}_Y$ is $\Omega(\mathcal{S}_Y) = \Omega$, with invariant factor map $\pi_Y = \pi|_Y$ (since $\pi(Y) = \Omega$).

Note also that (since $\pi(Y) = \Omega$)

$$\inf \{\varphi_Y(x,y) : (x,y) \in \mathcal{R}_Y, \; \varphi_Y(x,y) > 0\} = f(\pi x) \; \forall \, x \in Y.$$

Set also

$$m_Y(\cdot) = \frac{m(\cdot \cap Y)}{m(Y)}, \quad \overline{m}_Y = m_Y \circ \pi^{-1}, \quad \& \; h_Y = \frac{d\overline{m}_Y}{d\mu}.$$

New conditional probabilities $\{(m_Y)_z : z \in \Omega\}$ on $(X, \mathcal{B}(X))$ are induced as before by the invariant factor map $\pi_Y = \pi : Y \to \Omega(\mathcal{S}_Y) = \Omega$, and these are defined by

$$\int_A (m_Y)_z(B) d\overline{m}_Y(z) = m_Y(\pi^{-1}A \cap B), \quad (A \in \mathcal{B}(\Omega). \; B \in \mathcal{B}(X),$$

whence it follows that

$$(m_Y)_z(B) = \frac{m_z(B \cap Y)}{m_z(Y)} \quad (B \in \mathcal{B}(X), \; z \in \Omega).$$

For $A \in \mathcal{B}(\Omega)$,

$$\int_A h_Y \, d\mu = \int_A d\overline{m}_Y = \frac{m(\pi^{-1}A \cap Y)}{m(Y)} = \int_A \frac{m_z(Y)}{m(Y)} d\overline{m}(z) = \int_A \frac{m_z(Y)}{m(Y)} h(z) d\mu(z),$$

whence,

$$h_Y(z) = \frac{m_z(Y)}{m(Y)} h(z).$$

The reduction

Set

$$Z_1 = \{z \in \Omega : h'(\theta z) > h(z)\},$$
$$Z_2 = \{z \in \Omega : h'(\theta z) \le h(z)\}.$$

Choose $Y \in \mathcal{B}(X)$ and $Y' \in \mathcal{B}(X')$ such that

$$m_z(Y) = 1, \quad m'_{\theta z}(Y') = \frac{h(z)}{h'(\theta z)} \text{ for } z \in Z_1,$$

and

$$m_z(Y) = \frac{h'(\theta z)}{h(z)}, \quad m'_{\theta z}(Y') = 1 \text{ for } z \in Z_2.$$

It follows that for $z \in \Omega$,

$$h(z)m_z(Y) = h'(\theta z)m'_{\theta z}(Y'),$$

consequently:

$$m(Y) = m'(Y'),$$

$$h'_{Y'}(\theta z) = \frac{m'_{\theta z}(Y')}{m'(Y')}h'(\theta z)$$

$$= \frac{m_z(Y)}{m(Y)}h(z)$$

$$= h_Y(z),$$

and

$$\overline{m_Y} \circ \theta^{-1} = \overline{m'_{Y'}},$$

hence, for $k \in \mathbb{Z}$,

$$\frac{\overline{dm'_{Y'}U'^k}}{\overline{dm'_{Y'}}}(\theta z) = \frac{\overline{dm_Y U^k}}{\overline{dm_Y}}(z)$$

Lemma 4.1. *There is a measure preserving and measure space isomorphism* $\Phi : (Y, \mathcal{B}(Y), m_Y) \to (Y', \mathcal{B}(Y'), m'_{Y'})$ *with the following properties:*

(1) $\qquad\qquad\qquad \pi' \circ \Phi = \theta \circ \pi, \quad (x \in Y),$

(2) $\qquad\qquad\qquad (m'_{Y'})_{\theta z}(\Phi(A)) = (m_Y)_z(A), \quad (A \subset Y, z \in \Omega),$

(3) $\qquad\qquad\qquad (\Phi(x), \Phi(y)) \in \mathcal{R}'_{Y'} \text{ iff } (x, y) \in \mathcal{R}_Y,$

(4) $\qquad\qquad\qquad \varphi'_{Y'}(\Phi(x), \Phi(y)) = \varphi_Y(x, y), \quad ((x, y) \in \mathcal{R}_Y).$

Proof. This lemma can be thought of as a "relative" version of Dye's theorem ([Dye]) in that it establishes the existence of an orbit equivalence extending a given factor space isomorphism (conditions (1),(2), and (3)). Condition (4) will follow automatically as θ is an isomorphism of K-representations. The method of proof is to show that towers of $\mathcal{R}_Y$ can be copied as towers of $\mathcal{R}'_{Y'}$.

Copying Lemma ([K-W]). *Given any tower*

$$\xi = (\mathcal{P}, \mathcal{T}) \quad \mathcal{P} = \{E_\alpha : \alpha \in \Lambda\}, \ \mathcal{T} = \{e_{\alpha, \beta} : \alpha, \beta \in \Lambda\}$$

of $\mathcal{R}_Y$,
    there is a measure preserving and measure space isomorphism

$$\Phi : (Y, \mathcal{B}(Y), m_Y) \to (Y', \mathcal{B}(Y'), m'_{Y'})$$

and a tower

$$\xi' = (\mathcal{P}', \mathcal{T}') \quad \mathcal{P}' = \{E'_\alpha : \alpha \in \Lambda\}, \quad \mathcal{T}' = \{e'_{\alpha,\beta} : \alpha, \beta \in \Lambda\}$$

of $\mathcal{R}'_Y$, satisfying:

(1) $\qquad\qquad\qquad\qquad \pi' \circ \Phi = \theta \circ \pi$

(2) $\qquad\qquad\qquad\qquad \Phi(E_\alpha) = E'_\alpha$

(3) $\qquad\qquad\qquad\qquad \Phi e_{\beta,\alpha} = e'_{\beta,\alpha} \Phi \quad \text{on} \quad E_\alpha$

(4) $\qquad \pi'(e'_{\beta,\alpha} \circ \Phi(x)) = U'^{-n} \pi'(\Phi(x)) \quad \text{if} \quad \pi(e_{\beta,\alpha} x) = U^{-n} \pi(x),$

where $n \in \mathbb{Z}$.

Proof of the Copying Lemma

Let $\alpha, \beta \in \Lambda$. Then $\pi(e_{\beta,\alpha} x)$ is of the form:

$$\pi(e_{\beta,\alpha} x) = U^{-n} \pi(x) \quad (x \in E_\alpha),$$

where $n = n(\beta, \alpha, x) \in \mathbb{Z}$.

Partition each set E_α into countable disjoint subsets $E_{\alpha,i}, i \geq 1$ so that

$$n(\beta, \alpha, x) = n(\beta, \alpha, i) \text{ (constant) for } x \in E_{\alpha,i},$$

and, for each $\alpha, \beta \in \Lambda$ and $i \geq 1$,

$$E_{\beta,i} = e_{\beta,\alpha}(E_{\alpha,i}).$$

Set

$$Y_i = \bigcup_\beta E_{\beta,i}.$$

Now we have a countable disjoint family of the restrictions $\xi_i = (\mathcal{P} \cap Y_i, \mathcal{T}|_{Y_i})$ of the tower ξ to the sets Y_i. Here, $\mathcal{P} \cap Y_i := \{E_\alpha \cap Y_i : \alpha \in \Lambda\}$, and $\mathcal{T}|_{Y_i} = \{e_{\gamma,\beta,i} := e_{\gamma,\beta}|_{Y_i} : \gamma, \beta \in \Lambda\}$.

As we'll copy each ξ_i individually, and disjointly, we'll drop the subscript i, and "assume" that

$$n(\beta, \alpha, x) = \text{constant} = n_{\beta,\alpha} \text{ on } E_\alpha.$$

So, $e_{\beta,\alpha} \in N[\mathcal{S}_Y]_* \cap [\mathcal{R}_Y]_*$, and

$$U^{-n_{\beta,\alpha}} \pi(E_\alpha) = \pi(E_\beta)$$

In order to facilitate notation, we'll denote for the rest of the proof of the copying lemma:

$$\overrightarrow{m_Y} = \nu, \ (m_Y)_z = \nu_z, \ \overrightarrow{m'_{Y'}} = \nu', \ (m'_{Y'})_{z'} = m'_{z'}, \ (z \in \Omega, \ z' \in \Omega').$$

We recall that $\nu \circ \pi^{-1} = \nu'$, and note Lemma 3.6 can now be written as:

Lemma 3.6'. If $R \in N[\mathcal{S}_Y]_* \cap [\mathcal{R}_Y]_*, E \subset Dom(R)$, then for a.e. $z \in \pi(Im(R))$,

$$\nu_z(RE) = \nu_{U^{-n(z)}z}(E) \cdot \frac{d\nu U^{-n(z)}}{d\nu}(z),$$

19

where $n(z)$ be as in Lemma 3.5.

Choose a finite partition $\{E'_\beta | \beta \in \Lambda\}$ of Y' so that

$$\nu'_{\theta z}(E'_\beta) = \nu_z(E_\beta), \quad (z \in \Omega).$$

Fix $\alpha \in \Lambda$ and take a measure preserving and measure space isomorphism $\Phi = \Phi_\xi : E_\alpha \to E'_\alpha$ such that $\pi'\Phi = \theta\pi$. Write $n_{\beta,\alpha} = n_\beta$. It follows that

$$\nu'_{\theta z}(\Phi(A)) = \nu_z(A), \quad \left(z \in \Omega, \ A \subset E_\alpha\right).$$

It follows from Lemma 3.6' that for a.e.$z \in \pi(E_\beta)$,

$$\nu_z(E_\beta) = \nu_z(e_{\beta,\alpha} E_\alpha) = \nu_{U^{n_\beta} z}(E_\alpha)\frac{d\nu U^{n_\beta}}{d\nu}(z),$$

or for a.e. $z \in E_\alpha$,

$$
\begin{aligned}
\nu_{U^{-n_\beta} z}(E_\beta) &= \nu_{U^{-n_\beta} z}(e_{\beta,\alpha}(E_\alpha)) \\
&= \nu_z(E_\alpha)\left(\frac{d\nu U^{-n_\beta}}{d\nu}(z)\right)^{-1}.
\end{aligned}
$$

By corollary 3.4, one can choose $R' \in N[S'_{Y'}]_*$ such that

(1) $\qquad \operatorname{Dom}(R') \subset E'_\alpha$

(2) $\qquad \nu'_{z'}(\operatorname{Dom}(R')) > 0 \quad$ if and only if $\quad m'_{z'}(E'_\alpha) > 0 \quad$ a.e.z

(3) $\qquad \pi'(R'x') = U'^{-n_\beta}\pi'(x'), \quad (x' \in \operatorname{Dom}(R')).$

We now claim that for a.e.$z' \in \pi'(E'_\beta)$

$$\nu'_z(\operatorname{Im}(R')) \leq \nu'_z(E'_\beta).$$

To see this, we notice that $\pi'(\operatorname{Im}(R')) = \pi'(E'_\beta)$. By Lemma 3.6', for a.e.$z' \in \pi'(E'_\beta)$

$$
\begin{aligned}
\nu'_z(\operatorname{Im} R') &= \nu'_{U'^{n_\beta} z'}(\operatorname{Dom} R')\left(\frac{d\nu' U'^{-n_\beta}}{d\nu'}(U'^{n_\beta} z')\right)^{-1} \\
&\leq \nu'_{U'^{n_\beta} z'}(E'_\alpha)\frac{d\nu' U'^{-n_\beta}}{d\nu'}(z') \\
&= \nu_{U^{n_\beta} z}(E_\alpha)\frac{d\nu U^{n_\beta}}{d\nu}(z) \\
&= \nu_z(E_\beta) \\
&= \nu'_{z'}(E'_\beta)
\end{aligned}
$$

where $z = \theta^{-1}(z')$.

For a.e.$z' \in \pi'(E'_\beta)$ define $d = d_\beta = d_\beta(z') \geq 1$ by

$$d_\beta = [\frac{\nu'_{z'}(E'_\beta)}{\nu'_{z'}(\operatorname{Im} R')}] = \max\{k \in \mathbb{N} : k \leq \frac{\nu'_{z'}(E'_\beta)}{\nu'_{z'}(\operatorname{Im} R')}\}.$$

Applying Hopf-equivalence, we obtain $g'_1, g'_2, \cdots, g'_d \in [S'_{Y'}]_*$ satisfying:

(1) $\qquad \operatorname{Dom}(g'_i) \subset \operatorname{Im}(R'), \ \operatorname{Im}(g'_i) \subset E'_\beta \ (i \geq 1).$

(2) $\qquad$ The subsets $\operatorname{Im}(g'_i)$'s are disjoint.

(3) $\qquad \nu'_{U'^{-n_\beta} z'}(\operatorname{Dom}(g'_i)\triangle \operatorname{Im}(R')) = 0 \ (1 \leq i \leq d).$

(4) $\qquad \nu'_{U'^{-n_\beta} z'}(\operatorname{Dom}(g'_{d+1})) = \nu'_{U'^{-n_\beta} z'}(E'_\beta) - d \cdot \nu'_{U'^{-n_\beta} z'}(\operatorname{Im} R').$

where $z' \in \pi'(E'_\alpha)$ and $d = d_\beta(U'^{-n_\beta}z')$. Then obviously,

$$\bigcup_i \mathrm{Im}\,(g'_i) = E'_\beta.$$

We are going to show that for a.e. $z' \in \pi'(E'_\alpha)$

(1) $$[\frac{\nu'_{z'}(E'_\alpha)}{\nu'_{z'}(\mathrm{Dom}\,(R'))}] = d$$

(2) $$\nu'_{z'}(R'^{-1}(\mathrm{Dom}\,(g'_{d+1}))) = \nu'_{z'}(E'_\alpha) - d \cdot \nu'_{z'}(\mathrm{Dom}\,(R'))$$

where $d = d_\beta(U'^{-n_\beta}z')$.

Let $z' \in \pi'(E'_\alpha)$. The first is obtained from

$$
\begin{aligned}
\frac{\nu'_{U'^{-n_\beta}z'}(E'_\beta)}{\nu'_{U'^{-n_\beta}z'}(\mathrm{Im}\,(R'))}
&= \frac{\nu'_{U'^{-n_\beta}z'}(E'_\beta)}{\nu'_{z'}(\mathrm{Dom}\,(R'))} \cdot \frac{d\nu'U'^{-n_\beta}}{d\nu'}(z') \quad \left(\text{use Lemma 3.6'}\right) \\
&= \frac{\nu_{U^{-n_\beta}z}(E_\beta)}{\nu_{z'}(\mathrm{Dom}\,(R'))} \cdot \frac{d\nu U^{-n_\beta}}{d\nu}(z) \\
&= \frac{\nu_z(E_\alpha)}{\nu'_{z'}(\mathrm{Dom}\,(R'))} \\
&= \frac{\nu'_{z'}(E'_\alpha)}{\nu'_{z'}(\mathrm{Dom}\,(R'))}.
\end{aligned}
$$

The second is that if $z' \in \pi'(E'_\alpha)$ and $d = d_\beta(U'^{-n_\beta}z')$ then

$$
\begin{aligned}
&\nu'_{z'}(R'^{-1}(\mathrm{Dom}\,(g'_{d+1}))) \\
&= \nu'_{U'^{-n_\beta}z'}(\mathrm{Dom}\,(g'_{d+1}))\frac{d\nu'U'^{-n_\beta}}{d\nu'}(z') \\
&= \left(\nu'_{U'^{-n_\beta}z'}(E'_\beta) - d \cdot \nu'_{U'^{-n_\beta}z'}(\mathrm{Im}\,(R'))\right)\frac{d\nu'U'^{-n_\beta}}{d\nu'}(z') \\
&= \nu'_{z'}(E'_\alpha) - d \cdot \nu'_{z'}(\mathrm{Dom}\,(R')).
\end{aligned}
$$

Thus, if $z' \in \pi(E'_\alpha)$ and $d = d_\beta(U'^{-n_\beta}z')$ then

$$
\begin{aligned}
&\nu'_{z'}(R'^{-1}(\mathrm{Dom}\,(g'_{d+1}))) \\
&= \nu'_{z'}(E'_\alpha) - [\frac{\nu'_{z'}(E'_\alpha)}{\nu'_{z'}(\mathrm{Dom}\,(R'))}] \cdot \nu'_{z'}(\mathrm{Dom}\,(R')).
\end{aligned}
$$

Therefore, using Hopf-equivalence by $[\mathcal{S}'_{\gamma'}]_*$, we obtain $\rho'_i \in [\mathcal{S}'_{\gamma'}]_*$ satisfying the following conditions:

(1) $\mathrm{Dom}\,(\rho'_i) \subset \mathrm{Dom}\,(R'), \quad \mathrm{Im}\,(\rho'_i) \subset E'_\alpha$.

(2)   The sets $\mathrm{Im}\,(\rho'_i)$ are disjoint.

(3)   If $z' \in \pi'(E'_\alpha)$ and $d = d_\beta(U'^{-n_\beta}z')$ and $1 \leq i \leq d$ then $\nu'_{z'}(\mathrm{Dom}\,(\rho'_i)\triangle\mathrm{Dom}\,(R')) = 0 \quad (1 \leq i \leq d)$.

(4) $\nu'_{z'}(\mathrm{Dom}\,(\rho'_{d+1})\triangle R'^{-1}(\mathrm{Dom}\,(g'_{d+1}))) = 0$.

Then, obviously

$$\bigcup_i \mathrm{Im}\,(\rho'_i) = E'_\alpha.$$

Using g_i's and ρ_i's, let us define a map $e'_{\beta,\alpha} : E'_\alpha \to E'_\beta$ by setting for $x' \in \mathrm{Im}(\rho'_i)$,

$$e'_{\beta,\alpha}x' = g'_i \cdot R' \cdot \rho'^{-1}_i(x').$$

Obviously,

(1)
$$e'_{\beta,\alpha} \in N[\mathcal{S}'_{Y'}]_*,$$

(2)
$$\mathrm{Dom}\,(e'_{\beta,\alpha}) = E'_\alpha, \quad \mathrm{Im}\,(e'_{\beta,\alpha}) = E'_\beta$$

(3)
$$\pi'(e'_{\beta,\alpha}x') = U'^{-n_\beta}\pi'(x'), \quad (x' \in E'_\alpha)$$

Extend $\Phi : E_\alpha \to E'_\alpha$ by setting for each $\beta \in \Lambda$,

$$\Phi = e'_{\beta,\alpha} \circ \Phi \circ e_{\alpha,\beta}.$$

Set

$$
\begin{aligned}
e'_{\alpha,\beta} &= e'^{-1}_{\beta,\alpha}, \\
e'_{\beta,\epsilon} &= e'_{\beta,\alpha}e'_{\alpha,\epsilon}, \\
\xi' &= (\mathcal{P}', \mathcal{T}') \text{ where} \\
\mathcal{P}' &= \{E'_\beta : \beta \in \Lambda\} \text{ and} \\
\mathcal{T}' &= \{e'_{\beta,\epsilon}|\beta, \epsilon \in \Lambda\}.
\end{aligned}
$$

We have constructed the tower ξ', and completed the proof of the copying lemma. $\square$

By hyperfiniteness, $\mathcal{B}(Y)$ is generated by a sequence of towers of $\mathcal{R}_Y$, and each $\mathcal{R}_Y$-orbit is a countable increasing union of finite orbits by towers. So, in order to complete the proof of lemma 4.1, we apply the copying lemma to a refinement of ξ' in Y', which approximates $\mathcal{B}(Y')$ and $\mathcal{R}'_{Y'}$ orbits with some fixed precision, obtaining a refinement of ξ in Y, and continue this procedure back and forth as before. In the limit we obtain $\Phi : Y \to Y'$ satisfying conditions (1),(2), and (3) of the lemma. As mentioned above, condition (4) follows automatically. $\square$

We complete the proof of theorem 2 by extending the domain of definition of Φ. Set

$$Z_{2,n} = \{z \in Z_2 | nh'(\theta z) \leq h(z) < (n+1)h'(\theta z)\}, \quad n \geq 1.$$

Choose a countable partition $\{F_{n,i}|n \geq 1, 1 \leq i \leq n\}$ of the set $X \backslash Y$ satisfying

(1) The subsets $\pi(F_{n,i})$ are disjoint and

$$\bigcup_{i=1}^{n} \pi(F_{n,i}) = Z_{2,n}$$

(2)

$$
\begin{aligned}
m_z(F_{n,i}) &= \frac{h'(\theta z)}{h(z)} m_z(Y), \quad (i < n, z \in Z_{2,n}) \\
m_z(F_{n,n}) &= \frac{h(z) - nh'(\theta z)}{h(z)} \quad (z \in Z_{2,n})
\end{aligned}
$$

For each n and for each $1 \leq i \leq n$, a partial transformation $\alpha_{n,i} \in [S]_*$ with the domain $F_{n,i}$ and the image which is a subset of Y is defined. Set

$$G_{n,i} = \text{Im}\,(\alpha_{n,i}), \quad G'_{n,i} = \Phi(G_{n,i}).$$

We recall $m(Y) = m'(Y')$ and choose a countable partition $\{F'_{n,i}\}$ of $X'\backslash Y'$ such that

$$m'(F'_{n,i}) = m'(G'_{n,i})$$

Then, by Hopf-equivalence, we obtain partial transformations $\alpha'_{n,i} \in [R']_*$ with domain $F'_{n,i}$ and image $G'_{n,i}$.

We obtain a measure preserving isomorphism from X onto X' by extending the previous Φ by setting

$$\Phi(x) = \alpha'^{-1}_{n,i} \circ \Phi \circ \alpha_{n,i}(x) \quad (x \in F_{n,i}).$$

Finally let us define a measurable function $\eta(x)$ which cancels the difference of φ and φ' on the set Y^c. Set

$$
\begin{aligned}
\eta(x) &= 0 \quad (x \in Y), \\
\eta(x) &= \varphi'(\alpha'^{-1}_{n,i} \circ \Phi \circ \alpha_{n,i}(x), \; \Phi \circ \alpha_{n,i}(x)) \quad (x \in F_{n,i}).
\end{aligned}
$$

Then, it is easily checked that

$$\varphi(x,y) - \varphi'(\Phi(x), \Phi(y)) = \eta(y) - \eta(x).$$

□

Concluding Remarks

(1) The possible lack of $g \in [R]$ satisfying $\pi(gx) = U\pi x$ (as pointed out in section 3) makes the proof of theorem 2 more difficult. If the hyperfinite equivalence relation R admits a σ-finite, infinite invariant measure, then such g exist, and a simplification of our proof may establish uniqueness (up to cohomology via orbit equivalence) of a cocycle with a given free, conservative action. Indeed, in this type II_∞ setup, Bezugly and Golodets [B-G] obtained this result using Krieger's cohomology lemma and discrete decomposition theorem, which use the existence of g. However in our setup, the invariant measure is finite and so, the proof must be rigid.

(2) As a consequence of this, it is shown in [B-G] that $\varphi\times\text{Id}$ and $\varphi'\times\text{Id}$ are cohomologous up to orbit equivalence. Here, $\varphi\times\text{Id}$ is the cocycle on the product equivalence relation $R \times R_{\mathbb{Z}}$ of type II_∞ where

$$((x,i), \; (y,j)) \in R \times R_{\mathbb{Z}} \; \text{iff} \; (x,y) \in R, \; \text{and} \; i, \; j \in \mathbb{Z}$$

and where

$$(\varphi \times \text{Id})((x,i), \; (y,j)) = \varphi(x,y) \quad ((x,y) \in R, \; \text{and} \; i, \; j \in \mathbb{Z}).$$

Theorem 2 refines this result.

(3) When R is of type III, then the tensor product cocycle $(\varphi, \log(\rho))$ taking values in $\mathbb{R} \times \mathbb{R}$ where ρ is the Radon-Nikodym cocycle, and φ has a given action is unique up to cohomology via orbit equivalence. In this setup, the copying lemma will be proved by replacing S by $\text{Ker}(\varphi)\cap\text{Ker}(\log(\rho))$. This is established by a different method in [B-G].

References

[B-G]   S.I.Bezugly, V.Ya.Golodets, *Weak equivalence and the structures of cocycles of an ergodic automorphism,* Publ. RIMS, Kyoto Univ. **27**, (1991), 577-625

[C-F-W]   A. Connes, J. Feldman, B. Weiss, *An amenable equivalence relation is generated by a single transformation,* Ergod. Th. and Dynam. Sys. **1**, (1981), 431-450

[Dye]   H.A.Dye, *On groups of measure preserving transformations I,* Amer. J. Math. **81**, (1959), 119-159

[Fe-Mo]   J. Feldman and C.C. Moore, *Ergodic equivalence relations, cohomology, and von-Neumann algebras, I,* Trans. Amer. Math. Soc. **234**, (1977), 289-324.

[F]   H. Furstenberg, *Strict ergodicity and transformations of the torus,* American Journal of Math. **83**, (1961), 573-601.

[G-S1]   V.Ya.Golodets and S.D.Sinel'shchikov, *Existence and uniqueness of cocycles of an ergodic automorphism with dense ranges in amenable groups,* Kharkov preprint, Inst. Low Temp. Phys. & Engin. (1983).

[G-S2]   V.Ya.Golodets and S.D.Sinel'shchikov, *Classification and structure of cocycles of amenable ergodic equivalence relations,* Kharkov preprint, Inst. Low Temp. Phys. & Engin. (1993).

[Ham1]   T.Hamachi, *The normalizer group of an ergodic automorphism of type III and the commutant of an ergodic flow,* J. Funct. Anal. **40**, (1981), 387-403.

[Ham2]   T.Hamachi, *A measure theoretical proof of Connes-Woods theorem on AT-flows,* Pacific J. Math. **154**, (1992), 67-85.

[H-O1]   T.Hamachi and M.Osikawa, *Ergodic groups of automorphisms and Krieger's theorems,* Seminar on Mathematical Science of Keio Univ. **3**, (1981), 1-113

[H-O2]   T.Hamachi and M.Osikawa, *Computation of the associated flows of $ITPFI_2$ factors of type III_0,* Geometric methods in Operator algebras, ed. by H.Araki and E.G.Effros, Pitman Research Notes in Math. Series, **123**, (1983), 196-210.

[Ka]   A. B. Katok, *The special representation theorem for multi-dimensional group actions,* Asterisque, **49**, (1977), 117-140.

[Katz]   Y. Katznelson, *Lectures on orbit equivalence,* mimeographed notes, Orsay (1980).

[K-W]   Y. Katznelson, B. Weiss, *The classification of non-singular actions revisited,* Ergod. Th. and Dynam. Sys. **11**, (1991), 333-348.

[Kre]   U.Krengel, *Darstellungssätze für Strömungen und Halbströmungen,* Math. Ann. **182**, (1969), 1-39.

[Kri]   W. Krieger, *On ergodic flows and isomorphism of factors,* Math. Annalen, **223**, (1976), 19-70.

[Ku]   I. Kubo, *Quasi-flows,* Nagoya Math. J. **35**, (1969).

[Ma]   G.W. Mackey, *Ergodic theory and virtual groups,* Math. Ann. **166**, (1966), 187-207.

[P]   W. Parry, *Compact Abelian groups extensions of discrete dynamical systems* Z. Wahrsch. u.v.w Geb. **13**, (1969), 95-113.

[Sch] K. Schmidt, *Cocycles of Ergodic Transformation Groups,* Lect. Notes in Math. Vol. 1, MacMillan Co. of India (1977).

KOKSMA'S INEQUALITY AND GROUP EXTENSIONS OF KRONECKER TRANSFORMATIONS

Jon Aaronson[1], Mariusz Lemańczyk[2],
Christian Mauduit[3], and Hitoshi Nakada[4]

[1]School of Mathematical Sciences, Tel Aviv University
   69978 Tel Aviv, Israel
[2]Institute of Mathematics, Nicholas Copernicus University
   ul. Chopina 12/18, 87-100 Toruń, Poland
[3]Laboratoire de Mathématiques Discrètes
   Case 930, 163 avenue de Luminy
   13288 Marseille Cedex 9, France
[4]Dept. Math., Keio University, Hiyoshi 3-14-1 Kohoku
   Yokohama 223, Japan

ABSTRACT. We consider methods of establishing ergodicity of group extensions, proving that a class of cylinder flows are ergodic, coalescent and non-squashable. A new Koksma-type inequality is also obtained.

Introduction

We study locally compact group extensions of Kronecker transformations.

Let X be a compact monothetic group with Haar probability measure $m = m_X$, and G a locally compact metric group with Haar measure m_G. Let T be an ergodic translation on X, (called a *Kronecker* transformation) and set $\mu = m \times m_G$.

For $\varphi : X \longrightarrow G$ measurable (called a *cocycle*), consider the *skew product* (or G-*extension*) which is the measure preserving transformation $T_\varphi : (X \times G, \mu) \longrightarrow (X \times G, \mu)$ defined by

$$T_\varphi(x, g) = (Tx, \varphi(x)g).$$

Recall from [Aa1] that a measure preserving transformation $\tau : (Y, \nu) \longrightarrow (Y, \nu)$ is called *squashable* if $\exists Q \ni \quad Q\tau = \tau Q \quad$ and $\nu Q^{-1} = c\nu$ for certain $c \neq 1$. It follows from [Aa2, Th3.4] that if the group G is countable, and has no arbitrarily large finite normal subgroups (*e.g.* $G = \mathbb{Z}^k \times \mathbb{Q}^l$) then no ergodic G-extension is squashable.

Aaronson and Lemańczyk would like to thank Keio university for hospitality provided while this research was done. Lemańczyk's research was partly supported by KBN grant 512/2/91.

Most of the results in this paper are for the case $G = \mathbb{R}$. It is an open problem to decide if there is a conservative, ergodic, squashable $\mathbb{R}$-extension of a Kronecker transformation. Almost all of our results are in the other direction, showing that certain $\mathbb{R}$-extensions are nonsquashable.

We consider product-type cocycles for odometers in §1, obtaining conditions for ergodicity, nonsquashability, and coalescence (q.v.) Essentially the same ideas can be used in the context of [Kw-Le-Ru2] to obtain analytic *cylinder flows* (i.e. $\mathbb{R}$-extensions of rotations of the circle) which are ergodic, nonsquashable, and coalescent (see §4). We show in §5 that if $\varphi : \mathbb{T} \longrightarrow \mathbb{R}$ is $C^{1+\delta}$ then for a residual set of irrational rotations T, the cocycle is conservative and ergodic. We improve some recent results by D. Pask (in §6) [Pa1],[Pa2] on the ergodicity of cylinder flows also proving the non-squashability in this case.

One of our tools is a new Koksma-type inequality in $L^2(\mathbb{T})$ for functions whose Fourier coefficients are of order $O(1/n)$ (see §2) with possible speeds of convergence for smooth functions and irrational rotations admitting a speed of approximations by rationals (see §3).

The authors would like to thank E. Lesigne for a discussion on the proof of Theorem 5.1.

§1 Coalescence of group extensions, and ergodicity of product type cocycles

A non-singular transformation is called *coalescent* if all nonsingular commuting with it transformations are invertible. To begin this section, we study the form of non-singular transformations commuting with an ergodic, group extension of a Kronecker transformation.

Suppose that T is an ergodic measure-preserving transformation of the probability space $(X, \mathcal{B}, m)$; let $(G, \mathcal{T})$ be an abelian, locally compact, second countable, topological group $(\mathcal{T} = \mathcal{T}(G)$ denotes the family of open sets in the topological space $G)$, and let $\varphi : X \to G$ be a cocycle.

Let $T_\varphi : (X \times G, \mu) \longrightarrow (X \times G, \mu)$,

$$T_\varphi(x, g) = (Tx, \varphi(x)g)$$

be ergodic (this implies that G has to be amenable [Zim]), where T is a Kronecker transformation on X, and $\varphi : X \longrightarrow G$ is a cocycle.

Proposition 1.1. *Suppose that $Q : X \times G \longrightarrow X \times G$ is non-singular and $QT_\varphi = T_\varphi Q$. Then there exist a translation S of X, and a continuous group homomorphism $w : G \longrightarrow G$ which is non-singular in the sense that $m_G \circ w^{-1} \sim m_G$ and a measurable map $f : X \longrightarrow G$ such that*

$$Q(x, h) = (Sx, f(x)w(h)) \quad \text{for each } x \in X, \, h \in G.$$

Proof. Write $Q = (S, F)$, where $S : X \times G \longrightarrow X$ and $F : X \times G \longrightarrow G$. We have

$$S \circ T_\varphi = T \circ S \quad \& \quad F \circ T_\varphi = (\varphi \circ S) \cdot F.$$

Let $U : X \times G \longrightarrow X$ be defined by $U(x,h) = x^{-1}S(x,h)$, then $U \circ T_\varphi = U$, hence by ergodicity of T_φ, $U(x,h) = x_1$, and $S(x,g) = Sx = xx_1 = x_1x$. Therefore

$$FT_\varphi(x,h) = \varphi(Sx)F(x,h).$$

Denote $\sigma_g(x,h) = (x,hg)$ and note that for each $g \in G$, $\sigma_g T_\varphi = T_\varphi \sigma_g$. Hence

$$\begin{aligned}
\left(F^{-1} \cdot (F \circ \sigma_g)\right) \circ T_\varphi(x,h) &= F(T_\varphi(x,h))^{-1} F(T_\varphi(x,hg)) \\
&= \left(\varphi(Sx)F(x,h)\right)^{-1} \varphi(Sx)F(x,hg) \\
&= \left(F^{-1}F \circ \sigma_g\right)(x,h),
\end{aligned}$$

whence there exists $w : G \longrightarrow G$ such that $F^{-1}(F \circ \sigma_g) = w(g)$ for each $g \in G$. It follows that w is a measurable homomorphism (and hence continuous).

Set $\phi(x,h) = F(x,h)w(h)^{-1}$. By the above, $\phi \circ \sigma_g = \phi$ for each $g \in G$ whence there exists a measurable $f : X \longrightarrow G$ such that $\phi(x,h) = f(x)$ a.e., and

$$Q(x,g) = (Sx, f(x)w(g)).$$

To see that $w : G \to G$ is non-singular, note that $\mu \circ S_f^{-1} = \mu$, and since $QT_\varphi = T_\varphi Q$, $\exists c > 0$ such that $\mu \circ Q^{-1} = c\mu$. Moreover

$$\tilde{w} := \mathrm{Id} \times w = S_f^{-1} \circ Q$$

whence $\mu \circ \tilde{w}^{-1} = c\mu$, and $m \circ w^{-1} = cm$. $\qquad\square$

Remarks

(1) If T is an invertible, ergodic probability preserving transformation and φ an ergodic cocycle, and $Q(x,g) = (Sx, F(x,g))$ is non-singular, and commutes with T_φ, then Q has the above form.

(2) If $w : G \to G$ is non-singular and measurable, then w is continuous, and onto. To see this, note that $w(G)$ is a m_G-measurable subgroup of G, whence

$$\begin{aligned}
\exists x \notin w(G) \quad &\Rightarrow \quad xw(G) \subset G \setminus w(G) \\
&\Rightarrow \quad m(w(G)) = m(xw(G)) \leq m(G \setminus w(G)) = 0.
\end{aligned}$$

(3) If G is such that any continuous group non-singular homomorphism is 1-1 (e.g. $G = \mathbb{Z}^k \times \mathbb{Q}^l \times \mathbb{R}^m$) then any ergodic G-extension of a Kronecker transformation is coalescent. For coalescence of other group extensions, see theorem 1.5 below.

(4) In case $G = \mathbb{R}$ a skew product T_φ is squashable iff it commutes with a Q of form $Q(x,t) = (Sx, ct + \psi(x))$, where $|c| \neq 1$, or, in other words, $c\varphi - \varphi \circ S$ is a coboundary for some $|c| \neq 1$ and S a translation of X.

Next, we turn to methods of proving ergodicity of group extensions.

As in [Sch], the *essential values* of φ are defined as those group elements $a \in G$ with the property that

$$\forall \, A \in \mathcal{B}_+, \; U \in T(G) \text{ with } a \in U; \; \exists \, n \geq 1 \ni m(A \cap T^{-n} A \cap [\varphi^{(n)} \in U]) > 0$$

where $\varphi^{(n)}(x) = \varphi(T^{n-1}x) \cdot \ldots \cdot \varphi(x)$, $n \geq 1$.

The collection of essential values of φ is denoted by $E(\varphi)$. It is shown in [Sch] that $E(\varphi)$ is a closed subgroup of G, and is the collection of *periods* for T_φ-invariant functions:

$$E(\varphi) = \{a \in G : f(x, y + a) = f(x, y) \text{ a.e. } \forall \, f \circ T_\varphi = f \text{ measurable}\}.$$

In particular, T_φ is ergodic iff $E(\varphi) = G$. Also,

Lemma 1.2 [Sch]. *For any compact set K which is disjoint from $E(\varphi)$ there is a Borel set B, $\mu(B) > 0$, such that for each integer $m > 0$ we have*

$$\mu(B \cap T^{-m} B \cap [\varphi^{(m)} \in K]) = 0.$$

Definition. A sequence $q_n \in \mathbb{N}$ $(n \geq 1)$, $q_n \uparrow \infty$ is called a *rigidity time* for the probability preserving transformation T if $T^{q_n} \overset{\mathcal{U}(L^2(m))}{\longrightarrow}$ Id. Here $\mathcal{U}(L^2(m))$ denotes the collection of unitary operators on $L^2(m)$. Note that if T is a translation on the compact group X with Haar measure m then $T^{q_n} \overset{\mathcal{U}(L^2(m))}{\longrightarrow}$ Id iff $T^{q_n} \overset{X}{\longrightarrow}$ Id.

Lemma 1.3. *Suppose that $K \subset \mathbb{R}$ is compact, and that $\{q_n\}$ is a rigidity time for T such that*

$$\forall \, A \in \mathcal{B}_+, \; \liminf_{n \to \infty} m(A \cap [\varphi^{(q_n)} \in K]) > 0,$$

then

$$K \cap E(\varphi) \neq \emptyset.$$

Proof. Follows immediately from Lemma 1.2. $\qquad\qquad\qquad\qquad\qquad\qquad\qquad$ □

Let

$$D(\varphi) = \{a \in G : \exists \, q_n \to \infty, \; T^{q_n} \overset{\mathcal{U}(L^2(m))}{\longrightarrow} \text{Id} \ni \forall \; n_k \to \infty, \, a \in \{\varphi^{(q_{n_k})}\}'_{k \geq 1} \text{ a.e.}\}.$$

See also proofs of ergodicity in [Aa2, §4].

Proposition 1.4

$$D(\varphi) \subset E(\varphi).$$

Proof. Suppose that $y \in D$, and $T^{q_n} \to \text{Id}$, $y \in \{\varphi^{(q_{n_k})} : k \geq 1\}'$ a.e. $\forall \, n_k \to \infty$, then

$$\forall \, A \in \mathcal{B}_+ \; y \in U \in T(G), \; \exists \, \delta > 0 \ni \liminf_{n \to \infty} m(A \cap [\varphi^{(q_n)} \in U]) \geq \delta,$$

because if there were no such $\delta > 0$ we could choose $y \in U \in \mathcal{T}(G)$, and a subsequence q_{n_k}, $(k \geq 1)$ satisfying $m(A \cap [\varphi^{(q_{n_k})} \in U]) < 1/2^n$ and use the Borel-Cantelli lemma to get a contradiction to the definition of $y \in D(\varphi)$. Hence, since $T^{q_n} \longrightarrow \mathrm{Id}$, $\liminf_{n \to \infty} m(A \cap T^{-q_n} A \cap [\varphi^{(q_n)} \in U]) > \frac{\delta}{2} \ \forall \ n$ large, and therefore $y \in E(\varphi)$. $\qquad\square$

Set
$$\widetilde{D}(\varphi) = \{a \in G : \exists \ q_n \ni T^{q_n} \overset{\mathcal{U}(L^2(m))}{\longrightarrow} \mathrm{Id}, \ \& \ \varphi^{(q_n)} \to a \text{ a.e.}\}.$$

Clearly $\widetilde{D}(\varphi) \subset D(\varphi)$.

Theorem 1.5. *Assume that T is an ergodic translation. If $Gp(\widetilde{D}(\varphi))$ is dense in G, then T_φ is ergodic, and*

$$Q : X \times G \to X \times G \text{ nonsingular}, \ QT_\varphi = T_\varphi Q \ \Rightarrow \ Q(x, g) = (Sx, g + f(x))$$

where $ST = TS$ and $f : X \to G$ is measurable.
 In particular, such a T_φ is coalescent, and non-squashable.

Proof. By the previous proposition, T_φ is ergodic. We know from proposition 1.1 that

$$Q : X \times G \to X \times G \text{ nonsingular}, \ QT_\varphi = T_\varphi Q \ \Rightarrow \ Q(x, g) = (Sx, w(g) + f(x))$$

where $ST = TS$, $f : X \to G$ is measurable, and $w : G \to G$ is a continuous nonsingular homomorphism. It follows that

$$w(\varphi) - \varphi \circ S = f - f \circ T,$$

whence

$$\widetilde{D}(w(\varphi) - \varphi \circ S) = \{0\}.$$

However, if $a \in \widetilde{D}(\varphi)$, and

$$q_n \to \infty, \ T^{q_n} \overset{\mathcal{U}(L^2(m))}{\longrightarrow} \mathrm{Id}, \ \& \ \varphi^{(q_n)} \to a \text{ a.e.},$$

then

$$w(\varphi^{(q_n)}) - \varphi^{(q_n)} \circ S \to w(a) - a \text{ a.e.}$$

whence $w(a) - a \in \widetilde{D}(w(\varphi) - \varphi \circ S) = \{0\}$ and $w(a) = a \ \forall \ a \in \widetilde{D}(\varphi)$ and hence $\forall \ a \in G$.
$\square$

Set
$$C(\varphi) = \{a \in G : \underset{T^q \overset{\mathcal{U}(L^2(m))}{\longrightarrow} \mathrm{Id}, \ q \neq 0}{\liminf} 1_U(\varphi^{(q)}) = 1 \text{ a.e.} \forall \ a \in U \in \mathcal{T}(G)\}.$$

It is not hard to show that (for T Kronecker)

$$E(\varphi) \subset C(\varphi) \subset \widetilde{E}(\varphi)$$

where $\widetilde{E}(\varphi) :=$

$$\{a \in G : \forall \ I \in \mathcal{T}(X), \ a \in U \in \mathcal{T}(G) \ \exists \ n \geq 1 \ni m(I \cap T^{-n} I \cap [\varphi^{(n)} \in U]) > 0\}.$$

A popular misconception in the subject for the case $G = \mathbb{R}$ ([Con, proposition 1] [He-La1, lemma 3]) seems to have been that $C(\varphi) \subset E(\varphi)$.

This latter claim is wrong. A counterexample for a Kronecker transformation is given in example 1.7 (below). An analogous example for the case $G = \mathbb{T}$ was given in [Furst]. See [Or, proposition 1] for a related method of proving ergodicity not based on the above.

The rest of this section is devoted to

Cocycles of product type for an odometer

For $a_n \in \mathbb{N}$, $(n \in \mathbb{N})$, set $\Omega := \prod_{n=1}^{\infty}\{0, \ldots, a_n - 1\}$ equipped with the addition

$$(\omega + \omega')_n = \omega_n + \omega'_n + \epsilon_n \quad \text{mod } a_n$$

where $\epsilon_1 = 0$ and

$$\epsilon_{n+1} = \begin{cases} 0 & \omega_n + \omega'_n + \epsilon_n < a_n \\ 1 & \omega_n + \omega'_n + \epsilon_n \geq a_n. \end{cases}$$

Clearly, Ω equipped with the product discrete topology, is a compact Abelian topological group (called an *odometer group*), with Haar measure

$$m = \prod_{n=1}^{\infty} (\frac{1}{a_n}, \ldots, \frac{1}{a_n}).$$

Also if $\tau = (1, 0, \ldots)$ then $\Omega = \overline{\{n\tau\}}_{n \in \mathbb{Z}}$ whence $x \mapsto Tx(:= \tau + x)$ (called an *odometer transformation*) is ergodic.

A cocycle of *product type* is a measurable function $\varphi : \Omega \to G$ (where G is an Abelian topological group) of form

$$\varphi(\omega) = \sum_{n=1}^{\infty} (b_n(T\omega) - b_n(\omega))$$

where $b_n(\omega) = \beta_n(\omega_n)$, where $\beta_n : \{0, \ldots, q_n - 1\} \longrightarrow G$ (notice that $T\omega$ differs from ω only in finitely many places whenever $\omega \neq -\tau$, so φ is well-defined except for one point).

Set $q_1 = 1$, $q_{n+1} = \prod_{k=1}^{n} a_k$, then

$$(q_n\tau)_k = \begin{cases} 1 & k = n \\ 0 & k \neq n, \end{cases}$$

whence

$$T^{q_n}\omega = (\omega_1, \ldots, \omega_{n-1}, \tilde{\tau}_n + (\omega_n, \ldots))$$

where

$$\tilde{\tau}_n = (1, 0, \ldots) \in \prod_{k=n}^{\infty} \{0, \ldots, a_k - 1\}.$$

Note that

$$\varphi^{(k)}(\omega) := \sum_{j=0}^{k-1} \varphi(T^j\omega) \stackrel{!}{=} \sum_{n=1}^{\infty} [b_n(T^k\omega) - b_n(\omega)],$$

whence

$$\varphi^{(q_k)}(\omega) = \sum_{n=1}^{\infty}[b_n(T^{q_k}\omega) - b_n(\omega)]$$

$$= \sum_{n=0}^{\ell_k(\omega)-1}[\beta_{k+n}(0) - \beta_{k+n}(a_{k+n} - 1)]$$

$$+ \beta_{k+\ell_k(\omega)}(\omega_{k+\ell_k(\omega)} + 1) - \beta_{k+\ell_k(\omega)}(\omega_{k+\ell_k(\omega)}),$$

where

$$\ell_k(\omega) = \min\{n \geq 0 : \omega_{k+n} < a_{k+n} - 1\}.$$

We begin by considering cocycles of form

$$\beta_n(k) = k\lambda_n(:= \underbrace{\lambda_n + \cdots + \lambda_n}_{k \text{ times}}), \quad \text{for } 0 \leq k \leq a_n - 1, \quad \text{where } \lambda_n \in G.$$

Proposition 1.6. *If* $r_n \in I\!N$ *and* $\sum_{n=1}^{\infty} \frac{r_n}{a_n} < \infty$, *then*

$$\{k\lambda_n : n \geq 1, \ 1 \leq k \leq r_n\}' \subset \widetilde{D}(\varphi).$$

Proof. From the condition on $\{r_n\}_{n \in I\!N}$, for a.e. $\omega \in \Omega$

$$\exists \, N_\omega \in I\!N \ni \omega_n < a_n - r_n - 1 \ \forall \, n > N_\omega,$$

whence $\forall \, n \geq N_\omega, \ 0 \leq k \leq r_n$,

$$\varphi^{(kq_n)}(\omega) = \sum_{j=1}^{k} \varphi^{(q_n)}(T^{(j-1)q_n}\omega)$$

$$= \sum_{j=0}^{k-1}\left(\beta_n(\omega_n + j + 1) - \beta_n(\omega_n + j)\right) \quad (\because k < r_n)$$

$$= k\lambda_n$$

and if $k_\nu \lambda_{n_\nu} \to a$, then for a.e. $\omega \in \Omega$,

$$\varphi^{(k_\nu \cdot q_{n_\nu})} \approx k_\nu \lambda_{n_\nu} \to a \text{ a.e},$$

and $a \in \widetilde{D}(\varphi)$. □

Theorem 1.5, and Proposition 1.6 facilitate easy constructions of conservative, ergodic, coalescent, non-squashable G-extensions of odometers.

Example 1.7. There is a continuous $I\!R$-valued cocycle of product type which is a coboundary, and satisfies

$$\overline{\mathrm{Gp}}(C(\varphi)) = I\!R.$$

Proof. Assume that $\sum_{n=1}^{\infty} \frac{1}{a_n} < +\infty$, $a_n \geq 3$. Let

$$\varphi(\omega) = \sum_{n=1}^{\infty}(b_n(T\omega) - b_n(\omega))$$

where, as before, $b_n(\omega) = \beta_n(\omega_n)$. Set $\beta_{2n+1} \equiv 0$, and

$$\beta_{2n}(k) = \begin{cases} \frac{1}{n} & k = 1, \\ 0 & \text{else.} \end{cases}$$

By Borel-Cantelli lemma, since $\mu\{\omega : \omega_{2n} = 1\} = \frac{1}{a_n}$, $\varphi = \psi \circ T - \psi$ with

$$\psi = \sum_{n=1}^{\infty} b_n.$$

Note that $\varphi(-\tau) = 0$ (where $-\tau = (a_1 - 1, a_2 - 1, \ldots)$). For $\omega \neq -\tau$, $\ell(\omega) < \infty$

$$\begin{aligned}
\varphi(\omega) &= \sum_{n=0}^{\ell(\omega)-1} [\beta_n(0) - \beta_n(a_n - 1)] \\
&\quad + \beta_{\ell(\omega)}(\omega_{\ell(\omega)} + 1) - \beta_{\ell(\omega)}(\omega_{\ell(\omega)}) \\
&= \beta_{\ell(\omega)}(\omega_{\ell(\omega)} + 1) - \beta_{\ell(\omega)}(\omega_{\ell(\omega)}),
\end{aligned}$$

since $\beta_n(0) - \beta_n(a_n - 1) = 0$, whence

$$|\varphi(\omega)| \leq \frac{2}{\ell(\omega)}$$

and the continuity of φ is ensured.

For a.e. $\omega \in \Omega$, $\exists\, n_\omega$ such that $2 < \omega_n < a_n - 2 \; \forall\, n > n_\omega$. Set

$$\kappa_n(\omega) = a_{2n} - \omega_{2n}$$

for $n > \frac{n_\omega}{2}$. Clearly $\kappa_n(\omega) q_{2n} \tau \xrightarrow{\Omega} 0$.

Moreover, for $n > \frac{n_\omega}{2}$,

$$(T^{jq_{2n}}\omega)_{2n} = \begin{cases} \omega_{2n} + j & 0 \leq j \leq \kappa_n(\omega) - 1, \\ 0 & j = \kappa_n(\omega), \end{cases}$$

$$(T^{jq_{2n}}\omega)_{2n+1} = \begin{cases} \omega_{2n+1} & 0 \leq j \leq \kappa_n(\omega) - 1, \\ \omega_{2n+1} + 1 & j = \kappa_n(\omega), \end{cases}$$

and

$$(T^{jq_{2n}}\omega)_k = \omega_k \quad \forall\; 0 \leq j \leq \kappa_n(\omega),\ k \neq 2n, 2n+1;$$

whence

$$\begin{aligned}
\varphi^{((\kappa_n(\omega)+1)q_{2n})}(\omega) &= \sum_{k=1}^{\infty} \left(b_k(T^{(\kappa_n(\omega)+1)q_{2n}}\omega) - b_k(\omega) \right) \\
&\quad \sum_{k=1}^{\infty} \left(\beta_k((T^{(\kappa_n(\omega)+1)q_{2n}}\omega)_k) - \beta_k(\omega_k) \right) \\
&= \beta_{2n}((T^{(\kappa_n(\omega)+1)q_{2n}}\omega)_{2n}) - \beta_{2n}(\omega_{2n}) \\
&= \beta_{2n}(1) = \frac{1}{n}.
\end{aligned}$$

We use the fact that

$$\forall\, y > 0,\ N \geq 1,\ \exists\, N < n_k(N) \uparrow \infty \ni \sum_{k=1}^{\infty} \frac{1}{n_k(N)} = y.$$

Now, for fixed ω, y, and $N > \frac{n_\omega}{2}$ choose m_N such that

$$\left| \sum_{k=1}^{m_N} \frac{1}{n_k(N)} - y \right| < \frac{1}{N}$$

and set

$$Q_m^{(N)}(\omega) = \sum_{k=1}^{m} (\kappa_{n_k(N)} + 1)(\omega) q_{2n_k(N)}, \ \& \ Q_N = Q_N(\omega) := Q_{m_N}^{(N)}(\omega).$$

It follows that $Q_N T \xrightarrow{\Omega} 0$ whence $T^{Q_N} \xrightarrow{\mathcal{U}(L^2(m))}$ Id. On the other hand,

$$\varphi^{(Q_N)}(\omega) = \sum_{k=1}^{m_N} \varphi^{((\kappa_{n_k}+1)q_{2n_k})}(T^{Q_{k-1}(N)}\omega) = \sum_{k=1}^{m_N} \frac{1}{n_k(N)} \longrightarrow y.$$

Thus $C(\varphi) \supset \mathbb{R}_+$. With some minor adjustments, $C(\varphi) = \mathbb{R}$ can be arranged. $\qquad \square$

§2 Homogeneous Banach spaces and Koksma inequalities

Definition. By a *pseudo-homogeneous* Banach space on $\mathbb{T}$ we mean a Banach space $(B, \|\cdot\|_B)$ satisfying

(1) $B \subseteq L^1(\mathbb{T})$, and $\|\cdot\|_B \geq \|\cdot\|_1$,

(2) If $f \in B$ and $t \in \mathbb{T}$ then $f_t \in B$, and $\|f_t\|_B = \|f\|_B$, where $f_t(x) = f(x-t), x \in \mathbb{T}$.

A pseudo-homogeneous Banach space on $\mathbb{T}$ is called *homogeneous* if $t \mapsto f_t$ is continuous $\mathbb{T} \longrightarrow B, \forall\, f \in B$.

The following properties of pseudo-homogeneous Banach spaces are either contained in, or can be easily deduced from [Katzn, chapter 1]:

(1) there exists the largest homogeneous Banach subspace B_h contained in B defined by
$$B_h = \{f \in B : t \mapsto f_t \text{ is continuous } \mathbb{T} \to B\};$$

(2) the space B_h is the closure of trigonometric polynomials belonging to B (this is because B_h is homogeneous and hence if $f \in B_h$ and $g \in C(\mathbb{T})$ then the convolution of these two functions is an element of B_h);

(3) if $f \in B$ then $f \in B_h$ iff for each $n \in \mathbb{Z}$ such that $\hat{f}(n) \neq 0$ there exists $g \in B_h$ such that $\hat{g}(n) \neq 0$.

Suppose now that B is a Banach space and T is an isometry on it. Assume also that zero is the only fixed point of T. We say that for an $x \in B$ *the ergodic theorem holds* if

$$B - \lim_{n \to \infty} \frac{1}{n} \sum_{j=0}^{n-1} T^j x = 0.$$

The set of all elements of B for which the ergodic theorem holds is denoted by $ET(B,T)$. An element $x \in B$ is said to be a *(B-)coboundary* if $x = y - Ty$ for some $y \in B$ (called a *transfer* element). The following theorem is a version of the Mean Ergodic Theorem:

Theorem 2.1 (von Neumann). *An element $x \in ET(B,T)$ iff x belongs to the closure of the subspace of B-coboundaries.*

Suppose now that B is a pseudo-homogeneous Banach space on $\mathbb{T}$ (only functions with zero mean are considered). Let T denote an irrational translation by α, then T acts as an isometry on B. Note that if P is a trigonometric polynomial from B then P is a coboundary, in fact we have $P = Q - Q \circ T$, where Q is another trigonometric polynomial, hence $P, Q \in B_h$. This proves

Corollary 2.2

$$B_h \subset ET(B,T).$$

Let

$$\alpha = [0; a_1, a_2, \ldots]$$

be the continued fraction expansion of α. The positive integers a_n are called the *partial quotients* of α. Put

$$q_0 = 1, \quad q_1 = a_1, \quad q_{n+1} = a_{n+1} q_n + q_{n-1} \quad p_0 = 0, \quad p_1 = 1, \quad p_{n+1} = a_{n+1} p_n + p_{n-1}.$$

The rationals p_n/q_n are called the *convergents* of α and the inequality

$$\left| \alpha - \frac{p_n}{q_n} \right| < \frac{1}{q_n q_{n+1}}$$

holds. A *denominator* q_n is said to be a *Legendre* denominator if $\left| \alpha - \frac{p_n}{q_n} \right| < \frac{1}{2q_n^2}$. We'll sometimes denote the set of Legendre denominators of α by $\mathcal{L}(\alpha)$.

Note that if $q \in \mathcal{L}(\alpha)$ is a Legendre denominator then

(2.1) $$\| j\alpha - j'\alpha \| > \frac{1}{2q} \quad \text{whenever} \quad 0 \le j \ne j' \le q - 1.$$

Here, for $t \in \mathbb{R}$,

$$\| t \| = d(t, \mathbb{Z}) = \min_{n \in \mathbb{Z}} |n - t|.$$

We recall that one of any two consecutive denominators of an irrational α must be a Legendre denominator i.e. ($\forall \alpha \notin \mathbb{Q}$, $n \ge 1$), $\{q_n, q_{n+1}\} \cap \mathcal{L}(\alpha) \ne \emptyset$.

Let B be a pseudo-homogeneous Banach space on $\mathbb{T}$. We say that *Koksma's inequality* holds for the pair (B,T) provided that there exists a positive sequence

$\tilde{D}_N = \tilde{D}_N(\alpha)$, $N \geq 1$, satisfying $\tilde{D}_{q_n} = O(1/q_n)$ where $\{q_n\}$ is the sequence of denominators of α and

(2.2) $$\left\| \frac{1}{N} f^{(N)}(\cdot) - \int_0^1 f(t)dt \right\|_{L^1} \leq \|f\|_B \tilde{D}_N(\alpha) \quad \forall\, f \in B,$$

where $f^{(N)}(x) = \sum_{j=0}^{N-1} f(T^j x)$, $x \in \mathbb{T}$. For the classical cases where Koksma inequality is satisfied for functions with bounded variation or Lipschitz continuous functions we refer to [Ku-Ni], chapter 2.

The proposition below (essentially due to M. Herman, [He], p.189) will play a role in the proofs of ergodicity of certain cylinder flows.

Proposition 2.3. *If Koksma's inequality is satisfied for the pair (B,T) then for each $f \in B_h$ with $\int_0^1 f(t)\,dt = 0$ we have*

$$\lim_{n \longrightarrow \infty} f^{(q_n)} = 0 \quad in \ L^1(\mathbb{T}).$$

Proof. Denote by B_0 the subspace of B consisting of functions with zero mean. Then define a map $S : B_0 \longrightarrow l^\infty$ by

$$Sg = (\|g^{(q_n)}\|_{L^1})_{n \geq 1}.$$

Note that by the Koksma inequality, S is well-defined and continuous. Hence, the set $S^{-1}(c_0)$ is closed as c_0 is a closed subspace of l^∞. Each coboundary $f = h - hT$, $h \in B$ is in $S^{-1}(c_0)$ since for each function $u \in L^1(\mathbb{T})$ we have

(2.3) $$uT^{q_n} \longrightarrow u \quad in \ L^1(\mathbb{T}).$$

It follows from this, theorem 2.1 and corollary 2.2, that

$$B_h \subset ET(B,T) = \overline{\{h - h \circ T : h \in B\}} \subset S^{-1}(c_0).$$

$\square$

We will now pass to a proof of Koksma's inequality in the space $B = O(1/n)$ (of functions whose Fourier coefficients are of order $O(1/n)$), where the norm is defined as $\|f\|_B = \|f\|_{L^1} + \sup_{n \neq 0} |n\hat{f}(n)|$. If $\{x_1, \ldots, x_N\}$ is a finite set of points from $[0, 1)$ then by *discrepancy* $D_N = D_N(x_1, \ldots, x_N)$ we mean

$$D_N = \sup_{x < y}\{|\frac{\#\{1 \leq j \leq N\, x_j \in [x, y)\}}{N} - (y - x)|\}.$$

Lemma 2.4.
$$\sup_x \#\{1 \leq j \leq N\, x_j \in [x, x + \frac{1}{N})\} \leq ND_N + 1.$$

Proof. For an arbitrary $x \in [0, 1)$,

$$|\frac{\#\{1 \leq j \leq N\, x_j \in [x, x + \frac{1}{N})\}}{N} - (x + \frac{1}{N} - x)| \leq D_N,$$

whence the assertions follows immediately. □

Lemma 2.5. *There exists* $C > 0$ *such that*
$(\forall m \geq 1)(\forall a \geq 1)(\forall x_1, \ldots, x_{m-1} \in [0,1))$ *if in each interval of length* $\frac{1}{m}$: *there are at most* a *points of the form* x_i *then* $\sum_{\{i : x_i \in (\frac{1}{2m}, 1 - \frac{1}{2m})\}} \frac{1}{\|x_i\|^2} \leq Cam^2$.

Proof. Denote by I the set of those $1 \leq i \leq m-1$ so that $x_i \in (\frac{1}{2m}, 1 - \frac{1}{2m})$. Then define a map $i \mapsto j(i)$, $i \in I, 1 \leq j(i) \leq m-1$, by

$$(2.4) \qquad \left| x_i - \frac{j(i)}{m} \right| \leq \frac{1}{2m}.$$

Since $\|x_i\| > \frac{1}{2m}$,

$$(2.5) \qquad \frac{1}{2} \leq \frac{\|x_i\|}{\|\frac{j(i)}{m}\|} \leq 2.$$

Note that if k is in the image of the function j then

$$\# j^{-1}(k) \leq a$$

by our assumption and (2.4). Hence by (2.5)

$$\sum_{i \in I} \frac{1}{\|x_i\|^2} \leq 2a \sum_{k \in \operatorname{Im} j} \frac{1}{\|k/m\|^2} \leq 4a \sum_{i=1}^{m-1} \frac{1}{(i/m)^2} = Cam^2.$$

□

Combining this with Lemma 2.4, we obtain

Corollary 2.6. *Under the conditions of lemma 2.5,*

$$\sum_{i \in I} \frac{1}{\|x_i\|^2} \leq C(mD_m + 1)m^2,$$

where I *is the same as in the proof of Lemma 2.5.*

Now, suppose that $f \in O(\frac{1}{n})$,

$$f(x) = \sum_{k=-\infty}^{\infty} \hat{f}_k e^{2\pi i k x}.$$

We have

$$f^{(m)}(x) = \sum_{i=0}^{m-1} f(x + i\alpha) = f^{(m)}(x) = \sum_{k=-\infty}^{\infty} \hat{f}_k \frac{e^{2\pi i k m \alpha} - 1}{e^{2\pi i k \alpha} - 1} e^{2\pi i k x}.$$

Theorem 2.7 (Koksma's Inequality in $O(\frac{1}{n})$)*. There is a constant* $K > 0$ *such that if we denote*

$$\tilde{D}_m = \sqrt{K \left(\sum_{k \in A_m} \frac{1}{k^2} + (mD_m + 1)(\|m\alpha\|^2 + \frac{1}{m^2}) \right)}$$

then $\forall\, f \in O(\frac{1}{n})$,

$$\|\frac{1}{m}\sum_{i=0}^{m-1} f(\cdot + i\alpha) - \int_0^1 f(t)dt\|_{L^2}^2 \le \|f\|^2 O(\frac{1}{n})\tilde{D}_m,$$

where
$D_m = D_m(0, \alpha, 2\alpha, \dots, (m-1)\alpha)$, and $A_m = \{0 \le j \le m-1 : 0 < \|j\alpha\| < \frac{1}{2m}\}$.
Moreover,

$$\tilde{D}_{q_n} = O(1/q_n).$$

Proof. Without loss of generality we will assume that $\int_0^1 f(t)\,dt = 0$ and it is enough to prove that

$$(2.6)\ \ \|f^{(m)}\|_{L^2}^2 \le C_2\|f\|^2 O(\frac{1}{n})(m^2\sum_{k\in A_m}\frac{1}{k^2} + C(mD_m + 1)m^2\|m\alpha\|^2 + C_3(mD_m + 1)),$$

where C_2, C, C_3 are some absolute constants. Since f is real,

$$\|f^{(m)}\|_{L^2}^2 \le 2C_1\sum_{k=1}^{\infty}|\hat{f}_k|^2\frac{\|km\alpha\|^2}{\|k\alpha\|^2} = C_2(S_1 + S_2),$$

where

$$S_1 = \sum_{k=1}^{m-1}\frac{|\hat{f}_k|^2\|km\alpha\|^2}{\|k\alpha\|^2}, \quad S_2 = \sum_{k=m}^{\infty}\frac{|\hat{f}_k|^2\|km\alpha\|^2}{\|k\alpha\|^2}.$$

Now,

$$S_1 = \sum_{k=1}^{m-1}\frac{|\hat{f}_k k|^2\|km\alpha\|^2}{k^2\|k\alpha\|^2} \le \|f\|^2 O(\frac{1}{n})\sum_{k=1}^{m-1}\frac{\|km\alpha\|^2}{k^2\|k\alpha\|^2} = \|f\|^2 O(\frac{1}{n})(S_{11} + S_{12}),$$

where

$$S_{11} = \sum_{k\in A_m}\frac{\|km\alpha\|^2}{k^2\|k\alpha\|^2}, \quad S_{12} = \sum_{k\notin A_m}\frac{\|km\alpha\|^2}{k^2\|k\alpha\|^2}.$$

We have, $S_{11} \le m^2\sum_{k\in A_m}\frac{1}{k^2}$, and $S_{12} \le \|m\alpha\|^2\sum_{k\notin A_m}\frac{1}{\|k\alpha\|^2}$.
By Corollary 2.6,

$$S_{12} \le \|m\alpha\|^2 C(mD_m + 1)m^2.$$

We pass now to estimate S_2. We have

$$S_2 = \sum_{k=m}^{\infty}\frac{|\hat{f}_k|^2\|km\alpha\|^2}{\|k\alpha\|^2} = \sum_{p=1}^{\infty}\sum_{r=0}^{m-1}\frac{|\hat{f}_{pm+r}|^2\|(pm+r)m\alpha\|^2}{\|(pm+r)\alpha\|^2} \le$$

$$\|f\|^2 O(\frac{1}{n})\sum_{p=1}^{\infty}\frac{1}{p^2}\sum_{r=0}^{m-1}\frac{\|(pm+r)m\alpha\|^2}{m^2\|(pm+r)\alpha\|^2} \le$$

$$\frac{1}{m^2}\|f\|^2 O(\frac{1}{n})\sum_{p=1}^{\infty}\frac{1}{p^2}\sum_{r=0}^{m-1}\min(m^2, \frac{1}{\|pm\alpha + r\alpha\|^2}).$$

Denote $x = pm\alpha$. In the interval $(-\frac{1}{2m}, \frac{1}{2m}) = [0, \frac{1}{2m}) \cup [1 - \frac{1}{2m}, 1)$ (mod 1) we have at most $mD_m + 1$ points of the form $x + r\alpha$ because $D_m = D_m(x, x+\alpha, \ldots, x+(m-1)\alpha)$. By Corollary 2.6 we thus have

$$S_2 \leq \frac{1}{m^2}\|f\|^2_{O(\frac{1}{n})} \sum_{p=1}^{\infty} \frac{1}{p^2}((mD_m + 1)m^2 + C(mD_m + 1)m^2) \leq C_3\|f\|^2_{O(\frac{1}{n})}(mD_m + 1).$$

To complete the proof we have to show that the sequence $\{q_n \check{D}_{q_n}\}$ is bounded. But classically, $D_{q_n} = O(1/q_n)$ and also $q_n\|q_n\alpha\|$ is bounded. Now, note that in the interval $M_n = [0, \frac{1}{2q_n}) \cup [1 - \frac{1}{2q_n}, 1)$ we can have at most one point of the form $j\alpha$, where $j = 1, \ldots, q_n - 1$. Moreover, $|j\alpha - j\frac{p_n}{q_n}| < \frac{j}{q_n q_{n+1}}$, so if $j\alpha \in M_n$ then we must have $\frac{j}{q_n q_{n+1}} > \frac{1}{2q_n}$. In particular, $j > q_n/2$, so $\sum_{k \in A_{q_n}} \frac{1}{k^2} = O(1/q_n^2)$.

$\square$

Now, proceeding as in the proof of Proposition 2.3, we obtain the following extension of the main result from [Le-Ma]

Corollary 2.8. *If* $f \in o(\frac{1}{n})$, $\int_0^1 f(t)\,dt = 0$ *and* $\{q_n\}$ *is the sequence of all denominators of* α *then*

$$\|f^{(q_n)}\|_{L^2} \longrightarrow 0.$$

§3 Speed of approximation in Koksma's Inequality for spaces $O(1/a(n))$

Assume that $a : \mathbb{N} \longrightarrow \mathbb{R}^+$ satisfies

$$(3.1) \qquad\qquad a(k) \geq k,$$

$$(3.2) \qquad a(pm + r) \geq a(p)a(m), \quad \text{for arbitrary} \quad p, m \geq 1, r = 0, \ldots, m - 1.$$

We will now concentrate on a pseudo-homogeneous Banach space $B = O(1/a(n))$ of functions

$$f(x) = \sum_{k=-\infty}^{\infty} \hat{f}_k e^{2\pi i k x},$$

with $\hat{f}_k = O(1/a(k))$. The norm is defined as

$$\|f\|_{O(1/a(n))} = \|f\|_{L^1} + \sup_{n \neq 0} |a(n)\hat{f}_n|.$$

Notice that in this case $B_h = o(1/a(n))$ the subspace of functions whose Fourier coefficients are of order $o(1/a(n))$. Keeping the notation from the proof of Theorem 2.7 and proceeding as before we obtain that

$$S_1 \leq \|f\|^2_{O(1/a(n))}(S_{11} + S_{12}),$$

where

$$S_{11} = m^2 \sum_{k \in A_m} \frac{1}{a(k)^2},$$

and by (3.1)

$$S_{12} \leq \|m\alpha\|^2 \sum_{k \notin A_m} \frac{k^2}{a(k)^2} \frac{1}{\|k\alpha\|^2} \leq \|m\alpha\|^2 m^2 (D_m m + 1) \cdot C.$$

In view of (3.2),

$$S_2 \leq \|f\|^2 O_{(1/a(n))} \sum_{p=1}^{\infty} \frac{1}{a(p)^2} \sum_{r=0}^{m-1} \frac{\|(pm+r)m\alpha\|^2}{a(m)^2 \|(pm+r)\alpha\|^2} \leq$$

$$\frac{1}{a(m)^2} \|f\|^2 O_{(1/a(n))} m^2 C_4 (m D_m + 1) \sum_{p=1}^{\infty} \frac{1}{a(p)^2} \leq (\frac{m}{a(m)})^2 \|f\|^2 O_{(1/a(n))} (m D_m + 1) C_5.$$

For a function $a(\cdot)$ satisfying (3.1) and (3.2) denote

$$I(a) = \{\alpha \in [0,1) \setminus \mathbb{Q} : \liminf_{q \to \infty, \, q \in \mathcal{L}(\alpha)} a(q)\|q\alpha\| < \infty\}.$$

Lemma 3.1. If $f = gT - g$, $g \in O(1/a(n))$, $\alpha \in I(a)$ and $q_{n_k} \in \mathcal{L}(\alpha)$ with $a(q_{n_k})\|q_{n_k}\alpha\| = O(1)$, then

$$\|f^{(q_{n_k})}\|_{L^2} = o(\frac{q_{n_k}}{a(q_{n_k})}).$$

Proof. All we need to show is that $\sum_{s=1}^{\infty} |\hat{g}_s|^2 \|q_{n_k} s\alpha\|^2 = o((\frac{q_{n_k}}{a(q_{n_k})})^2)$. We have

$$\sum_{s=1}^{\infty} |\hat{g}_s|^2 \|q_{n_k} s\alpha\|^2 \leq \|g\|^2 O_{(1/a(n))} \left(\sum_{s=1}^{q_{n_k}-1} \frac{\|q_{n_k} s\alpha\|^2}{a(s)^2} + \sum_{s=q_{n_k}}^{\infty} \frac{\|q_{n_k} s\alpha\|^2}{a(s)^2} \right) \leq$$

$$\|g\|^2 O_{(1/a(n))} (q_{n_k} \|q_{n_k}\alpha\|^2 + q_{n_k} \sum_{p=1}^{\infty} \frac{1}{(a(p)a(q_{n_k}))^2}) =$$

$$\|g\|^2 O_{(1/a(n))} (\frac{q_{n_k}}{a(q_{n_k})^2} a(q_{n_k})^2 \|q_{n_k}\alpha\|^2 + \frac{q_{n_k}}{a(q_{n_k})^2} \sum_{p=1}^{\infty} \frac{1}{a(p)^2}) = o((\frac{q_{n_k}}{a(q_{n_k})})^2).$$

$\square$

Corollary 3.2. If $f \in O(1/a(n))$, $\int_0^1 f(t)\,dt = 0$ and $\alpha \in I(a)$ and $q_{n_k} \in \mathcal{L}(\alpha)$ with $a(q_{n_k})\|q_{n_k}\alpha\| = O(1)$, then

$$\|f^{(q_{n_k})}\|_{L^2} \leq const.\|f\|_{O_{(1/a(n))}} \frac{q_{n_k}}{a(q_{n_k})}.$$

Moreover, if in addition $f \in o(\frac{1}{a(n)})$ then

(3.3) $$\|f^{(q_{n_k})}\|_{L^2} = o(\frac{q_{n_k}}{a(q_{n_k})}).$$

Proof. Since (3.3) is satisfied for all coboundaries by Lemma 3.1, the mechanism described in the proof of Proposition 2.3 works well. The map S is defined as $Sf = (\frac{a(q_{n_k})}{q_{n_k}} \|f^{(q_{n_k})}\|_{L^2})_{k \geq 1}$.

$\square$

Suppose now that $a(n) = \frac{1}{n^t}$ for certain natural number $t \geq 1$. Hence $I(a) =: I(t)$ is the set of those irrationals α for which $(q_{n_k}^t \|q_{n_k}\alpha\|)$ is bounded for certain subsequence of Legendre denominators of α.

Corollary 3.3. *If $f \in o(\frac{1}{n^t})$, $\int_0^1 f d\lambda = 0$ then for an arbitrary $\alpha \in I(t)$ and $q_{n_k} \in \mathcal{L}(\alpha)$ with $q_{n_k}^t \|q_{n_k}\alpha\| = O(1)$, we have*

(i) $\|f^{(q_{n_k})}\|_{L^2} = o(\frac{1}{q_{n_k}^{t-1}})$,

(ii) *the sequence $(q_{n_k}^t)$ is a rigidity time for α and*

$$\lim_{k \to \infty} f^{(q_{n_k}^t)} = 0 \quad in \quad L^2(\mathbb{T}).$$

Proof. It is enough to notice that $f^{(q_{n_k}^t)} = f^{(q_{n_k} q_{n_k}^{t-1})}$ and that $\|f^{(q_{n_k} q_{n_k}^{t-1})}\|_{L^2} \leq q_{n_k}^{t-1} \|f^{(q_{n_k})}\|_{L^2}$. $\qquad\square$

§4 Constructions of ergodic analytic cylinder flows

Constructions which are known of ergodic cylinder flows are rather based on some irregularities in the smoothness of the cocycle (e.g. [He-La1], [He-La2], [Pa1], [Pa2], [Ba-Me1], [Ba-Me2]). Below, we will show a new method coming from [Kw-Le-Ru2] for constructing analytic cylinder flows which are ergodic.

Assume that $Tx = x + \alpha$, where $\alpha = [0; a_1, a_2, \ldots]$. From the continued fraction expansion of α we obtain, for each n, two Rokhlin towers $\xi_n, \overline{\xi}_n$ whose union coincides with the whole circle. For n even

$$\xi_n = \{[0, \{q_n\alpha\}), T[0, \{q_n\alpha\}), \ldots, T^{(a_{n+1}q_n + q_{n-1})-1}[0, \{q_n\alpha\})\},$$

$$\overline{\xi}_n = \{[\{q_{n+1}\alpha\}, 1), \ldots, T^{q_n-1}[\{q_{n+1}\alpha\}, 1)\}.$$

Given a subsequence $\{n_k\}$ of natural numbers we will denote

$$I_k = [0, \{a_{2n_k+1}q_{2n_k}\alpha\}), \quad J_t^k = T^{(t-1)q_{2n_k}}(0, \{q_{2n_k}\alpha\}],$$

$t = 1, \ldots, a_{2n_k+1}$. Notice that

$$I_k = \bigcup_{t=1}^{a_{2n_k+1}} J_t^k,$$

and

(4.1) $$|J_1^k| < \frac{1}{a_{2n_k+1}q_{2n_k}}.$$

We will recall here a notion of an a.a.c.c.p. (almost analytic cocycle construction procedure) from [Kw-Le-Ru2] which is to construct a real 1-periodic cocycle $\tilde{\varphi} : \mathbb{R} \longrightarrow \mathbb{R}$ such that in its $\mathbb{R}$–cohomology class (for certain α) there is an analytic cocycle.

An a.a.c.c.p. is given by a collection of parameters as follows. We are given a sequence $\{M_k\}$ of natural numbers and an array $\{(d_{k,1}, \ldots, d_{k,M_k})\}$, $d_{k,i} \in \mathbb{R}$ satisfying for each k

$$\sum_{i=1}^{M_k} d_{k,i} = 0.$$

Denote $D_k = \max_{1 \leq i \leq M_k} |d_{k,i}|$. Choose a sequence $\{\varepsilon_k\}$ of positive real numbers satisfying

$$\sum_{k=1}^{\infty} \sqrt{\varepsilon_k} M_k < +\infty,$$

$$\sum_{k=1}^{\infty} \varepsilon_k < 1,$$

$$\varepsilon_k < \frac{1}{D_k^2}, \quad k = 1, 2, \ldots.$$

Finally, we are given $A > 1$ completing the parameters of the a.a.c.c.p.

We say that this a.a.c.c.p. *is realized over an irrational number α with continued fraction expansion* $[0; a_1, a_2, \ldots]$ and convergents $p_n/q_n, n \geq 1$ if there exists a strictly increasing sequence $\{n_k\}$ of natural numbers such that

$$A^{N_k} \frac{D_k M_k \|P_k\|_{\mathcal{F}}}{a_{2n_k+1} q_{2n_k}} < \frac{1}{2^k}$$

and

$$\frac{D_k \|P_k'\|_{\infty}}{a_{2n_k+1} q_{2n_k}} < \sqrt{\varepsilon_k},$$

where $\{P_k\}$ is a sequence of "bump" real trigonometric polynomials, *i.e.*

(i) $\int_0^1 P_k(t)dt = 1$,
(ii) $P_k \geq 0$,
(iii) $P_k(t) < \varepsilon_k$ for each $t \in (\eta_k/2, 1)$,

where the η_k's are chosen in such a way that

(4.2) $$4 M_k \eta_k < \frac{\varepsilon_k}{q_{2n_k}}$$

and N_k is the degree of P_k. Finally, $a_{2n_k+1} > 1$ and

(4.3) $$\frac{1}{a_{2n_k+1} q_{2n_k}} < \frac{1}{2}\eta_k$$

Using the above parameters define a cocycle

$$\varphi = \sum_{k=1}^{\infty} \varphi(k)$$

as follows. In view of (4.2),(4.3) (and (4.1)), in the interval $I_k = [0, \{a_{2n_k+1} q_{2n_k} \alpha\})$ we can choose $w_{k,1}, \ldots, w_{k,M_k}$ to be consecutive pairwise disjoint intervals of the same length contained between η_k and $2\eta_k$ such that each $w_{k,i}$ consists of say e_k consecutive subintervals J_t^k, where e_k is an odd number. Let $J_{s_{k,i}}^k$ be the central subinterval in $w_{k,i}$ and now define

$$\varphi(k)(x) = \begin{cases} d_{k,i} & \text{if } x \in J_{s_{k,i}}^k \\ 0 & \text{otherwise.} \end{cases}$$

Note that the $\varphi(k)$'s have disjoint supports so φ is well defined.

As proved in [Kw-Le-Ru2]

(A) The set of α's over which an a.a.c.c.p. is realized is a G_δ and dense subset of the circle.

(B) If an a.a.c.c.p. is realized over α then there exists an analytic cocycle $f : \mathbb{T} \longrightarrow \mathbb{R}$ which is α–cohomologous to φ.

We will need an additional property of an a.a.c.c.p. which is not explicitly formulated in [Kw-Le-Ru2]. Namely,

(4.4) $\varphi|_{T^s I_k}$ is constant for $s = 1, \ldots, q_{2n_k} - 1$, & $\displaystyle\sum_{s=1}^{q_{2n_k}-1} \varphi|_{T^s I_k} = 0$

which is Lemma 3 from [Kw-Le-Ru1].

Example 4.1. There is an a.a.c.c.p. with $\mathrm{Gp}(\widetilde{D}(\varphi)) = E(\varphi) = \mathbb{Z}\lambda$.

Proof. Assume that $\lambda \in \mathbb{R}$ is given. We will assume that an a.a.c.c.p. satisfies the following additional requirements:

$$a_{2n_k+1} = M_k r_k + N_k,$$

with $0 \le N_k < r_k$ and both M_k, r_k tending to infinity. We put $d_{k,1} = 0, d_{k,i} = \lambda$ for $i = 2, \ldots, M_k - 1$ and $d_{k,M_k} = -(M_k - 1)\lambda$. In the definition of φ_k we require that $\varphi_k|_{J^k_{ir_k+1}} = d_{k,i}$ for $i = 0, \ldots, M_k - 1$ and zero for all others subintervals J^k_t, $k \ge 1$.

Notice that $E(\varphi) \subset \mathbb{Z}\lambda$ because the values of φ are from the group $\mathbb{Z}\lambda$. It is then enough to show that $\lambda \in \widetilde{D}(\varphi)$. Define

$$X_k = \bigcup_{s=0}^{q_{2n_k}-1} \bigcup_{t=r_k+1}^{(M_k-1)r_k} T^s J^k_t.$$

By our definition of φ and a basic property of an a.a.c.c.p. (see (4.4)) we have $\varphi^{(M_k r_k)}(x) = \lambda$ for all $x \in X_k$. It is clear also that $M_k r_k$ is a rigidity time for T. Therefore $\lambda \in \widetilde{D}(\varphi)$.   □

Example 4.2. An a.a.c.c.p. with $\overline{\mathrm{Gp}(\widetilde{D}(\varphi))} = \mathbb{R}$.

This is an obvious modification of the previous construction. We divide the sequence $\{n_k\}$ into two disjoint subsequences say $\{n^i_k\}_k$ ($i = 1, 2$) and repeat the previous construction for rationally independent $\lambda_1, \lambda_2 \in \mathbb{R}$, with the sequences $\{n^i_k\}$, $i = 1, 2$. From the previous arguments we find $\lambda_1, \lambda_2 \in \widetilde{D}(\varphi)$. The group generated by λ_1, λ_2 is dense in $\mathbb{R}$ and the advertised condition is attained.

Remark It follows from proposition 1.5 that the cocycles of example 4.2 are ergodic, coalescent, and nonsquashable.

§5 Ergodicity of smooth cylinder flows. Generic point of view

Suppose that $f : \mathbb{T} \to \mathbb{R}$ is smooth. We shall prove that under certain assumptions, the set of those irrational translations for which the corresponding cylinder flow is ergodic is residual. For similar results see [Kr], [Ka].

Assume that $f(x) = \sum_{n=-\infty}^{\infty} b_n e^{2\pi i n x}$ with zero mean is in $A(\mathbb{T})$, that is its Fourier transform is absolutely summable. Put $f_m(x) = f(x) + f(x + \frac{1}{m}) + \ldots + f(x + \frac{m-1}{m}) = m \sum_{l=-\infty}^{\infty} b_{lm} e^{2\pi i l m x}$, $m = 1, \ldots$.

Theorem 5.1. *Suppose that there exist an infinite subsequence $\{q_n\}$ and a constant $C > 0$ such that*

(1) $q_n \sum_{l=-\infty}^{\infty} |b_{l q_n}| \leq C \|f_{q_n}\|_{L^2}$, $n = 1, 2, \ldots$,

(2) $0 < \|f_{q_n}\|_{L^2} \to 0$,

then there exists a dense G_δ set of irrational numbers α such that the corresponding cylinder flow T_f, $Tx = x + \alpha$ is ergodic.

Proof. We will need the following

Lemma 5.2. *Given $C > 0$ there exist positive numbers K, L, M such that $0 < K < 1 < L$, $0 < M < 1$ and for each $h \in L^4(\mathbb{T})$ if $\|h\|_4 \leq C \|h\|_2$, then*

$$\mu\{x \in \mathbb{T} : K\|h\|_2 \leq |h(x)| \leq L\|h\|_2\} > M$$

We will prove the lemma later. Denote

$$g_n(x) = q_n \sum_{l=-\infty}^{\infty} b_{l q_n} e^{2\pi i l x}.$$

In view of (1) we have that

(5.1) $$g_n(q_n x) = f_{q_n}(x), \quad x \in \mathbb{T}$$

and

$$\|g_n\|_{L^\infty} \leq q_n \sum |b_{l q_n}| \leq C\|g_n\|_{L^2},$$

in particular, $\|g_n\|_4 \leq C\|g_n\|_2$. Hence by Lemma 5.2

$$\mu\{x \in \mathbb{T} : K\|g_n\|_2 \leq |g_n(x)| \leq L\|g_n\|_2\} > M.$$

By (2) we have $\|g_n\|_2 = \|f_{q_n}\|_2 \to 0$.

Let $\{D_n\}$ be a family of pairwise disjoint closed intervals, $D_n = [c_n, d_n]$, with

$$d_n/c_n = 100\frac{L}{K} \quad \text{and} \quad d_n \to 0.$$

Assume that $\{D_n'\}$ is a sequence of the above intervals with the property that each D_n repeats infinitely many times in $\{D_n'\}$.

Now, fix n, that is we have the interval D_n'. Choose a natural number k_n so that for some natural s_n

$$[s_n K\|g_{k_n}\|_{L^2}, s_n L\|g_{k_n}\|_{L^2}] \subset \tilde{D}_n',$$

where $\tilde{D}_n'$ is a strict subinterval of D_n'. This gives us a subsequence $\{k_n\}$. For it we have that

$$\mu\{x \in \mathbb{T} : |s_n g_{k_n}(x)| \in \tilde{D}_n'\} \geq M.$$

From this and (5.1) we obtain that for each interval I of length being a multiple of $\frac{1}{q_{k_n}}$

(5.2) $$\mu\{x \in I : |s_n f_{q_{k_n}}(x)| \in \tilde{D}'_n\} \geq M|I|.$$

We will also use the following lemma whose proof is contained in [Kw-Le-Ru2].

Lemma 5.3. *Given an infinite set $\{Q_n\}$ of natural numbers and a positive real valued function $\delta = \delta(Q_n)$ the set*

$$\mathcal{A} = \{\alpha \in [0,1) : \#\{n : \exists\, P_n \ni \frac{P_n}{Q_n} \text{ a convergent of } \alpha, \& \left|\alpha - \frac{P_n}{Q_n}\right| < \delta(Q_n)\} = \infty\}$$

is a dense G_δ.

Let us fix r. So we have infinitely many $n = n(r)$ with $D'_n = D_r$. Consider now those α which are approximated by $\frac{p_{k_{n(r)}}}{q_{k_{n(r)}}}$ so well to have

$$\|s_{n(r)}q_{k_{n(r)}}\alpha\| \to 0$$

and

(5.3) $$\mu\{x \in I : |f^{(s_{n(r)}q_{k_{n(r)}})}(x)| \in D'_{n(r)}\} \geq \frac{M}{2}|I|$$

for each interval I with $|I| = \frac{t}{q_{k_{n(r)}}}, t = 1, \ldots, q_{k_{n(r)}}$ (remember that we know the modulus of continuity of f and that

$$\sum_{i=0}^{s-1}\left(\sum_{j=0}^{q-1} f(x + \frac{j}{q}) - \sum_{k=0}^{q-1} f(x + iq\alpha + k\alpha)\right) =$$

$$\sum_{i=0}^{s-1}\left(\sum_{k=0}^{q-1} f(x + k\frac{p}{q}) - f(x + iq\alpha + k\alpha)\right) \leq \sum_{i=0}^{s-1}\sum_{k=0}^{q-1} \omega(f, iq\alpha + k(\alpha - \frac{p}{q})),$$

where $\gcd(p,q) = 1$, $p = p_{k_{n(r)}}, q = q_{k_{n(r)}}$ and $\omega(f,h)$ stands for the modulus of the continuity of f; now given s, q the size of the above quantity depends on the distance between α and $\frac{p}{q}$.)

In view of Lemma 5.3 we have a G_δ and dense subset of α, say Y_r, for which (5.3) holds true for an infinite subsequence of $\{q_{k_{n(r)}}\}$. Finally take

$$Y = \bigcap_{r=1}^{\infty} Y_r$$

which is G_δ and dense. If we take $\alpha \in Y$ then for each r we have an infinite subsequence $n(\alpha)$ such that

$$\mu\{x \in I : |f^{(s_{n(\alpha)}q_{k_{n(\alpha)}})}(x)| \in D'_{n(\alpha)}\} \geq \frac{M}{2}|I|$$

for each interval I with $|I| = \frac{t}{q_{k_{n(r)}}}$ and $D'_{n(\alpha)} = D_r$.

It remains to prove that if $Tx = x + \alpha$, where $\alpha \in Y$ then the cylinder flow T_f is ergodic. Suppose that $E(f) = \lambda \mathbb{Z}$. Choose r so big to have that the compact set $K_r := D_r \cup (-D_r)$ is disjoint with $\lambda \mathbb{Z}$. By Lemma 1.2 there exists a Borel set B, with $\mu(B) > 0$ such that for all $m \geq 1$

$$(5.4) \qquad \mu(B \cap T^{-m} B \cap \{x \in \mathbb{T} : f^{(m)}(x) \in K_r\}) = 0.$$

If $m = s_n q_{k_n}$, $n = n(\alpha)$, then $\mu(B \triangle T^{s_n q_{k_n}} B) \to 0$ since $s_n q_{k_n}$ is a rigidity time for T. If y is a density point of B then for an interval I of length t/q_{k_n} containing y we will have $\mu(B \cap I) > (1 - \frac{M}{4})|I|$. Hence a subset A_n of B of measure at least $\frac{M}{4}\mu(B)$ has the property that $f^{(s_n q_{k_n})}(x) \in K_r$ whenever $x \in A_n$. This contradicts (5.4). □

Proof of Lemma 5.2. It is enough to consider the case $\|h\|_2 = 1$. Take two real numbers K, L satisfying $0 < K < 1 < L$. From Tchebycheff inequality we have

$$\mu\{|h| \leq L\} \geq \mu\{\|h|^2 - 1| \leq L^2 - 1\} \geq 1 - \operatorname{Var}(|h|^2)(L^2 - 1)^{-2} \geq 1 - (C^4 - 1)(L^2 - 1)^{-2}.$$

On the other hand, from Cauchy-Schwartz inequality

$$1 = \int_{\{|h|>K\}} h^2 + \int_{\{|h|\leq K\}} h^2 \leq (\int h^4)^{1/2}(\mu\{|h| > K\})^{1/2} + K^2;$$

whence $\mu\{|h| > K\} \geq (1 - K^2)^2/C^4$. Now to have the conclusion of the lemma it is enough to choose $\varepsilon > 0$, put $M = 1/C^4 - 2\varepsilon$, then find K small enough to have $(1 - K^2)^2/C^4 > M + \varepsilon$ and finally select L sufficiently big to have $(C^4 - 1)(L^2 - 1)^{-2} < \varepsilon$. □

Remarks

As shown in [Kw-Le-Ru2], the assumptions of Theorem 5.1 are satisfied for each zero mean function $f \in C^{1+\delta}(\mathbb{T})$, $\delta > 0$ which is not a trigonometric polynomial. Recall that a subset $E \subset \mathbb{Z}$ is called *of type* $\Lambda(2)$ if for every $q \geq 2$ there exists a constant $C = C(q, E)$ such that for every function $h \in L^q(\mathbb{T})$ we have $\|h\|_q \leq C\|h\|_2$ whenever $\operatorname{supp}(\hat{h}) \subset E$. For instance, each lacunary subset is of that type ([Katzn], Chapter 5.). Now, if $f \in L^2(\mathbb{T})$ with the absolutely summable Fourier transform has the property that the support of its Fourier transform is an *infinite* $\Lambda(2)$ type set and moreover that $\hat{f}(n) = o(1/n)$ then the assumptions of Theorem 5.1 are also satisfied.

§6 Ergodicity of a class of cylinder flows

This section will be devoted to a generalization of a result of Pask [Pa1].

A function $f : \mathbb{T} \to \mathbb{R}$ is called *piecewise linear (piecewise absolutely continuous)* if there are points $x_1 < x_2 < \ldots < x_K$ such that f restricted to $[x_j, x_{j+1})$ is linear (absolutely continuous), $j = 1, 2, \ldots$ (mod K). Denote by d_j the jump of the values of f at x_j. It is clear that if f is piecewise absolutely continuous then

$$\int_0^1 f'(t)\,dt = \sum_{j=1}^{K} d_j.$$

Lemma 6.1. *Suppose that* $f : \mathbb{T} \to \mathbb{R}$, $\int_0^1 f(t)\,dt = 0$ *is piecewise linear, and* $\sum_{j=1}^K d_j \neq 0$, *then for each irrational number* α *the corresponding cylinder flow* T_f *is ergodic.*

Proof. There is no loss of generality in assuming that $\sum_{j=1}^K d_j > 0$. Since f' is Riemann integrable, the ergodic theorem holds uniformly, so

$$\frac{1}{q} \sum_{j=0}^{q-1} f'(x + j\alpha) \to \int_0^1 f'(t)\,dt > 0$$

uniformly in x. Hence, we can find two constants $0 < C_1 < C_2$ such that for all q sufficiently large,

$$(6.1) \qquad\qquad C_1 q \leq f^{(q)'}(x) \leq C_2 q \; \forall \, x \in \mathbb{T}.$$

On the other hand, $f^{(q)}$ is still piecewise linear with the discontinuity points of the form $x_i + j\alpha$, with the jump at it equal to d_i, where $i = 1, \ldots, K$, $j = 0, \ldots, q-1$. Substitute from now on $q = q_n$ a Legendre denominator of α. Take the division of the circle given by the points of the form $x_i + j\alpha$. It may happen that for $i \neq i'$ we will have for some $j \neq j'$ that $x_i + j\alpha = x_{i'} + j'\alpha$. This gives rise to a partition, say ξ_n, of the circle into closed-open subintervals. Consequently the number of atoms in ξ_n is not bigger than $K q_n$. Note that no subinterval in ξ_n can be longer than $1/q_n$, so ξ_n is tending to the point partition. Let us call a subinterval in ξ_n *long* if its length is at least $\frac{1}{100 K q_n}$. Hence there must exist a constant $D = D(K) > 0$ such that for all $n \geq 1$ the number of long subintervals is at least $D q_n$. Finally, by the classical Koksma inequality, we have

$$|f^{(q_n)}(x) - f^{(q_n)}(y)| \leq \mathrm{Var}(f) \quad \text{for all } x, y \in \mathbb{T}.$$

Suppose now that $E(f) = \mathbb{Z}\lambda$. Choose a very small $\varepsilon = \varepsilon(\lambda, \mathrm{Var}(f), C_1, C_2, D) > 0$ and let

$$K = \{r \in [-2\,\mathrm{Var}(f), 2\,\mathrm{Var}(f)] : \mathrm{dist}(r, \mathbb{Z}\lambda) \geq \varepsilon\}.$$

It is clear that K is compact. If ε is small enough, in view of (6.1) and (6.2), there exists a constant $F > 0$ such that for each long subinterval of ξ_n there exists a subset with measure at least $F\frac{1}{q_n}$ such that for each x from this subset we have $f^{(q_n)}(x) \in K$. It is now sufficient to apply Lemma 1.3 to obtain an obvious contradiction to $K \cap E(f) = \emptyset$. $\qquad\square$

It is clear that the arguments from the above proof persist if instead of a piecewise continuous function we consider a function $g = f + h$, where f is piecewise linear with $\int_0^1 f'(t)\,dt \neq 0$, h is integrable, $\int_0^1 f\,dt = \int_0^1 h\,dt = 0$ and $h^{(q_n)}$ is tending to zero in measure along the sequence of Legendre denominators of α. In particular, because of Proposition 2.3, we have proved the following

Theorem 6.2. *Let B be a homogeneous Banach space on $\mathbb{T}$ and T an irrational translation. If for the pair (B, T) the Koksma inequality holds true then for each cocycle $g = f + h$, where f is piecewise linear with $\int_0^1 f'(t)\,dt \neq 0$, $h \in B_h$, $\int_0^1 f\,dt = \int_0^1 h\,dt = 0$ the corresponding cylinder flow T_f is ergodic.*

In particular (see Corollary 2.8)

Corollary 6.3. *Suppose that* $g = f + h$ *where* f *is piecewise linear with* $\int_0^1 f'(t)\,dt \neq 0$, *and* $\hat{h}(n) = o(1/n)$, $\int_0^1 f\,dt = \int_0^1 h\,dt = 0$ *then for each irrational translation* T *the corresponding cylinder flow* T_f *is ergodic.*

Remarks 1. Assume as in [Pa1] that $g : \mathbb{T} \to \mathbb{R}$ is piecewise absolutely continuous, with $\int_0^1 g'(t)\,dt \neq 0$ and $\int_0^1 g(t)\,dt = 0$. Denote by $x_1, \ldots, x_K$ the discontinuity points and let d_j be the jump at x_j. Take any piecewise linear function f with the same discontinuity points and the same jumps as g; in particular $\int_0^1 f'(t)\,dt \neq 0$. By adding a constant if necessary we can assume that $\int_0^1 f(t)\,dt = 0$. Define $h = g - f$. We have that h has zero mean and is absolutely continuous. Now, the result from [Pa1] directly follows from Corollary 6.3.

2. Notice that if g is of the form as in Corollary 6.3 then for each $\beta \in \mathbb{T}, c \neq 1$ the cocycle $g(\cdot + \beta) - cg(\cdot)$ is still of the same form, hence ergodic. We have proved that all ergodic cocycles from Corollary 6.3 are not squashable. In particular, piecewise absolutely continuous cocycles with a nonzero sum of the jumps are not squashable.

3. Using our result on the speed in Koksma's inequality (see Corollary 3.3) and the technique from [Pa2], we can slightly improve the main result of that paper by requiring that the functions from this paper can be modified by those whose Fourier coefficients are of order $o(\frac{1}{n^t})$ with an additionally remark that all those cocycles are not squashable.

References

[Aa1] J. Aaronson, *The asymptotic distributional behaviour of transformations preserving infinite measures*, J. d'Analyse Math., **39**, (1981), 203-234.

[Aa2] ——— , *The intrinsic normalising constants of transformations preserving infinite measures*, J. d'Analyse Math., **49**, (1987), 239-270.

[Ba-Me1] L. Bagget, K. Merrill, *Smooth cocycles for an irrational rotation*, preprint.

[Ba-Me2] L. Bagget, K. Merrill, *On the cohomological equation of a class of functions under irrational rotation of bounded type*, preprint.

[Con] J.P.Conze, *Ergodicite d'un flot cylindrique*, Bull. Soc. Mat. de France, **108**, (1980), 441-456.

[Furst] II.Furstenberg, *Strict ergodicity and transformations of the torus*, Amer. J. Math., **83**, (1961), 573-601.

[He] M. Herman, *Sur la conjugaison différentiable des difféomorphismes du cercle à des rotations*, Publ. Mat. IHES, **49**, (1979), 5-234.

[He-La1] P.Hellekalek, G.Larcher, *On ergodicity of a class of skew products*, Israel J. Math., **54**, (1986), 301-306.

[He-La2] P.Hellekalek, G.Larcher, *On Weyl sums and skew products over irrational rotations*, Th. Comp. Sc., Fund. St., **65**, (1989), 189-196.

[Ka] A.B. Katok, *Constructions in Ergodic Theory*, preprint.

[Katzn] Y. Katznelson, *An Introduction to Harmonic Analysis*, Dover Publ. INC., New York (1967).

[Kr] A. Krygin, *Examples of ergodic cascades*, Math. Notes USSR, **16**, (1974), 1180-1186.

[Kw-Le-Ru1] J. Kwiatkowski, M. Lemańczyk, D. Rudolph, *Weak isomorphisms of measure-preserving diffeomorphisms*, Israel J. Math. (1992), 33-64.

[Kw-Le-Ru2] _____ , *A class of real cocycles having an analytic coboundary modification*, preprint.

[Ku-Ni]   L.Kuipers, H.Niederreiter, *Uniform Distribution of Sequences*, Wiley,(1974).

[Le-Ma]   M. Lemańczyk, Ch. Mauduit, *Ergodicity of a class of cocycles over irrational rotations*, Bull. London Math. Soc., to appear.

[Or]   I. Oren, *Erdodicity of cylinder flows arising from irregularities of distribution*, Israel J. Math., **44**, (1983), 127-138.

[Pa1]   D.A. Pask, *Skew products over the irrational rotation*, Israel J. Math., **69**, (1990), 65-74.

[Pa2]   D.A. Pask, *Ergodicity of certain cylinder flows*, Israel J. Math., **76**, (1991), 129-152.

[Sch]   K. Schmidt, *Cocycles of Ergodic Transformation Groups*, Lect. Notes in Math. Vol. 1, Mac Millan Co. of India (1977).

[Zim]   R. Zimmer, *Amenable ergodic group actions and an application to Poisson boundaries of random walks*, J.Funct. Anal., **27**, (1978), 350-372.

NORMALITY TO DIFFERENT BASES

Gavin Brown

The University of Adelaide
G.P.O. Box 498
Adelaide, 5001
South Australia

1. INTRODUCTION

For a positive integer $s(> 1)$, we say that the real number x is normal to base s or *s-normal* if the sequence $(s^n x)_{n-1}^{\infty}$ is uniformly distributed modulo one. Some 80 years ago Borel showed that almost all real numbers are normal to all bases. Some 40 years ago Steinhaus asked whether 2-normality coincides with 3-normality and Cassels answered the question in the negative (see [3]).

The definitive on integer bases was established by Wolfgang Schmidt in 1962, [4], and is the following:

> Let $s, t(> 1)$ be positive integers. The properties of s-normality and t-normality coincide if and only if there exist integers a, b such that $s^a = t^b$. Otherwise there are infinitely many s-normal (resp. t-normal) numbers which fail to be t-normal (resp. s-normal).

I would like to discuss two extensions of this result. The first of these concerns itself with multi-dimensional bases, essentially defined by integer matrices, and is joint work with William Moran, [1]. We have extended Schmidt's result to commuting ergodic integer matrices on the m-torus. The second tackles the question of non-integer bases and describes joint work with William Moran and Andy Pollington, [2]. The particular result upon which attention will be focussed is our demonstration that there are numbers normal to base 10 which fail to be normal to base $\sqrt{10}$.

In order to explain the new methods used it will be useful to sketch another proof of Schmidt's basic one-dimensional result. This will be done in Section 3 after a brief discussion of some links with ergodic theory.

Algorithms, Fractals, and Dynamics
Edited by Y. Takahashi, Plenum Press, New York, 1995

2. BACKGROUND AND STATEMENT OF RESULTS

We say than an $m \times m$ integer matrix S is *ergodic* if S is invertible and has no root of unity as an eigenvalue. A vector x in $\mathbf{T}^m$ is *S-normal* if the sequence $(S^n x)_{n=1}^{\infty}$ is uniformly distributed on the m-torus. S-normality of x is equivalent to the statement that

$$\lim_{N \to \infty} \frac{1}{N} \sum_{n=1}^{N} f(S^n x) = \int f(y) dm(y), \qquad (1)$$

for all continuous f on $\mathbf{T}^m$, where m is Haar measure. For ergodic S, m almost all vectors are S-normal.

It is as well to note, in passing, that Weyl's criterion corresponds to the choice of exponentials in (1) viz.

$$\lim_{N \to \infty} \frac{1}{N} \sum_{n=1}^{N} \exp(2\pi i k S^n x) = 0, \quad k \in \mathbf{Z}^n, \ k \neq 0. \qquad (2)$$

Schmidt showed that invertible $m \times m$ matrices with rational entries, whose eigenvalues are algebraic numbers but not roots of unity, have the property that there exists some integer d such that dS^n has integer entries for every n. Accordingly we may enlarge the class of ergodic matrices and use (2) to characterize S-normality. Schmidt extended his one-dimensional theorem to the case of commuting ergodic matrices S and T, under the hypothesis that all eigenvalues of S have modulus strictly greater than one. See [5]. The latter condition relates *inter alia* to convergence criteria for explicit base S expansions. It has been removed in [1] so we quote:

Theorem 1(Brown & Moran). *Let S, T be commuting ergodic matrices. S-normality coincides with T-normality if and only if there exist integers a, b such that $S^a = T^b$. Otherwise there are infinitely many vectors which are S-normal (resp. T-normal) but not T-normal (resp. S-normal).*

The difficult part of the proof is the demonstration that S and T are rationally related when S-normality and T-normality coincide. That is where the hypothesis that $ST = TS$ is used and we conjecture that it may be removed.

The 'easy' part of the proof of Theorem 1 depends on the fact that S-normality coincides with S^k-normality. Even in the one-dimensional case this is not entirely obvious and, indeed, the corresponding property is typically violated in the non-integer case which we are about to discuss. It may therefore be helpful to sketch the proof of the following lemma.

Lemma 1. *Let $s(> 1)$ be a positive integer. Then x is normal to base s if and only if x is normal to base s^2.*

Proof sketch Suppose that x is normal to base s^2 then so is sx by the Weyl criterion, (2). But then both

$$\frac{1}{N} \sum_{n=1}^{N} \exp(2\pi i k s^{2n} x), \quad \frac{1}{N} \sum_{n=1}^{N} \exp(2\pi i k s^{2n+1} x)$$

tend to zero as $N \to \infty$ for every $k \in \mathbf{Z}$, $k \neq 0$. A simple average demonstrates Weyl's criterion for normality to base s.

Suppose next that x is normal to base s and let f be an arbitrary non-negative continuous function on the circle. From (1) we see that

$$\varlimsup_{N \to \infty} \frac{1}{N} \sum_{n=1}^{N} f(S^{2n}x) \leq 2 \int f \, dm,$$

(Throw away the terms involving odd powers of s)

and hence that every weak $*$ limit point of ρ of $\rho_N = \frac{1}{N} \sum_{n=1}^{N} \delta(s^{2n}x)$ ($\delta(u)$ is the point probability at u) is absolutely continuous with respect to m. ρ is clearly s^2-invariant so we have

$$\int f \, d\rho = \int \frac{1}{N} \sum_{n=1}^{N} f(s^{2n}y) d\rho(y) \to \int \int f \, dm \, d\rho = \int f \, dm,$$

and $\rho = m$. Thus x is s^2-normal.

In studying normality to non-integer bases we encounter a problem concerning iteration of the operation involved. For example, in general,

$$\{\sigma^2 x\} \neq \{\sigma\{\sigma x\}\},$$

where $\{\}$ denotes fractional part. Because of this, there are very few positive results concerning passage between different non-integer bases and Moran, Pollington and I have recently established a compendium of cases there results which are true in the integer case do not transfer. Before stating a selection of these let me note two results which do transfer.

Lemma 2. *Every number normal to base $\sqrt{10}$ is normal to base 10.*

Proof. In the second half of the proof of Lemma 1 replace s by $\sqrt{10}$. Because we worked throughout with s^2 there is no difficulty !

Theorem 2 (Moran & Pollington). *Let σ, τ be real numbers greater than one. Then*

$$\log \sigma / \log \tau \in \mathbb{Q} \Rightarrow \sigma\text{-normality} \not\equiv \tau\text{-normality}.$$

On the other hand we have proved:

Theorem 3 (Brown,Moran & Pollington). *If θ (> 1) is an algebraic number such that θ-normality implies θ^k-normality for some integer k greater than one, then either $\theta^r \pm \theta^{-r} \in \mathbb{N}$ for some r in $\mathbb{N}$ or $\theta = s^{1/j}$, for some $j \in \mathbb{N}$. Conversely if $\theta^r \pm \theta^{-r} \in \mathbb{N}$, for some r in $\mathbb{N}$ or $\theta = s^{1/j}$, for some j in $\mathbb{N}$, then θ-normality implies θ^k-normality for every k in $\mathbb{N}$. Moreover if θ is not an integer then there exists a positive integer k such that θ^k normality does not imply θ-normality.*

The only part of this result which will be discussed further here is the statement that there exist numbers which are normal to base 10 but not normal to base $\sqrt{10}$.

3. BASIC RESULT

In this section we will fix two positive integers s, t such that $\log s / \log t$ is not rational and show the existence of s-normal numbers which are not t-normal. By Baker's estimates for linear forms in logarithms there is some constant B such that

$$|n \log s - m \log t| > B^{-\log N}, \tag{3}$$

whenever $0 < \max(|m|, |n|) \le N$. We replace s, t by powers s^K, t^K to ensure

$$\min(s, t) \ge \max(B^2, 3) \tag{4}$$

without disturbance to (3) or to the sets of s-normality and t-normality.

Define the measure μ as the Riesz product

$$d\mu = \lim_{N \to \infty} \prod_{n=1}^{N} (1 + \cos 2\pi t^n x) dm(x), \tag{5}$$

noting that μ is a probability measure whose Fourier transform vanishes off words of the form

$$\sum_{j=1}^{J} \varepsilon_j t^j \quad \varepsilon_j \in \{0, 1, -1\},$$

where it is characterized by the formula

$$\mu^{\wedge}\left(\sum \varepsilon_j t^j\right) = \left(\frac{1}{2}\right)^{\sum |\varepsilon_j|}. \tag{6}$$

It follows, in particular, that

$$\frac{1}{N} \sum_{n=1}^{N} \exp(2\pi i t^n x) \to \frac{1}{2} \quad (\mu \text{ a.e.}),$$

(almost everywhere convergence uses the low correlation property of Riesz products) and Weyl's criterion for t-normality (cf.(2)) is violated.

It remains to show that μ almost all numbers are s-normal, and it is worth remarking that this general proof method follows Cassels's original solution of Steinhaus's problem. Cassels (and Schmidt afterwards) chose a measure μ for which the typical number was not t-normal. In Cassels's case $t = 3$ and μ is essentially the natural measure on the Cantor middle third set. The next step is to demonstrate s-normality by estimating the transform of μ, and it is in this process that Riesz products confer technical advantages. Following Davenport, Erdös and LeVeque, we use the criterion

$$\sum_{N=1}^{\infty} N^{-3} \sum_{k=1}^{N} \sum_{j=1}^{k-1} |\mu^{\wedge}(rs^k - rs^j)| < \infty \Rightarrow s\text{-normality} \quad \mu \text{ a.e.} \tag{7}$$

Let's write $t(N) = \log N$ and denote by G_N the number of $k \le N$ such that for some j with $0 \le j \le k - t(N)$ and some ε_i, there is a solution of

$$r(s^k - s^j) = \sum_{i=1}^{m} \varepsilon_i t^i, \quad \varepsilon \in \{0, 1, -1\}. \tag{8}$$

Using (6) crudely to count non-zero entries, we deduce from (7) that s-normality holds μ a.e. provided

$$\sum_{N=1}^{\infty} N^{-3}(NG_N + N_t(N)) < \infty. \tag{9}$$

To estimate G_N we rewrite (8), with $\eta_i = \varepsilon_{m-i}$ and assuming $\varepsilon_m = 1$,

$$s^k/t^m = r^{-1}(1 - s^{j-k})^{-1}\left(1 + \sum_{i=1}^{m-1} \eta_i t^{-1}\right). \tag{10}$$

Suppose we fix the first $[\log N]$ choices of η_i and consider pairs $(k, m), (k', m')$ appearing in (10). It is simple to deduce from (4) that

$$|(m - m')\log t - (k - k')\log s| = O(B^{-2}\log N),$$

and this violates (3). It follows that G_N is no greater that $3^{\log N} = N^{\log 3}$, and the result is established (providing we agreed to interpret log as $\log_{10}$ throughout!).

4. MULTI-DIMENSIONAL CASE

Because there is an integer d such that $(dT^n)_{n=1}^{\infty}$ has integer entries, we are able to construct Riesz products of the type

$$d\mu(x) = \lim_{N\to\infty} \prod_{n=1}^{N}(1 + \cos 2\pi a d T_x^n)dm(x),$$

where a is an m-vector with integer entries and the Haar measure is m-dimensional.

It is quite plausible that we may follow a route analogous to the way of Section 3 and end by counting solutions to matrix Diophantine equations. This is, in fact, the case. Nevertheless it is necessary to develop some complicated *ad hoc* arguments to handle the linear algebra, so let us agree to omit that discussion. Full details appear in [1].

5. BASE ROOT TEN

Our object here is to demonstrate the existence of numbers which are normal to base 10 but not normal to base $\sqrt{10}$. In using Section 3 as a model we have, therefore, $s = 10, t = \sqrt{10}$. It is necessary to modify (5) using a kernel function, so we write

$$d\mu(x) = \lim_{N\to\infty} \prod_{n=1}^{N}\left(1 + \cos 2\pi(\sqrt{10})^{2n+1}x\right)(1 - \cos x)x^{-2}dm(x). \tag{11}$$

This standard device produces a probability measure μ, the graph of whose transform is a set of triangles having bases of width 2 centered at the points $\sum_{n=1}^{N}\varepsilon_i(\sqrt{10})^{2i+1}$, $\varepsilon_i \in \{0, \pm 1\}$ and heights $2^{-\sum|\varepsilon_i|}$. An argument similar to that used before shows

$$\liminf_{N} \frac{1}{N}\sum_{n=1}^{N}\exp\left(2\pi i(\sqrt{10})^n x\right) \geq \frac{1}{4} \quad \mu \text{ a.e.} \tag{12}$$

55

and Weyl's criterion is violated for $\left((\sqrt{10})^n x\right)$. Thus μ almost all numbers fail to be normal to base $\sqrt{10}$, and it remains to check that μ almost all numbers *are* normal to base 10.

Once more we use criterion (7) but the crude count of one for each non-zero transform entry and the simple choice of $t(N) = \log N$ no longer suffice. Choose, in fact,

$$t(N) = N(\log N)^{-2} \tag{13}$$

and let $g(n,m)$ be the number of non-zero coefficients $\varepsilon_i \in \{0, \pm 1\}$ in an inequality of the form

$$\left| r(10^n - 10^m) - \sum_{i=1}^{k} \varepsilon_i 10^i \sqrt{10} \right| < 1 \tag{14}$$

with $m \le n - t(n)$. Set $g(n,m) = \infty$, if no such inequality as (14) holds.

By the Erdös-Davenport-LeVeque criterion, (7), it suffices to show that

$$\sum_{N=1}^{\infty} N^{-3} \sum_{m < n \le N} 2^{-g(n,m)} < \infty. \tag{15}$$

It is certainly the case that (15) holds unless (14) has infinitely many solutions for pairs m, n. Accordingly we assume that $r\sqrt{10}$ has a decimal expansion of the form

$$r\sqrt{10} = \sum_{i=-K}^{\infty} \delta_i 10^{-i}, \quad \delta_i \in \{0, \pm 1\}. \tag{16}$$

Next we define

$$h(n) = \sum_{i \le t(n)} |\delta_i|, \tag{17}$$

and consider, for large n, the rearrangement of (14) given by

$$\left| \left(\sum_{i=-K}^{\infty} \delta_i 10^{-i} \right) (1 - 10^{m-n}) - \sum_{i=1}^{k} \varepsilon_i 10^{i+1-n} \right| < 10^{-n}. \tag{18}$$

Note that, for large n,

$$\sum_{m < n \le N} 2^{-g(n,m)} \le \sum_{n=1}^{N} n 2^{-h(n)} \le N^2 2^{-h(N)}. \tag{19}$$

Now (15) may be replaced by the simple criterion

$$\sum_{N} N^{-1} 2^{-h(N)} < \infty, \tag{20}$$

and (20) certainly holds if $h(n) > \alpha \log n$ for large n and some positive constant α.

In the case which remains to be considered we may now assume that there exist infinitely many j such that $\delta_j \ne 0$ but $\delta_{j+i} = 0$, for $k = 1, 2, \cdots, 2j$. Let us write

$$10^j r\sqrt{10} = 10^j \sum_{k=-K}^{j} \delta_i 10^{-i} + R = A + R,$$

where A is an integer, $R \ne 0, |A| < 10^{j+1} r, R \ge 10^{-2j}$. This forces

$$10^{2j+1} r^2 = A^2 + 2AR + R^2,$$

where, for large j, $2AR + R^2$ is non-zero non-integer. This is impossible and the theorem is proved.

References

1. G.Brown & W.Moran, *Schmidt's conjecture on normality for commuting matrices*, Inventions Math, to appear.
2. G.Brown, W.Moran & A.Pollington, *Normality to non-integer bases*, to appear.
3. J.W.S. Cassels, *On a problem of Steinhaus about normal numbers*, Colloq. Math. **7**, 1959, 95-101.
4. W.M.Schmidt, *Über die normalität von zahlen zu verschiedenen basen*, Acta Math. **7**, 1962, 299-301.
5. W.M.Schmidt, *Normalität bezüglich matrizen*, J. für die Riene u. Angewandte Math. **2314/5**, 1964, 227-260.

SPIRALS AND PHASE TRANSITIONS

Yves Dupain[1], Michel Mendes France[1], and Claud Tricot[2]

[1] Départment de Mathématiques
Université Bordeaux I, F - 34405 Talence Cedex France
[2] Départment de Mathématiques Appliqueés
Ecole Polytechnique, CP 6079 succ.A Montréal (Québec)
Canada H 3C 3A 7

Abstract. By a stretch of imagination we shall identify spirals with systems of interacting particles. Mimicking the formalism of Statistical Mechanics we shall then discover that spirals go through a phase transition as the "temperature" increases. The inverse critical temperature coincides with the box dimension of the spiral. The article is a restatement of previous joint work [1].

§1 Integral geometry and statistical mechanics

A straight line D is completely determined in the plane by the knowledge of its distance p from the origin and by the angle w of the x axis with its normal; $p \geq 0, 0 \leq w < 2\pi$, see figure 1.

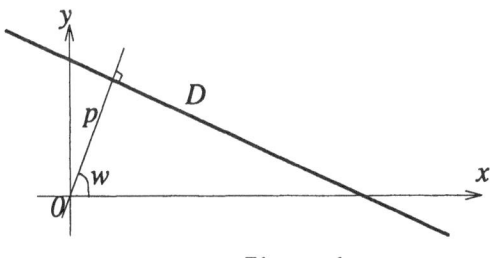

Figure 1

Let Γ be a rectifiable curve in the plane. Let $E = E(\Gamma)$ be the set of straight lines that intersect Γ. In the p, w plane the set E is compact. On E we define the probability measure

$$dP = dp \, dw / \text{meas}(E)$$

where meas(E) is the Lebesgue measure of E. Denote by $E_n(n \geq 1)$ the set of $D \in E$ which intersect Γ in n points exactly. The probability that E cuts Γ in exactly n points is then

$$P_n = \int_{E_n} dP.$$

A classical and beautiful result of Steinhaus ([2] and [3]) states the expectation of the number of intersection points is

$$\sum_{n=1}^{\infty} n P_n = \frac{2L}{C}$$

where L is the length of the curve Γ and where C is its perimeter *i.e.* the length of the boundary of its convex hull.

A random straight line has more chances to fall within the largest E_n, say E_{n_0}. We assume for simplicity that n_0 is unique.

The energy of E_n is defined as $H_n = \log(1/P_n)$ so that the energy is minimal at n_0. From now on integers are called states (of energy) and n_0 is the "fundamental state".

At this point we introduce an order-disorder parameter β defined on (β_0, ∞) for some $\beta_0 \in (0, 1]$, which we identify with the inverse temperature T^{-1} (T could also be thought of as vibrational energy). The formalism of Statistical Mechanics tells us that the probability of being in state n at temperature $T = 1/\beta$ is

$$P_n(\beta) = \frac{e^{-\beta H_n}}{Z(\beta)} = \frac{P_n^\beta}{Z(\beta)}$$

where

$$Z(\beta) = \sum_{m=1}^{\infty} e^{-\beta H_m}$$

is the partition function. β_0 is chosen so that $Z(\beta)$ converge for all $\beta > \beta_0$.

Observe that

$$P_n(\infty) = \lim_{T \searrow 0} P_n(\beta) = \begin{cases} 1 & \text{if } E_n \text{ is the fundamental state} \\ 0 & \text{if not} \end{cases}$$

At temperature 0, the fundamental state has probability 1. Notice also $P_n(1) = P_n$. The expected state at temperature $T = \beta^{-1}$ is

$$\begin{aligned} \overline{N}(\beta) &= \sum_{n=1}^{\infty} n P_n(\beta) \\ &= \frac{1}{Z(\beta)} \sum_{n=1}^{\infty} n P_n^\beta. \end{aligned}$$

In particular

$$\overline{N}(1) = \frac{2L}{C}.$$

Now we suppose that the curve Γ can have infinite length but that its convex hull stays bounded. For example this would be the case of the spiral $\rho = (\theta + 1)^{-1}, \theta \geq 0$ (polar coordinates). $\overline{N}(1)$ is then infinite. There may be a critical value β_c such that for all $\varepsilon > 0$

$$\overline{N}(\beta_c - \varepsilon) = +\infty \quad \text{and} \quad \overline{N}(\beta_c + \varepsilon) < +\infty.$$

If Γ has finite length $\beta_c \leq 1$. If however Γ has infinite length $\beta_c \geq 1$. In this case we shall link β_c to the box dimension or Bouligand dimension of Γ.

2. The box dimension

Let Γ be any bounded subset of the plane. For $\varepsilon > 0$ define the ε-sausage of Γ

$$\Gamma(\varepsilon) = \{M \in \mathbf{R}^2 \,|\, \text{dist}\,(M,\Gamma) \leq \varepsilon\}.$$

The dimension (box or Bouligand) of Γ is defined as

$$\Delta = 2 - \limsup_{\varepsilon \searrow 0} \frac{\log(\text{meas }\Gamma(\varepsilon))}{\log \varepsilon}.$$

The reader will easily convince himself that the dimension of a finite set of points is 0, that the dimension of a straight segment is 1 and that the dimension of a disc is 2. A bounded set has dimension $\Delta \in [0,2]$. A bounded curve has dimension $\Delta \in [1,2]$.

Here are three examples that show that for any $\Delta \in [1,2]$ there exists a spiral Γ with dimension Δ.

Example 1 Let $\alpha > 0$ be given. The spiral

$$\rho = (1+\theta)^{-\alpha}$$

has dimension

$$\Delta = \begin{cases} \dfrac{2}{1+\alpha} & \text{if } \alpha \leq 1 \\ 1 & \text{if } \alpha > 1. \end{cases}$$

Example 2 The spiral

$$\rho = 1/\log(\theta + 2)$$

has dimension 2.

Example 3 The spiral

$$\rho = e^{-\theta}$$

has dimension 1.

Theorem *Let f be a real function defined for $\theta \geq 0$ and suppose that for large θ f is convex and decreases to 0 as θ increases to infinity. Let Γ be the spiral defined by its polar equation*

$$\rho = f(\theta)$$

and suppose furthermore that for large θ, Γ is locally convex. Then its dimension Δ and the critical inverse temperature β_c are linked by the formula

$$\Delta = \max\{1, \beta_c\}$$

In particular, if Γ has infinite length then $\Delta = \beta_c$.

In example 1, $\beta_c = 2/(1+\alpha)$ for all $\alpha > 0$. In example 2, $\beta_c = 2$ and finally in example 3, $\beta_c = 0$.

The theorem a rewording of a previous result which we published ten years ago [1]. The concept of Steinhaus dimension that we introduced then is exactly that of the critical β_c discussed here. For the proof of the above Theorem 1 we refer to our earlier paper.

References

1. Y. Dupain, M. Mendes France, C. Tricot. *Dimensions des spirales,* Bull. Soc. Math. Fr. **111**, (1983), 193-201.
2. H. Steinhaus. *Sur la portée pratique et théorique de quelque théorèms sur la measure des ensembles de droites,* Comptes rendus du 1er congrés des Mathématiciens des pays Slaves, (1930), 353-354.
3. H. Steinhaus. *Length, Shape and Area,* Colloquium Mathematicum, **3**, (1954), 1-13.

ON AN ORDERING OF DYNAMICS OF HOMEOMORPHISMS

Koichi Hiraide

Department of Mathematics
Faculty of Science
Ehime University
Matsuyama 790, Japan

As is well known, an important part of dynamics of a homeomorphism (or a diffeo-morphism) is on an invariant set, which is certainly not a manifold, such as recurrent set, nonwandering set and chain recurrent set. In many cases, the restriction of the map to such an invariant set possesses expansivity (or sensitive dependence on initial conditions, see Devaney [D] for the definition). For instance, from a result of Shub [Sh] we see that a diffeomorphism of a closed manifold can be C^0-approximated by an Axiom A diffeomorphism which is expansive on the chain recurrent set.

In this paper we will discuss a one-parameter family of homeomorphisms on a compact manifold and, by making use of expansivity, give a basic framework in order to study global perturbation of dynamics of a homeomorphism within its isotopy class. Also, we will state some problems relevant to an ordering of dynamics of homeomorphisms.

As a special case we will investigate a one-parameter family of homeomorphisms starting at a hyperbolic infra-nilautomorphism, and give a partial answer to the prob-lems. The results seem to imply that we can define a partial order in a natural way among dynamics of homeomorphisms (or diffeomorphisms) in a large class.

Let $f : X \to X$ be a homeomorphism of a metric space with a metric d. We say that f is *expansive* if there is a constant $e > 0$, called an *expansive constant* for f, such that for $x, y \in Y$ if $x \neq y$ then $d(f^n(s), f^n(y)) > e$ for some integer $n \in \mathbb{Z}$.

Let X be a compact metric space and let $C(X)$ denote the set of all continuous maps of X endowed with the C^0-topology, i.e. the topology induced by the metric $d(f, g) = \max\{d(f(x), g(x))\}$. For a homeomorphism $f : X \to X$ we denote as $C(f)$ the set of continuous maps h of X commuting with f, i.e. $f \circ h = h \circ f$. Then the identity map and any iteration of f belong to $C(f)$. We take an interest in the fact that $C(f)$ is discrete in $C(X)$ if f is expansive (see Lemma 1.3 of §1 stated after). This fact will lead us to consider a monodromy map $C(f) \to C(f)$ along a loop of a one-parameter family of homeomorphisms from f to f.

In §1 we will discuss a one-parameter family of homeomorphisms starting at at home-omorphism f with the property that its restriction $f_{|\Lambda}$ to some invariant set Λ is expansive, and in such a general setting define a monodromy map $C(f_{|\lambda}) \to C(f_{|\lambda})$ along the one-parameter family if it is a loop. The monodromy maps will have a connection with an ordering of dynamics of homeomorphisms in a topological sense. In the last part of §1 we will state two problems on the monodromy maps.

In §§2 and 3 we will put the case such that the invariant set Λ is an infra-nilmanifold and the restriction $f_{|\Lambda}$ is hyperbolic infra-nilautomorphism, and give an affirmative answer to the problems in the case.

Let N be a connected simply connected nilpotent Lie group with a left invariant Riemannian metric, and denote as $\mathrm{Aff}(N)$ the group of transformations of N generated by all left translations and by all automorphisms from N onto itself. Let $\Gamma \subset \mathrm{Aff}(N)$ be a group which acts freely and discretely on N. Then Γ is the semi-direct product of a discrete subgroup of N and finite group of automorphisms of N (see Auslander [A]). If the quotient N/Γ is compact, then it is called an *infra-nil-manifold*. Each automorphism $N \to N$ which respects Γ induces a map $N/\Gamma \to N/\Gamma$. Such a map is called an *infra-nilautomorphism*.

As a corollary of Theorem 2.1 of §2 stated after we will obtain the following Theorem 0.2, which makes sure that every hyperbolic infra-nilautomorphism is minimum in its isotopy class. Compare with Theorem 2.2 Franks [Fr].

Theorem 0.2. *Let $A : N/\Gamma \to N/\Gamma$ be a hyperbolic infra-nilautomorphism and denote as $\mathrm{Iso}(A)$ the isotopy class of A. Then for $h \in C(A)$ there is a unique continuous map $H_h : \mathrm{Iso}(A) \to \mathrm{Hom}(h)$ with $H_h(A) = h$ such that $A \circ H_h(f) = H_h(f) \circ f$ for all $f \in \mathrm{Iso}(A)$, where $\mathrm{Hom}(h)$ denotes the homotopy class of h. Moreover, the following properties hold:*

(1) if $h, h' \in C(A)$ and $h \neq h'$, then $H_h(f) \neq H'_h(f)$ for all $f \in \mathrm{Iso}(A)$,

(2) if a continuous map $g : N/\Gamma \to N/\Gamma$ satisfies $A \circ g = g \circ f$ for some $f \in \mathrm{Iso}(f)$, then there is $h \in C(A)$ such that $g = H_h(f)$.

Notice that a continuous map of a closed manifold is surjective if it is homotopic to a homeomorphism.

In the final section we will mention a result on the inverse of the above theorem.

Remark 0.3 Let $\mathcal{H}(X)$ denote the set of homeomrphisms of X onto itself. Then $\mathcal{H}(X)$ is a group and $C(f) \cap \mathcal{H}(X)$ is a subgroup of $\mathcal{H}(X)$. Such a subgroup is called the *centralizer* of f. Let $\mathcal{A}(\mathbb{T}^n)$ denote the set of all Anosov diffeomorphisms of an n-torus $\mathbb{T}^n$ Palis-Yoccoz [P-Y2] showed that for some open and dense subset of $\mathcal{A}(M)$ if a diffeomorphism f belongs to its subset then the centralizer $C(f) \cap \mathcal{D}(\mathbb{T}^n)$ is trivial, i.e. $C(f) \cap \mathcal{D}(M) = \{f^n : n \in \mathbb{Z}\}$. Here $\mathcal{D}(\mathbb{T}^n)$ denotes the set of all diffeomorphisms. See also Palis [P] and Palis-Yoccoz [P-Y1]. On the other hand, it might be proved that for any $f \in \mathcal{A}(\mathbb{T}^n)$ the centralizer $C(f) \cap \mathcal{H}(\mathbb{T}^n)$ has comparatively a rich algebraic structure determined by fixed points of f, which depends only on its isotopy class.

§1. Semi-conjugacy maps and those monodromy

As above, let X be a compact metric space and denote as $C(X)$ the set of all continuous maps of X. For $h \in C(X)$ the path connected component of h in $C(X)$ is the homotopy class of h, which is denoted as $\mathrm{Hom}(h)$. It is well known that $C(X)$ is locally contractible if X is a compact manifold. This implies that each homotopy class is open and closed in $\mathcal{C}(X)$.

Let $\mathcal{H}(X)$ denote the topological group of all homeomorphisms of X onto itself. Given $f \in \mathcal{H}(X)$, the isotopy class of f is defined as the path connected component of f in $\mathcal{H}(X)$. We denote it as $\mathrm{Iso}(f)$. It is known that the homeomorphism group of a compact topological manifold is locally contractible (see Edwards-Kirby [E-K] and Černavskii [Ce]). Hence, if X is a compact topological manifold then $\mathcal{H}(X)$ is locally path connected and each isotopy class is open and closed in $\mathcal{H}(X)$.

A path in $\mathcal{H}(X)$ is called an isotopy or a path of homeomorphisms.

Let 2^X denote the family of non-empty closed subset of X. Then 2^X is a compact metric space under the Hausdorff metric (see Kuratowski [K]). Let Y be a metric space with a metric d'. A map $f : Y \to 2^X$ is called *upper semi-continuous* if for $y \in Y$ and $\varepsilon > 0$ there is $\delta > 0$ such that $f(z) \subset U_\varepsilon(f(y))$ whenever $d'(y, z) < \delta$. Here $U_\varepsilon(f(y))$ denotes the ε-open neighborhood of $f(y)$ in X. It is easy to see that $f : Y \to 2^X$ is upper semi-continuous if and only if $\{(x, y) : x \in f(y), y \in Y\}$ is closed in $X \times Y$.

Let $\{\Lambda_t : t \in I\}$ be an upper semi-continuous family of points in 2^X, i.e. the map from I to 2^X defined by $t \mapsto \Lambda_t$ is upper semi-continuous. Here $I = [0, 1]$. For $t \in I$ let $f_t : \Lambda_t \to \Lambda_t$ be a homeomorphism. If the map $(x, t) \mapsto f_t(x)$ is a continuous map from $\{(x, t) : x \in \Lambda_t, t \in I\}$ to X, then we say that the family $\{f_t : t \in I\}$ is a *path of homeomorphisms* on $\{\Lambda_t\}$.

EXAMPLE 1.1. Let $f : X \to X$ be a homeomorphism. A sequence $\{x_i : a < i < b\}$ of points in X called a δ-*pseudo orbit* of f uf $d(f(x_i), x_{i+1}) < \delta$ for $i \in (a, b-1)$. For $x, y \in X$ and $\alpha > 0$, x is α-*related* to y (written $x \overset{\alpha}{\sim} y$) if there are α-pseudo orbits of f such that $x_0 = x, x_1, \cdots, x_k = y$ and $y_0 = y, y_1, \cdots, y_\ell = x$. If $x \overset{\alpha}{\sim} y$ for any $\alpha > 0$ then x is *related* to y (written $x \sim y$). The *chain recurrent set* of f, $CR(f)$ is $\{x \in X : x \sim x\}$. It is clear that $CR(f) = f(CR(f))$ and $\Omega(f) \subset CR(f)$ where $\Omega(f)$ is the nonwandering set. If $\{f_t : t \in I\}$ is a path of homeomorphisms of X, then

(1) $\{CR(f_t) : t \in I\}$ is an upper semi-continuous family,

(2) $\{f_{t|CR(f_t)} : t \in I\}$ is a path of homeomorphisms on $\{CR(f_t)\}$.

Indeed, suppose (1) is false. Then we can find $t_0 \in I$ and $\varepsilon_0 > 0$ such that for every $n > 0$ there is $t_n \in I$ with $|t_0 - t_n| < 1/n$ so that $CR(f_{t_n}) \not\subset U_{\varepsilon_0}(CR(f_{t_0}))$. Take $x_n \in CR(f_{t_n})$ such that $x_n \notin U_{\varepsilon_0}(CR(f_{t_0}))$. We may suppose that x_n converges to some $x_0 \in X$. Then $x_0 \notin CR(f_{t_0})$. Since $x_n \sim x_n$, for $\delta > 0$ there is a periodic δ-pseudo orbit $\{x_n = y_0, y_1 \cdots, y_k = x_n\}$ of f_{t_n}. Since $d(f_{t_0}, f_{t_n}) < \delta$ for large n, it follows that the sequence is a periodic 2δ-pseudo orbit of f_{t_0}, from which we have $x_0 \sim x_0$ for f_{t_0}, thus contradicting. Therefore, (1) holds. (2) is clear from the fact that $\{f_t\}$ is a path of homeomorphisms of X.

Let Λ be a closed set of X and $f : \Lambda \to \Lambda$ be a homeomorphism. As above, let $\{\Lambda_t : t \in I\}$ be an upper semi-continuous family and $\{f_t : t \in I\}$ be a path of homeomorphisms on $\{\Lambda_t\}$. A family $\{h_t : t \in I\}$ of continuous surjections $h_t : \Lambda_t \to \Lambda$ is said to be a *continuous family of semi-conjugacy maps* between f and $\{f_t\}$ if the following properties hold;

(1) the map $(x, t) \to h_t(x)$ is a continuous map from $\{(x, t) : x \in \Lambda_t, t \in I\}$ to Λ,

(2) $f \circ h_t = h_t \circ f_t$ for all $t \in I$.

Lemma 1.2. *Let f and $\{f_t : t \in I\}$ be as above. If f is expansive and $h : \Lambda_0 \to \Lambda$ is a continuous surjection satisfying $f \circ h = h \circ f_0$, then there exists at most one continuous family of semi-conjugacy maps $\{h_t : t \in I\}$ between f and $\{f_t\}$ such that $h_0 = h$.*

Proof: Let $\{h_t\}$ and $\{h_t'\}$ be continuous families of semi-conjugacy maps between f and $\{f_t\}$ satisfying $h_0 = h_0' = h$. If $\{h_t\} \neq \{h_t'\}$, then for $\varepsilon > 0$ there is $t \in I$ such that $h_t \neq h_t'$ and $d(h_t, h_t') < \varepsilon$, which contradicts the following Lemma 1.3. $\square$

Lemma 1.2. *Let (Z, d_Z) and (W, d_W) be metric spaces and let $f : Z \to Z$ and $g : W \to W$ be homeomorphisms. Let $h, h' : W \to Z$ are continuous maps satisfying $f \circ h = h \circ g$ and $f \circ h' = h' \circ g$ respectively. Suppose f is expansive and $e > 0$ is an expansive constant for f. If*

$$d(h, h') = \sup\{d_Z(h(y), h'(y)) : y \in W\} \leq e,$$

then $h = h'$.

Proof. This is easily checked as follows. If $h \neq h'$, then we have

$$
\begin{aligned}
d(h, h') &= \sup\{d_Z(h(y), h'(y)) : y \in W\} \\
&= \sup\{d_Z(h \circ g^n(y), h' \circ g^n(y)) : y \in W, x \in \mathbb{Z}\} \\
&= \sup\{d_Z(f^n \circ h(y), f^n \circ h'(y)) : y \in W\} > e
\end{aligned}
$$

from which the conclusion is obtained. $\square$

Let $\{f_t : t \in I\}$ and $\{f_t' : t \in I\}$ be paths of homeomorphisms on upper semi-continuous families $\{\Lambda_t\}$ and $\{\Lambda_t'\}$ respectively. An *inverse path* of $\{f_t\}$ is defined by

$$\overline{\{f_t\}} = \{f_{1-t} : t \in I\},$$

which is a path of homeomorphisms on $\{\Lambda_{1-t} : t \in I\}$. If $\Lambda_1 = \Lambda_0'$ and $f_1 = f_0'$, then we can define a *product* of $\{f_t\}$ and $\{f_t'\}$ by $\{f_t\} \cdot \{f_t'\} = \{f_t \cdot f_t' : t \in I\}$ where

$$
f_t \cdot f_t' = \begin{cases} f_{2t} & 0 \leq t \leq 1/2 \\ f_{2t-1}' & 1/2 \leq t \leq 1 \end{cases}
$$

Then $\{f_t\} \cdot \{f_t'\}$ is a path of homeomorphisms on an upper semi-continuous family $\{\Lambda_t \cdot \Lambda_t' : t \in I\}$ where

$$
\Lambda_t \cdot \Lambda_t' = \begin{cases} \Lambda_{2t} & 0 \leq t \leq 1/2 \\ \Lambda_{2t-1}' & 1/2 \leq t \leq 1 \end{cases}.
$$

Let $\Lambda_0 = \Lambda_0', \Lambda_1 = \Lambda_1'$ and $f_0 = f_0', f_1 = f_1'$. We say that $\{f_t\}$ and $\{f_t'\}$ is *homotopic* if there are an upper semi-continuous family $\{\Lambda_{t,s} : t, s \in I\}$ and a family $\{f_{t,s} : t, s \in I\}$ of homeomorphisms $f_{t,s} : \Lambda_{t,s} \to \Lambda_{t,s}$, with the property that the map from $\{(x, t, s) : x \in \Lambda_{t,s}, t, s \in I\}$ to X defined by $(x, t, s) \to f_{t,s}(x)$ is continuous, such that

$$
\begin{cases} \Lambda_{t,0} = \Lambda_t, f_{t,0} = f_t & (\forall t \in I) \\ \Lambda_{t,1} = \Lambda_t', f_{t,1} = f_t' & (\forall t \in I) \end{cases}
\qquad
\begin{cases} \Lambda_{0,s} = \Lambda_0, f_{0,s} = f_0 & (\forall s \in I) \\ \Lambda_{1,s} = \Lambda_1, f_{1,s} = f_1 & (\forall s \in I) \end{cases}
$$

Here $(\{\Lambda_{t,s}\}, \{ft, s\})$ is called a *homotopy* from $\{f_t\}$ to $\{f_t'\}$.

Lemma 1.4. *Let $\{f_t\}$ and $\{f'_t\}$ be homotopic and let $(\{\Lambda_{t,s}\},\{f_{t,s}\})$ denote a homotopy from $\{f_t\}$ to $\{f'_t\}$. Let $f : \Lambda \to \Lambda$ be a homeomorphism and suppose for each fixed $s \in I$ there is a continuous family $\{h_{t,s}\}$ of semi-conjugacy maps between f and $\{f_{t,s}\}$. If f is expansive and $h_{0,s}$ does not depend on s, then*

(1) *the map $(x,t,x) \to h_{t,s}$ is a continuous map from $\{(x,t,s) : x \in \Lambda_{t,s}, t, s \in I\}$ to X,*

(2) *$h_{1,s}$ does not depend on s.*

Proof: To show (1), it suffice to see that $h_{t,s}$ varies continuously with respect to parameters t and s. Let $e > 0$ be an expansive constant for f. We first prove the following statement.

(A) Suppose $h_{t,s}$ is discontinuous at a point (t_0, s_0). Then there exists a sequence $\{(t_i, s_i)\}$ converging to (t_0, s_0) such that $d(h_{t_0,s_0}(x_0), h_{t_i,s_i}(x_i)) > e$ for some $x_0 \in \Lambda_{t_0,s_0}$ and some sequence $\{x_i\}$ with $x_i \in \Lambda_{t_i,s_i}$ satisfying $d(x_0, x_i) \to 0$ as $i \to \infty$.

By discontinuity we can find a number $\varepsilon_0 > 0$ and a sequence $\{(t_i, s_i)\}$ converging to (t_0, s_0) such that $d(h_{t_0,s_0}(y_0), h_{t_i,s_i}(y_i)) > \varepsilon_0$ for some $y_0 \in \Lambda_{t_0,s_0}$ and some sequence $\{y_i\}$ with $y_i \in \Lambda_{t_i,s_i}$ satisfying $d(y_0, y_i) \to 0$ as $i \to \infty$. Since $f : \Lambda \to \Lambda$ is expansive, there is $N > 0$ such that if $a, b \in \Lambda$ and $d(a, b) > \varepsilon_0$ then $d(f^i(a), f^i(b)) > e$ for some i with $|i| \leq N$. Indeed, if such an N does not exist, then for $n \in \mathbb{Z}$ there is $a_n, b_n \in \Lambda$ with $d(a_n, b_n) > \varepsilon_0$ such that $d(f^i(a_n), f^i(b_n)) \leq e$ for all i with $|i| \leq n$. Since Λ is compact, we may suppose that $a_n \to a_0$, $b_n \to b_0$ as $n \to \infty$ for some $a_0, b_0 \in \Lambda$. Then $d(a_0, b_0) \geq \varepsilon_0$ and $d(f^i(a_0), f^i(b_0)) \leq e$ for all $i \in \mathbb{Z}$, thus contradicting.

Therefore, for each $i > 0$ we can find n with $|n| \leq N$ satisfing

$$
\begin{aligned}
e &< d(f^n \circ h_{t_0,s_0}(y_0), f^n \circ h_{t_i,s_i}(y_i)) \\
&= d(h_{t_0,s_0} \circ f^n_{t_0,s_0}(y_0), h_{t_i,s_i} \circ f^n_{t_i,s_i}(y_i)).
\end{aligned}
$$

By taking a subsequence we may suppose that such an n is constant. Put $x_0 = f^n_{t_0,s_0}(y_0)$ and $x_i = f^n_{t_i,s_i}(y_i)$. Then $x_0 \in \Lambda_{t_0,x_0}$ and $x_i \in \Lambda_{t_i,s_i}$. By continuity of $\{f_{t,s}\}$ we have $x_i \to x_0$ as $i \to \infty$. Thus (A) holds.

Let J be the set of points t in I such that $h_{t,s}$ is continuous at each point in $[0, t] \times I$. By the assumption we have that $h_{t,s}$ varies continuously with respect to t if s is fixed, and that $h_{0,s}$ does not depend on s, hence, $h_{0,s}$ is continuous with respect to s, From these facts together with (A), it follows that $0 \in J$, and so $J \neq \emptyset$. Use (A) again. Then, it is easy to see that J is open and closed. Therefore, (1) holds. Since $f_{1,s} = f_1 = f'_1$ and $f \circ h_{1,s} = h_{1,s} \circ f_{1,s}$ for all $s \in I$, from Lemma 1.2 together with (1) we obtain (2). $\square$

We are in a position to define monodromy for continuous families of semi-conjugacy maps.

Let M be a compact topological manifold. Let $\{\Lambda_t : t \in I\}$ be an upper semi-continuous family of points in 2^M and let $\{f_t : t \in I\}$ be a path of homeomorphisms on $\{\Lambda_t\}$. Suppose $\Lambda_0 = \Lambda_1$, $f_0 = f_1$ and f_0 is expansive. If a continuous family $\{h_t : t \in I\}$ of semi-conjugacy maps between f_0 and $\{f_t\}$ exists, then the correspondence $h_0 \mapsto h_1$ is said to be the *monodromy of semi-conjugacy maps* along a path of homeomorphisms $\{f_t : t \in I\}$. In this case we write $h_1 = \Phi(h_0)$. From Lemma 1.4 it follows that Φ

is invariant under homotopies between paths of homeomorphisms whenever continuous families of semi-conjugacy maps exist.

The questions are the following (MC1) and (MC2).

(MC1) *Is $\Phi(h_0)$ a homeomorphism when h_0 is identity map id of Λ_0 ?*

It is easy to see that (MC1) is equivalent to the assertion that for a homeomorphism $h_0 : \Lambda \to \Lambda$ belonging $C(f_0)$ there is a continuous family of semi-conjugacy maps $\{h_t : t \in I\}$ between f_0 and $\{f_t\}$ such that h_1 is also a homeomorphism.

(MC2) *Does one have $\Phi^n(id) = id$ for some integer $n > 0$?*

Let $\mathcal{E}(M)$ denote the class of all homeomorphisms $f : M \to M$ with the property that the restriction of f to the chain recurrent set $CR(f)$ is expansive. If (MC1) is affirmative for any closed path of homeomorphisms starting at an element in $\mathcal{E}(M)$, then we can define a partial order for the class $\mathcal{E}(M)$ in a natural way as follows. Write $f \leq g$ if for some isotopy $\{f_t : t \in I\}$ from f to g there is a continuous family of semi-conjugacy maps $\{h_t : t \in I\}$ between $f_{|CR(f)}$ and $\{f_{t|CR(f_t)} : t \in I\}$ such that h_0 is the identity map of $CR(f)$, and $f \sim g$ if, in addition, $h_1 : CR(g) \to CR(f)$ is a homeomorphism. Then the relation $\leq$ is a partial order on the identifying space of $\mathcal{E}(M)$.

To see this, we first show that $\sim$ is an equivalence relation. Clearly $f \sim f$ for all $f \in \mathcal{E}(M)$. If $f \sim g$, then for some isotopy $\{f_t : t \in I\}$ from f to g there is a continuous family of semi-conjugacy maps $\{h_t : t \in I\}$ between $f_{|CR(f)}$ and $\{f_{t|CR(f_t)} : t \in I\}$ such that h_0 is the identity map of $CR(f)$ and $h_1 : CR(g) \to CR(f)$ is a homeomorphism. Write $\{h'_t\} = \{h_1^{-1} \circ h_{1-t} : t \in I\}$. Then h'_0 is the identity map of $CR(g)$, $h'_1 = h_1^{-1}$, and $\{h'_t\}$ is a continuous family of semi-conjugacy maps between $g_{|CR(g)}$ and $\{f_{t|CR(f_t)}\}$. This shows $g \sim f$.

Let $f \sim g$ and $g \sim k$. Then there are an isotopy $\{f_t : t \in I\}$ from f to g and a continuous family $\{h_t\}$ of semi-conjugacy maps between $f_{|CR(f)}$ and $\{f_{t|CR(f_t)}\}$. Moreover, for some isotopy $\{g_t\}$ from g to k there is a continuous family $\{h'_t\}$ of semi-conjugacy maps between $g_{|CR(g)}$ and $\{g_{t|CR(g_t)}\}$. Here h_0 and h'_0 are the identity maps, and h_1 and h'_1 are homeomorphisms. Write $\{h''_t\} = \{h \cdot (h_1 \circ h'_1) : t \in I\}$ where

$$h_t \cdot (h_1 \circ h'_t) = \begin{cases} h_{2t} & 0 \leq t \leq 1/2 \\ h_1 \circ h'_{2t-1} & 1/2 \leq t \leq 1 \end{cases}$$

Then $h''_0 = h_0$, $h''_1 = h_1 \circ h'_1$ and $\{h''_t\}$ is a continuous family of semi-conjugacy maps between $f_{|CR(f)}$ and $\{f_{t|CR(f_t)}\} \cdot \{g_{t|CR(g_t)}\}$. Therefore, $f \sim k$.

Next, let us show that $\leq$ is a partial order on $\mathcal{E}(M)/\sim$. It is easy to show that $\leq$ satisfies reflexivity and antisymmetricity. To see transitivity of $\leq$, suppose $f \leq g$ and $g \leq f$. Then by the definition we can take an isotopy $\{f_t : t \in I\}$ from f to g and a continuous family $\{h_t\}$ of semi-conjugacy maps between $f_{|CR(f)}$ and $\{f_{t|CR(f_t)}\}$. Also, for some isotopy $\{g_t\}$ from g to f there is a continuous family $\{h'_t\}$ of semi-conjugacy maps between $g_{|CR(g)}$ and $\{g_{t|CR(g_t)}\}$. Here h_0 and h'_0 are the identity maps respectively. Write $\{h''_t\} = \{h_t \cdot (h_1 \circ h'_t) : t \in I\}$ as above. Then $\{h''\}$ is a continuous family of semi-conjugacy maps between $f_{|CR(f)}$ and $\{f_{t|CR(f_t)}\} \cdot \{g_{t|CR(g_t)}\}$, and by the assumption it follows that $h_1 \circ h'_1$ is a homeomorphism of $CR(f)$. Similarly, so is $h'_1 \circ h_1$. Hence $h_1 : CR(f) \to CR(g)$ is a homeomorphism, from which we obtain $f \sim g$.

§2. Perturbation of hyperbolic infra-nilautomorphisms

Let X be a compact connected locally connected metric space. As is well known, X is path connected and locally path connected. We say that X is *semilocally 1-connected* if for $x \in X$ there is a neighborhood U of x in X such that the inclusion $U \to X$ induces the zero map $\pi_1(U, x) \to \pi_1(X, x)$ of fundamental groups. If X is semilocally 1-connected, then it has the universal covering space (see Spanier [Sp] for example).

The following Theorem 2.1 is an extension of Theorem 2.2 of Franks [Fr].

Theorem 2.1. *Let S^1 be a circle $\mathbb{R}/\mathbb{Z}$ and $e : \mathbb{R} \to S^1$ denote the natural projection. Suppose X is semilocally 1-connected and let $p : X \to S^1$ be a fiber bundle with connected fiber. Let $f : X \to X$ be a homeomorphism which preserves each fiber for p, and for $t \in \mathbb{R}$ let $X_{e(t)} = p^{-1}(e(t))$ and $f_{e(t)} = f_{|X_{e(t)}}$. Suppose $A : N/\Gamma \to N/\Gamma$ is a hyperbolic infra-nilautomorphism and for some $t_0 \in \mathbb{R}$ there is a homomorphism $\phi : \pi_1(X_{e(t_0)}, b_0) \to \pi_1(N/\Gamma, c_0)$ such that the following diagram commutes:*

$$
\begin{array}{ccc}
\pi_1(X_{e(t_0)}, b_0) & \xrightarrow{\ f_{e(t_0)*}\ } & \pi_1(X_{e(t_0)}, f(b_0)) \\
\phi \downarrow & & \downarrow v_* \circ \phi \circ u_* \\
\pi_1(N/\Gamma, c_0) & \xrightarrow{\quad A_* \quad} & \pi_1(N/\Gamma, A(c_0))
\end{array}
$$

where u_* (resp. v_*) denote the induced isomorphism by a path u (resp. v) in X (resp. N/Γ) from b_0 (resp. $A(c_0)$) to $f(b_0)$ (resp. C_0). Then

(1) *there exists a unique continuous family $h_t : X_{e(t)} \to N/\Gamma$, $t \in \mathbb{R}$, of continuous maps with $h_{t_0}(b_0) = c_0$ such that $h_{t_0*} = \phi$ and $A \circ h_t = h_t \circ f_{e(t)}$ for all $t \in \mathbb{R}$,*

(2) *there exists a homeomorphism $T : N/\Gamma \to N/\Gamma$ homotopic to the identity map such that $T \in C(A)$ and $T \circ h_t = h_{t+1}$ holds for all $t \in \mathbb{R}$,*

(3) *if the following short exact sequence splits:*

$$
0 \longrightarrow \pi_1(X_{e(t_0)})/\mathrm{Ker}(\phi) \xrightarrow{\ i_*\ } \pi_1(X)/i_*(\mathrm{Ker}(\phi)) \xrightarrow{\ p_*\ } \pi_1(S^1) \longrightarrow 0.
$$

then the map T is the identity map, where i_* denotes the induced homomorphism by the inclusion.

From the above theorem we have that (MC1) and (MC2) stated in §1 are affirmative in a special case.

Indeed, by (2) it follows that $T \circ h_0 = h_1$ and T is a homeomorphism, which implies that (MC1) is true. Since $T \in C(A)$, if $q \in N/\Gamma$ is a fixed point then so is $T(q)$. Let $T' : N/\Gamma \to N/\Gamma$ be a homeomorphism homotopic to id and suppose $T' \in C(A)$ and $T(q) = T'(q)$. Then $T = T'$. This is easily checked as follows.

Let $\pi' : N \to N/\Gamma$ be the natural projection and let $\bar{A} : N \to N$ denote the automorphism which covers A. Choose $\bar{q} \in N$ such that $\pi(\bar{q}) = q$. Since $T, T' \in C(A)$ and $T(q) = T'(q)$, we can find lifts $\bar{T}, \bar{T}' : N \to N$ of T, T' such that $T(\bar{q}) = T'(\bar{q}), \bar{A} \circ \bar{T} = \bar{T} \circ \bar{A}$ and $\bar{A} \circ \bar{T}' = \bar{T}' \circ \bar{A}$. Since T and T' are homotopic to id, it is clear that T is homotopic to T', from which $D(T, T') = \sup\{D(\bar{T}(x), \bar{T}'(x)) : x \in N\}$ is finite where D denotes a left invariant Riemannian metric. By this fact and Lemma 1.3 together with Lemma 3.2 (2) stated after, it follows that $\bar{T} = \bar{T}'$.

Since the set of fixed points of A is a finite set, we have that $C(A) \cap \mathcal{H}(N/\Gamma) \cap \text{Hom}(id)$ is a finite group. This shows $T^n = id$ for some $n > 0$. On the other hand, $T^n \circ h_0 = h_n$ by (2). Thus (MC2) is also true.

From the above Theorem 2.1 we have the following corollary.

Corollary 2.2. *Let $A : N/\Gamma \to N/\Gamma$ be a hyperbolic infra-nilautomorphism and let $f_t : N/\Gamma \to N/\Gamma$, $t \in I$, be an isotopy starting at A. Then for $h \in C(A)$ there is a homotopy $h_t : N/\Gamma \to N/\Gamma$, $t \in I = [0,1]$, with $h_0 = h$ such that $A \circ h_t = h_t \circ f_t$ for all $t \in I$. Furthermore, if $f_0 = f_1$ then $h_1 = h = h_0$.*

Proof: Let $\{g_t : t \in I\} = \{f_t\} \cdot \overline{\{f_t\}}$ and define $g : N/\Gamma \times S^1 \to N/\Gamma \times S^1$ by $g(x,t) = (g_t(x),t)$. Here $S^1 = I/\{0,1\}$. Since $g_0 = g_1 = A$, by Theorem 2.1 there is a unique continuous map $H : N/\Gamma \times S^1 \to N/\Gamma$ such that $H_{|N/\Gamma \times \{0\}} = h$ and $A \circ H = H \circ g$, which shows the first part of the corollary. When $f_0 = f_1$, let us define $g : N/\Gamma \times S^1 \to N/\Gamma \times S^1$ by $g(x,t) = (f_t(x),t)$. Use Theorem 2.1 again, then there is a unique continuous map $H' : N/\Gamma \times S^1 \to N/\Gamma$ such that $H'_{|N/\Gamma \times \{0\}} = h$ and $A \circ H' = H' \circ g$. By uniqueness we obtain the second part of the corollary. $\square$

Proof of Theorem 0.2: For $f \in \text{Iso}(A)$ let $f_t, t \in I$, be an isotopy from A to f. Then, by the first part of Corollary 2.2 for $h \in C(A)$ there is a homotopy $h_t, t \in I$, with $h_0 = h$ such that $A \circ h_t = h_t \circ f_t$ and so we define $H_h : \text{Iso}(A) \to \text{Hom}(h)$ by $H_h(f) = h_1$. This is well-defined by the last part of Corollary 2.2. By this together with the fact that $\mathcal{H}(N/\Gamma)$ is locally contractible, we have that H_h is continuous. (1) of Theorem 0.2 follows from Lemma 1.2. Use Theorem 2.1 again, then (2) of Theorem 0.2 is easily checked in the same way as in the proof of Corollary 2.2. $\square$

§3. Proof of Theorem 2.1

To prove Theorem 2.1 we will apply the method in the author [Hi2]. First we prepare the following Lemma 3.1 and 3.2.

Lemma 3.1. *Let X be a compact metric space with a metric d and let $\bar{X}$ be a connected topological space. Suppose $p : \bar{X} \to X$ is covering map. If X is connected and locally connected, then there are a compatible metric ρ for $\bar{X}$ such that*

(1) all covering transformations for p are isometries under ρ,

(2) $(\bar{X}, \rho)$ is a complete metric space,

(3) for all $x \in \bar{X}$ and $r > 0$ the closed ball $\{y \in \bar{X} : \rho(x,y) \leq r\}$ is compact.

Proof: Since $p : \bar{X} \to X$ is a covering map, by Theorem 1 of Duvall-Husch [D-H] there exist a metric $\bar{d}$ for $\bar{X}$ and a constant $\delta_0 > 0$ satisfying the following properties:

(1)' all covering transformations for p are isometries under $\bar{d}$,

(2)' $(\bar{X}, \bar{d})$ is complete metric space,

(3)' for $0 < \delta \leq \delta_0$ and $x \in \bar{X}$ the restriction $p : U_\delta(x) \to U_\delta(p(x))$ is an isometry where $U_\delta(x) = \{y \in \bar{X} : \bar{d}(x,y) < \delta\}$ and $U_\delta(p(x)) = \{y \in \bar{X} : \bar{d}(p(x),y) < \delta\}$.

Fix $0 < \delta < \delta_0/2$. For $x, y \in \bar{X}$ let $\{x_i : 0 \leq i \leq \ell + 1\}$ be a δ-chain from x to y, i.e. $\bar{d}(x_i, x_{i+1}) \leq \delta$ for $0 \leq i \leq \ell$, and define ρ by

$$\rho(x, y) = \inf\{\sum_{i=0}^{\ell} \bar{d}(x_i, x_{i+1})\}$$

where the infimum is taken over all finite δ-chains from x to y. By the triangle inequality of $\bar{d}$ we have $\rho(x, y) \geq \bar{d}(x, y)$, from which ρ is a metric for $\bar{X}$. Clearly $\bar{d}(x, y) = \rho(x, y)$ if $\bar{d}(x, y) \leq \delta$. Thus ρ is a compatible metric and by $(2)'$ we obtain (2). From the construction of ρ together with $(1)'$ it is easy to see (1).

To show (3), let K be a compact subset of $\bar{X}$ and write $N_\delta(K) = \{y \in \bar{X} : \rho(y, K) \leq \delta\}$ where $\rho(y, K) = \min\{\rho(y, x) : x \in K\}$. If $\rho(y, x) \leq \delta$, then by the definition of ρ we have $\bar{d}(y, x) \leq \delta$, and hence $\bar{d}(y, x) = \rho(y, x)$, from which $N_\delta(K) = \{y \in \bar{X} : \bar{d}(y, K) \leq \delta\}$. Let $\{y_i\}$ be a sequence of points in $N_\delta(K)$. Then we can find a sequence $\{x_i\}$ of points in K such that $y_i \in \bar{B}_\delta(x_i)$ for all i. Here $\bar{B}_\delta(x_i) = \{z \in \bar{X} : \bar{d}(x_i, z) \leq \delta\}$. Since K is compact, clearly there is a subsequence $\{x_j\}$ converging to some $x_\infty \in K$. Hence $y_j \in \bar{B}_{2\delta}(x_\infty)$ if j is sufficiently large. Since $2\delta < \delta_0$, by $(3)'$ we have that $\bar{B}_{2\delta}(x_\infty)$ is compact, and so $\{y_j\}$ contains a subsequence $\{y_k\}$ converging to some y_∞. Since $\bar{d}(y_k, x_k) \leq \delta$, it follows that $y_\infty \in N_\delta(K)$. Therefore, $N_\delta(K)$ is compact. From the fact that $\bar{d}(x, y) = \rho(x, y)$ if $\rho(x, y) \leq \delta$, it is easy to see that $N_{2\delta}(K) = N_\delta(N_\delta(K))$, which shows that $N_{2\delta}(K)$ is also compact. By induction, so is $N_{n\delta}(K)$ for every $n \geq 1$. Let $x \in \bar{X}$ and $r > 0$, then we can find $n \geq 1$ such that $\{y \in \bar{X} : \rho(x, y) \leq r\} \subset N_{n\delta}(K)$. Therefore, (3) holds. $\qquad\square$

Lemma 3.2. *Let N be a connected simply connected nilpotent Lie group with a left invariant Riemannian metric D and let $A : N \to N$ be a hyperbolic automorphism. Then*

(1) *for $L > 0$ and $\varepsilon > 0$ there exists $J > 0$ such that if $D(A^i(x), A^i(y)) \leq L$ for all i wutg $|i| \leq J$, then $D(x, y) \leq \varepsilon$,*

(2) *for given $L > 0$, if $D(A^i(x), A^i(y)) \leq L$ for all $i \in \mathbb{Z}$, then $x = y$,*

(3) *for $L > 0$ there exists $\delta_L > 0$ such that for any L-pseudo orbit $\{x_i : i \in \mathbb{Z}\}$ of A there is a unique point $x \in N$ so that $D(A^i(x), x_i) \leq \delta_L$ for $i \in \mathbb{Z}$.*

Proof: Let $\mathcal{L}(N)$ be the Lie algebla of N and $dA : \mathcal{L}(N) \to \mathcal{L}(N)$ denote the linear automorphism induced by A. Since dA is hyperbolic by the assumption, $\mathcal{L}(N)$ splits into the direct sum $\mathcal{L}(N) = E^s \oplus E^u$ of subspaces E^s and E^u such that $dA(E^s) = E^s$ and $dA(E^u) = E^u$, and such that there are $c > 1$, $0 < \lambda < 1$ so that for all $n \geq 0$

$$\|dA^n(v)\| \leq c\lambda^n\|v\| \quad (\forall v \in E^s) \quad \text{and} \quad \|dA^{-n}(v)\| \leq c\lambda^n\|v\| \quad (\forall v \in E^u)$$

where $\| \ \|$ is the Riemannian metric. Let $\exp : \mathcal{L}(N) \to N$ denote the exponential map. Since N is simply connected and nilpotent, $\exp$ is a diffeomorphism and $dA = \exp^{-1} \circ A \circ \exp$ holds (see Hochschild [Ho]). Write $\mathcal{L}^\sigma(e) = \exp(E^\sigma)$. Here e denotes the identity of N. For $x \in N$ let $L^\sigma(x) = xL^\sigma(e)$. Since left translations are isometries under the metric D, it follows that for all $x \in N$

$$D(A^n(x), A^n(y)) \leq c\lambda^n D(x, y) \quad \text{for } y \in L^s(x),$$
$$D(A^{-n}(x), A^{-n}(y)) \leq c\lambda^n D(x, y) \quad \text{for } y \in L^u(x),$$

and hence the families $\mathcal{F}^\sigma = \{L_\sigma(x) : x \in N\}$ ($\sigma = s, u$) are consistent with the those of stable sets and unstable sets of A respectively. Moreover, for all $x, y \in N$, $L^s(x) \cap L^u(y)$ consists of exactly one point. Indeed, by the fact that $L^s(e) \cap L^u(e) = \{e\}$, it is easy to see that $L^s(x) \cap L^u(y)$ is at most one point. Since $L^s(e)$ and $L^u(e)$ intersect transversally, we can find $\delta > 0$ such that if x, y belong to a δ-neighborhood $U_\delta(e)$ then $L^s(x)$ intersects $L^u(y)$. Let x belong to the δ-neighborhood $U_\delta(L^u(e))$ of $L^u(e)$ then $x \in aU_\delta(e)$ for some $a \in L^u(e)$, and so $L^s(x)$ intersects $L^u(e)$. In the same way $L^s(x) \cap L^u(e) \neq \emptyset$ for $x \in U_\delta(U_\delta(L^u(e)))$. Notice $U_\delta(U_\delta(L^u(e))) = U_{2\delta}(L^u(e))$. By induction, we have the same result for $x \in U_{n\delta}(L^u(e))$ and $n > 0$. Since $\bigcup_{n>0} U_{n\delta}(L^u(e)) = N$, it follows that $L^s(x) \cap L^u(y) \neq \emptyset$ for all $x \in N$, from which $L^s(x) \cap L^u(y) \neq \emptyset$ for all $x, y \in N$.

For $x, y \in N$ denote as $\beta(x, y)$ the point in $L^s(x) \cap L^u(y)$. Since $\mathcal{F}^s$ and $\mathcal{F}^u$ are invariant under left translations, for $L > 0$ there is $\delta_L > 0$ such that diam$\{x, y, \beta(x, y)\} < \delta_L$ if $D(x, y) < L$. For given $\varepsilon > 0$ choose $J > 0$ such that $\delta_L c \lambda^J < \varepsilon$. Suppose $D(A^i(x), A^i(y)) \leq L$ for $-J \leq i \leq J$ and let $z_i = \beta(A^i(x), A^i(y))$. Then $D(z_J, A^J(y)) < \delta_L$. Since $z_J \in L^u(A^J(y))$, we have $D(z_0, y) = D(A^{-J}(z_J), A^{-J} \circ A^J(y)) \leq \delta_L c \lambda^N < \varepsilon$. Similarly $D(z_0, x) < \varepsilon$. Therefore $D(x, y) < 2\varepsilon$. This shows (1). Since ε is arbitrary, (2) holds.

To see (3), we first show that the above δ_L can be chosen as a number less than a polynomial of L. Let $u \in E^u$ and $v \in E^s$. Then by Campbell-Hausdorff formula we have

$$
\begin{aligned}
L^s(\exp(u)) &= \exp(u)\exp(E^s) = \exp(\eta(u, E^s)) \\
&= \exp(u + E^s + \tau(u, E^s)).
\end{aligned}
$$

Since N is nilpotent, we note that η and τ are determined by a polynomial (see Hochschild [Ho]). Similarly, $L^u(\exp(v)) = \exp(v + E^u + \tau(v, E^u))$. Denote as $\exp(w)$ the point in $L^s(\exp(u)) \cap L^u(\exp(v))$. Then there is a unique point $(x, y) \in E^s \times E^u$ such that

(B) $$w = u + x + \tau(u, x) = v + y + \tau(v, y),$$

and then

$$\exp(w) = \exp(u)\exp(x) = \exp(v)\exp(y)$$

Hence $\exp(-v)\exp(u) = \exp(y)\exp(-x)$, and we have

$$\eta_{|E^s \times E^u}(-v, u) = \eta_{|E^s \times E^u}(y, -x).$$

Since η is determined by a polynomial, letting $q = \eta_{|E^s \times E^u}(-v, u)$ we have that $\|q\|$ is bounded by a polynomial of $\|u\|$ and $\|v\|$. From the fact that $\eta : E^u \times E^s \to \mathcal{L}(N)$ are bijective, it follows that $\|x\|$ is bounded by a polynomial of $\|q\|$. Hence $\|x\|$ is less than a polynomial of $\|u\|$ and $\|v\|$. Combining this and (B), we obtain that $\|w\|$ has an upper bound expressed by a polynomial of $\|u\|$ and $\|v\|$. This show the above assertion.

Tow show (3), choose $k > 0$ with $\delta_L c \lambda^k < L/2$. Since A is uniformly continuous, there is $b > 0$ such that $D(A(x), A(y)) < bD(x, y)$. Take $M > 0$ such that $b^k M = L/2$. By the above result we may suppose that δ_L is a polynomial of L, from which $M \to \infty$ as $l \to \infty$.

Let $\{x_i\}$ be an M-pseudo orbit of A. Then $\{x_{ik}\}$ is an $L/2$-pseudo orbit of A^k. For fixed $i \geq 0$ let $z_i = \beta(x_{ik}, A^k(x_{(i-1)k})) < \delta_L$. Then $D(z_i, A^k(x_{(n-1)k})) < \delta_L$. Since $z_i \in$

$L^u(A^k(x_{(i-1)k}))$, we have $D(A^{-k}(z_i), x_{(i-1)k}) < \frac{L}{2}$, from which $D(A^{-k}(z_i), A^k(x_{(i-2)k})) < L$. Put $z_{i-1} = \beta(A^{-k}(z_i), A^k(x_{(i-2)k}))$. Then $D(A^{-k}(z_{i-1}), A^k(x_{(i-3)k})) < L$. Inductively, define $\{z_i, z_{i-1}, \cdots, z_1\}$ by $z_j = \beta(A^{-k}(z_{j+1}), A^k(x_{(j-1)k}))$ for $1 \le j \le i-1$, and put $z_0 = A^{-k}(z_1)$. Then $D(A^k(z_0), x_k) < \delta_L + L$. Since $D(z_1, A^{-k}(z_2)) < \delta_L$ and $z_1 \in L^s(A^{-k}(z_2))$,

$$D(A^{2k}(z_0), x_{2k}) < D(A^k(z_1), z_2) + D(z_2, x_{2k})$$
$$< L/2 + \delta_l + L/2 = \delta_L + L.$$

By induction we have $D(A^{jk}(z_0), x_{jk}) < \delta_L + L$ for $0 \le j \le i$. Since N is complete, we let z_0' and accumulation point of $\{z_0 : i > 0\}$. Then $D(A^{jk}(z_0'), x_{jk}) < \delta_L + L$ for all $j \ge 0$. Similarly, there is $w_0' \in N$ such that $D(A^{jk}(w_0'), x_{jk}) < \delta_L + L$ for all $j \le 0$. Put $x = \beta(z_0', w_0')$, then $D(A^{jk}(x), x_{jk}) < 2(\delta_L + L)$ for all $j \in \mathbb{Z}$. Therefore, $D(A^i(x), x_i) < 2b^k(\delta_L + L)$ for all $i \in \mathbb{Z}$. From this together with (2) we obtain (3). □

Proof of Theorem 2.1: Let $\pi : \bar{X} \to X$ be the universal covering map. Then we can find a continuous map $\bar{p} : \bar{X} \to \mathbb{R}$ such that the diagram

$$\begin{array}{ccc} \bar{X} & \xrightarrow{\ \bar{p}\ } & \mathbb{R} \\ {\scriptstyle \pi}\downarrow & & \downarrow{\scriptstyle e} \\ X & \xrightarrow{\ p\ } & S^1 \end{array}$$

commutes. Since $\mathbb{R}$ is contractible, it follows that $\bar{p} : \bar{M} \to \mathbb{R}$ is a trivial fiber bundle. For $t \in \mathbb{R}$ denote as $\bar{X}_t$ the fiber over t. For simplicity let $t_0 = 0$. Then for $i \in \mathbb{Z}$ we have $\pi(\bar{X}_i) = X_{e(0)}$ and each restriction $\pi : \bar{X}_i \to X_{e(0)}$ is the universal covering map.

Choose $\bar{b}_0 \in \bar{X}_0$ such that $\pi(\bar{b}_0) = b_0$. Since u is a path in X starting at b_0, we can find a lift $\bar{u}$ of u by π satisfying $\bar{u}(0) = \bar{b}_0$. Then $\bar{u}(1) \in \bar{X}_0$ and $\pi(\bar{u}(1)) = f(b_0)$. Hence there is a lift $\bar{f} : \bar{X} \to \bar{X}$ of f by π such that $\bar{f}(\bar{b}_0) = \bar{u}(1)$. Since f preserves each fiber for p, it follows that $\bar{f}(\bar{X}_t) = \bar{X}_t$ for all $t \in \mathbb{R}$. Put $\bar{f}_0 = \bar{f}_{|\bar{X}_0}$.

Let $\pi' : N \to N/\Gamma$ be the natural projection and choose $\bar{c}_0 \in N$ such that $\pi'(\bar{c}_0) = c_0$. Then there is a lift $\bar{A} : N \to N$ of A such that $\bar{A}(\bar{c}_0) = \bar{v}(0)$ where $\bar{v}$ is a lift of v by π' with $\bar{v}(1) = \bar{c}_0$.

Let $G(\pi)$ denote the covering transfomation group for π. Then $G(\pi)$ is isomorphic to $\pi_1(X, b_0)$ (cf. Spanier [Sp]). Denote as F the subgroup of $G(\pi)$ which corresponds to $\pi_1(X_{e(0)}, b_0)$. Since $X_{e(0)}$ is the fiber over $e(0)$ for the fiber bundle $p : X \to S^1$, $G(\pi)/F$ is isomorphic to $\mathbb{Z}$. Since $p_*(\pi_1(M_0, b_0)) = 0$, it follows that $\alpha(\bar{X}_t) = \bar{X}_t$ for all $\alpha \in F$ and $f \in \mathbb{R}$.

Let $\bar{f}_* : G(\pi) \to G(\pi)$ denote the induced homomorphism by $\bar{f}$. Then $\bar{f} \circ \beta = \bar{f}_*(\beta) \circ \bar{f}$ holds for all $\beta \in G(\pi)$. Since $f_*(\pi_1(X_{e(0)}, f(b_0)) = \pi_1(X_{e(0)}, f(b_0))$, we have $\bar{f}_*(F) = F$ and the restriction $\bar{f}_* : F \to F$ is consistent with the induced homomorphism $\bar{f}_{0*}$.

Since $\pi' : N \to N/\Gamma$ is the natural projection, Γ is the covering transformation group for π'. Let $\bar{A}_* : \Gamma \to \Gamma$ denote the induced homomorphism. As usual, we consider $\phi : \pi(X_{e(0)}, b_0) \to \pi_1(N/\Gamma, c_0)$ as a homomorphism from F to Γ. Since $A_* \circ \phi = (v_* \circ \phi \circ u_*) \circ f_{e(0)*}$ by the assumption, from the choice of $\bar{f}$ and $\bar{A}$ we have $\bar{A}_* \circ \phi = \phi \circ \bar{f}_*$ on F. Hence, letting $F(b) = \{\alpha(b) : \alpha \in F\}$ $(b \in \bar{X})$ and $\Gamma(c) = \{\ell(c) : \ell \in \Gamma\}$ $(c \in N)$, we have the following commutative diagram:

$$
\begin{array}{ccc}
F(\bar{b}_0) & \xrightarrow{\ \bar{f}_0\ } & F(\bar{f}_0(\bar{b}_0)) \\
{\scriptstyle \phi'}\downarrow & & \downarrow{\scriptstyle \phi''} \\
\Gamma(\bar{c}_0) & \xrightarrow[\ \bar{A}\ ]{} & \Gamma(\bar{A}(\bar{c}_0))
\end{array}
$$

where ϕ' and ϕ'' are defined by $\phi'(\alpha(\bar{b}_0)) = \phi(\alpha)(\bar{c}_0)$ and $\phi''(\alpha(\bar{f}_0(\bar{b}_0))) = \phi(\alpha)(\bar{A}(\bar{c}_0))$ for all $\alpha \in F$, respectively.

Since A is a hyperbolic infra-nilautomorphism, by Lemmas 3.2 it is easily checked that $\bar{A}$ possesses the following properties:

(P1) for $L > 0$ and $\varepsilon > 0$ there exists $J > 0$ such that if $D(\bar{A}^i(x), \bar{A}^i(y)) \leq L$ for all i with $|i| \leq J$, then $D(x, y) \leq \varepsilon$,

(P2) for given $L > 0$, if $D(\bar{A}^i(x), \bar{A}^i(y)) \leq L$ for all $i \in \mathbb{Z}$, then $x = y$,

(P3) for $L > 0$ there exists $\delta_L > 0$ such that for any L-pseudo orbit $\{x_i : i \in \mathbb{Z}\}$ of $\bar{A}$ there is a unique point $x \in N$ so that $D(\bar{A}^i(x), x_i) \leq \delta_L$ for $i \in \mathbb{Z}$.

where D denotes a left invariant Riemannian metric.

Indeed, let $\bar{A}'$ denote the automorphism of N which covers A. Then $\bar{A} = \theta \circ \bar{A}'$ for some $\theta \in \Gamma$. For $L > 0$ and $\varepsilon > 0$ let J be as in Lemma 3.2 (1). Suppose $D(\bar{A}^i(x), \bar{A}^i(y)) \leq L$ for $-L \leq i \leq L$. Notice that $\bar{A}^i = \theta_i \circ \bar{A}'^i$ for some $\theta_i \in \Gamma$. Since each θ_i is an isometry under D, we have $D(\bar{A}'^i(x), \bar{A}'^i(y)) \leq L$ and by Lemma 3.2(1), $D(x, y) \leq \varepsilon$. Thus (P1) holds. (P2) follows from (P1). To see (P3) let $L > 0$ and choose $\delta_L > 0$ as in Lemma 3.2(3). Let $\{x_i\}$ be an L-pseudo orbit of $\bar{A}$. Then for fixed $j \leq 0$ a one-sided sequence

$$
x_i,\ \theta(x_{j+1}),\quad \theta \circ \bar{A}_*(\theta)(x_{j+2}),\quad \theta \circ \bar{A}_*(\theta) \circ \bar{A}_*^2(\theta)(x_{j+3}),\ \cdots
$$

is an L-pseudo orbit of $\bar{A}'$. By Lemma 3.2(3) this sequence is δ_L-traced by a point z^j under $\bar{A}'$. Then we have $D(\bar{A}^i(z^j)x_{j+i}) < \delta_L$ for $i \geq 0$. Let x be an accumulation point of $\{\bar{A}^j(z^j) : j \leq 0\}$, then $D(\bar{A}^i(x), x_i) < \delta_L$ for all $i \in \mathbb{Z}$. Therefore (P3) holds.

Let d a metric for X and choose a metric ρ for $\bar{X}$ as in Lemma 3.1. Then the following property holds:

(P4) for $t \in \mathbb{R}$ there is $K_t > 0$ such that $\bar{X}_t \subset U_{K_t}(\bar{X}_0)$ where $U_{K_t}(\bar{X}_0) = \{x \in \bar{X} : \rho(x, \bar{X}_0) < K_t\}$.

For, let D be a compact covering domain for $\pi : \bar{X}_t \to X_{e(t)}$ and choose $K > 0$ such that $D \subset U_K(\bar{b}_0)$. Here $U_K(\bar{b}_0) = \{y \in \bar{X} : \rho(y, \bar{b}_0) < K\}$. Then $\cup_{\alpha \in F} \alpha(D) = \bar{X}_t$. Hence, by Lemma 3.1(1) we obtain (P4).

Let $x \in \bar{X}$ and choose $t \in \mathbb{R}$ such that $x \in \bar{X}_t$. Since $\bar{f}(\bar{X}_t) = \bar{X}_t$, the orbit $\{\bar{f}^i(x) : i \in\}$ is a subset of $\bar{X}_t$. By (P4) there is $K_t > 0$ such that for $i \in \mathbb{Z}$ there is $\alpha_i \in F$ such that $\alpha_i(\bar{b}_0) \in U_{K_t}(\bar{f}^i(x))$. Notice that $\alpha_i(\bar{b}_0) \in \bar{X}_0$ for all $i \in \mathbb{Z}$. Since $\bar{f}$ is uniformly continuous under ρ, we can find $K_t' > K_t$ such that $\rho(\bar{f}(x), \bar{f}(y)) < K_t'$ whenever $\rho(x, y) < K_t$. Put $B_t = \{z \in \bar{X} : \rho(\bar{b}_0, z) \leq K_t + K_t'\}$. Then B_t is compact by Lemma 3.2(3). Hence

$$
\Sigma = \{\alpha \in F : \alpha(B_t \cup \bar{f}(B_t)) \cap (B_t \cup \bar{f}(B_t)) \neq \emptyset\}
$$

is finite. Let $L = L_t = \max_{\alpha \in \Sigma} \rho(\phi(\alpha)(\bar{A}(\bar{c}_0)), \bar{c}_0)$. Then by (P3) there exists $\delta_L > 0$ such that any L-pseudo orbit of $\bar{A}$ is δ_L-traced by a unique point in N. Notice that L depends on t.

Claim 1. For $x \in \bar{X}$ let $\{\alpha_i : i \in \mathbb{Z}\}$ be as above. Then $\{\phi' \circ \alpha_i(\bar{b}_0) : i \in \mathbb{Z}\}$ is an L-pseudo orbit of $\bar{A}$.

Indeed, since

$$\bar{f}_*(\alpha_i)(\bar{f}(\bar{b}_0)) = \bar{f}_*(\alpha_i(\bar{f}(\bar{b}_0))) \in U_{K'_t}(\bar{f}^{i+1}(x)),$$

we have $\rho(\bar{f}_*(\alpha_i)(\bar{f}(\bar{b}_0)), \alpha_{i+1}(\bar{b}_0)) < K_t + K'_t$, and hence $\alpha_{k+1}^{-1} \circ \bar{f}_*(\alpha_i) \in \Sigma$, from which there is $\gamma_i \in \Sigma$ such that $\bar{f}_*(\alpha_i) = \alpha_{i+1} \circ \gamma_i$. Then we have

$$
\begin{aligned}
\bar{A} \circ \phi' \circ \alpha_i(\bar{b}_0) &= \phi'' \circ \bar{f} \circ \alpha_i(\bar{b}_0) \\
&= \phi'' \circ \bar{f}_*(\alpha_i)(\bar{f}(\bar{b}_0)) \\
&= \phi'' \circ \alpha_{i+1} \circ \gamma_i(\bar{f}(\bar{b}_0)) \\
&= \phi(\alpha_{i+1}) \circ \phi(\gamma_i) \circ \phi''(\bar{f}(\bar{b}_0)) \\
&= \phi(\alpha_{i+1}) \circ \phi(\gamma_i)(\bar{A}(\bar{c}_0))
\end{aligned}
$$

and so

$$
\begin{aligned}
D(\bar{A} \circ \phi' \circ \alpha_i(\bar{b}_0), \phi' \circ \alpha_{i+1}(\bar{b}_0)) & \\
&= D(\phi(\alpha_{i+1}) \circ \phi(\gamma_i)(\bar{A}(\bar{c}_0)), \phi(\alpha_{i+1})(\bar{c}_0)) \\
&= D(\phi(\gamma_i)(\bar{A}(\bar{c}_0)), \bar{c}_0) \leq L
\end{aligned}
$$

for all $i \in \mathbb{Z}$. Therefore, Claim 1 holds.

Let z_x be a δ_L-tracing point of $\{\phi' \circ \alpha_i(\bar{b}_0) : i \in \mathbb{Z}\}$.

Claim 2. z_x is independent of the choice of $\{\alpha_i\}$.

Indeed, for $i \in \mathbb{Z}$ let $\alpha'(\bar{b}_0) \in U_{K_t}(\bar{f}^i(x))$ for some $\alpha'_i \in F$. Then $\{\phi' \circ \alpha'_i(\bar{b}_0)\}$ is an L-pseudo orbit of $\bar{A}$ and there is a unique $z'_x \in N$ which is δ_L-tracing $\{\phi' \circ \alpha'_i(\bar{b}_0)\}$. Since $\rho(\alpha_i(\bar{b}_0), \alpha'_i(\bar{b}_0)) < K_t + K_t < K_t + K'_t$, we have $\alpha'_i = \alpha_i \circ \gamma_i$ for some $\gamma_i \in \Sigma$, and hence for $i \in \mathbb{Z}$

$$
\begin{aligned}
D(\bar{A}^i(z_x), \bar{A}^i(z'_x)) &\leq D(\bar{A}^i(z_x), \phi' \circ \alpha_i(\bar{b}_0)) + D(\phi' \circ \alpha_i(\bar{b}_0), \phi' \circ \alpha_i(\bar{b}_0)) \\
&\quad + D(\phi' \circ a'_i(\bar{b}_0), \bar{A}^i(z'_x)) \\
&\leq 2\delta_L + D(\phi(\alpha_i)(\bar{c}_0), \phi(\alpha_i) \circ \phi(\gamma_i)(\bar{c}_0) \\
&\leq 2\delta_L + D(\bar{c}_0, \phi(\gamma_i)(\bar{c}_0)).
\end{aligned}
$$

Therefore $z_x = z'_x$ by (P2).

Define map $\bar{H} : \bar{X} \to N$ by $x \mapsto z_x$. This is well-defined by Claim 2.

Let $x = \bar{b}_0$. Then, from the choice of $\bar{f}$ and $\bar{A}$ it is easily checked that $D(\phi \circ \alpha_i(c_0), \bar{A}^i(c_0)) < L = L_0$ for all i. This shows $\bar{H}(\bar{b}_0) = \bar{c}_0$.

For $x \in \bar{X}$ let $\{\alpha_i\}$ be as above. then we have for $i \in \mathbb{Z}$

$$
\begin{aligned}
D(\bar{A}^i \circ \bar{A} \circ \bar{H}, \bar{A}^i \circ \bar{H} \circ \bar{f}(x)) &\leq D(\bar{A}^{i+1} \circ \bar{H}(x), \phi' \circ \alpha_{i+1}(\bar{b}_0)) \\
&\quad + D(\phi' \circ \alpha_{i+1}(\bar{b}_0), \bar{A}^i \circ \bar{H}(\bar{f}(x))) \\
&\leq 2\delta_L,
\end{aligned}
$$

and hence $\bar{A} \circ \bar{H} = \bar{H} \circ \bar{f}$ holds on $\bar{X}$.

Let $x \in \bar{X}$ and $\{\alpha_i\}$ be as above. Since $\alpha_i(\bar{b}_0) \in U_{K_t}(\bar{f}^i(x))$ for all $i \in \mathbb{Z}$, for given $J > 0$ there is a neighborhood $U(x)$ of x in $\bar{X}$ such that $\alpha_i(\bar{b}_0) \in U_{K_t}(\bar{f}^i(y))$ for all $y \in U(x)$ and all i with $|i| \leq J$. Then $D(\bar{A}^i \circ \bar{H}(y), \bar{A}^i \circ \bar{H}(x)) < 2\delta_L$ for i with $|i| \leq J$. Choose J large enough, then $D(\bar{H}(x), \bar{H}(y))$ is small by (P1), which shows continuity of $\bar{H}$.

Claim 3. $\bar{H} \circ \alpha = \phi(\alpha) \circ \bar{H}$ for $\alpha \in F$.

Indeed, let $x \in \bar{X}$ and $\{\alpha_i\}$ be as above. Since $\alpha_i(\bar{b}_0) \in U_{K_t}(\bar{f}^i(x))$, it follows that $\bar{f}_*^i(\alpha) \circ \alpha_i(\bar{b}_0) \in U_{K_t}(\bar{f}^i(\alpha(x)))$, and so $D(\bar{A}^i \circ \bar{H} \circ \alpha(x), \phi' \circ \bar{f}_*^i(\alpha) \circ \alpha_i(\bar{b}_0)) < \delta_L$ for all $i \in \mathbb{Z}$. Since

$$
\begin{aligned}
\phi' \circ \bar{f}_*^i \circ \alpha_i(\bar{b}_0) &= \phi(\bar{f}_*^i(\alpha)) \circ \phi' \circ \alpha_i(\bar{b}_0) \\
&= \bar{A}_*^i(\phi(\alpha)) \circ \phi' \circ \alpha_i(\bar{b}_0),
\end{aligned}
$$

we have

$$
\begin{aligned}
&D(\bar{A}^i \circ \bar{H} \circ \alpha(x), \bar{A}^i \circ \phi(\alpha) \circ \bar{H}(x)) \\
&\quad \leq D(\bar{A}^i \circ \bar{H} \circ \alpha(x), \phi' \circ \bar{f}_*^i(\alpha) \circ \alpha_i(\bar{b}_0)) \\
&\qquad + D(\bar{A}_*^i(\phi(\alpha)) \circ \phi' \circ \alpha_i(\bar{b}_0), \bar{A}_*^i(\phi(\alpha)) \circ \bar{A}^i \circ \bar{H}(x)) \\
&\quad \leq \delta_L + D(\phi' \circ \alpha_i(\bar{b}_0), \bar{A}^i \circ \bar{H}(x)) \\
&\quad \leq 2\delta_L \qquad (\forall i \in \mathbb{Z})
\end{aligned}
$$

from which $\bar{H} \circ \alpha(x) = \phi(\alpha) \circ \bar{H}(x)$. Therefore, the conclusion is obtained.

By Claim 3 for each $t \in \mathbb{R}$ the restriction $\bar{H} : \bar{X}_t \to N$ can be projected to a continuous map $h_t : X_{e(t)} \to N/\Gamma$. Since $\bar{A} \circ \bar{H} = \bar{H} \circ \bar{f}$, we have that $A \circ h_t = h_t \circ f_{e(t)}$ for all $t \in \mathbb{R}$. Since $\bar{H}(\bar{b}_0) = \bar{c}_0$, clearly $h_0(b_0) = c_0$, and by Claim 3, $h_{0*} = \phi$ holds. Since H is continuous, it follows that $\{h_t : t \in \mathbb{R}\}$ is a continuous family.

To show uniqueness, let $\{h'_t\}$ be another continuous family such that $h'_0(b_0) = c_0$, $h'_{0*} = \phi$, and $A \circ h'_t = h'_t \circ f_{e(t)}$ for all $t \in \mathbb{R}$. Since $\pi : \bar{X} \to X$ is the universal covering map, there is a lift $\bar{H}' : \bar{X} \to N$ of the famly $\{h'_t\}$ such that $\bar{H}'(\bar{b}_0) = \bar{c}_0$. Then $\bar{A} \circ \bar{H}' = \bar{H}' \circ \bar{f}$. Since $h'_{0*} = \phi$, it follows that $\bar{H} \circ \alpha = \phi(\alpha) \circ \bar{H}$ for all $\alpha \in F$. By this and Claim 3 we have that $D(\bar{H}, \bar{H}') = \sup\{D(\bar{H}(x), \bar{H}'(x)) : x \in \bar{X}\}$ is finite, and hence $\bar{H} = \bar{H}'$ by Lemma 1.3 and (P2). (1) of Theorem 2.1 was proved.

Next we show (3). Let P denote the subgroup of F corresponding to $\text{Ker}(\phi)$. Since the sequence of (3) splits, we can choose $\beta \in G(\pi)$ with $\beta(\bar{X}_0) = \bar{X}_1$ such that $\bar{f} \circ \beta = \kappa \circ \beta \circ \bar{f}$ for some $\kappa \in P$. Take a lift $\bar{H}' : \bar{X} \to N$ of the continuous family $\{h_t\}$ such that $\bar{H}'(\beta(\bar{b}_0)) = \bar{c}_0$. Then $\bar{H}' \circ \beta = \bar{H}$. Hence

$$
\begin{aligned}
\bar{A} \circ \bar{H}' &= \bar{A} \circ \bar{H} \circ \beta \\
&= \bar{H} \circ \bar{f} \circ \beta \\
&= \bar{H} \circ \kappa \circ \beta \circ \bar{f} \\
&= \bar{H} \circ \beta \circ \bar{f} \\
&= \bar{H}' \circ \bar{f}.
\end{aligned}
$$

Let $\alpha \in F$. Then $\beta \circ \alpha = \kappa' \circ \alpha \circ \beta$ for some $\kappa' \in P$. This shows $\bar{H}' \circ \alpha = \phi(\alpha) \circ \bar{H}'$. Hence $\bar{H}' = \bar{H}$ by Lemma 1.3 and (P2). Since $\beta(\bar{X}_0) = \bar{X}_1$, it is clear that $\beta(\bar{X}'_t) = \bar{X}_{t+1}$ for all $t \in \mathbb{R}$, and therefore $h_t = h_{t+1}$ for all $t \in \mathbb{R}$. (3) of Theorem 2.1 was proved.

To show (2), take $\beta \in G(\pi)$ such that $\beta(\bar{X}_0) = \bar{X}_1$. Then there is a lift $\bar{H}' : \bar{X} \to N$ of the family $\{h_t\}$ such that $\bar{H}'(\beta(\bar{b}_0)) = \bar{c}_0$, i.e. $\bar{H}' \circ \beta = \bar{H}$. Let $\alpha = \bar{f}_*^{-1}(\beta) \circ \beta^{-1}$. Then $\alpha \in F$ and $\bar{A} \circ \bar{H}' = \bar{H}' \circ (\bar{f} \circ \alpha)$. Hence we have

(C) $$\bar{H}' \circ (\bar{f} \circ \alpha) = \gamma \circ \bar{H}' \circ \bar{f}$$

where $\gamma = \phi \circ \bar{f}_*(\alpha) \in \Gamma$.

On the other hand, since $\gamma \circ \bar{A}$ is a lift of A, it follows that $\gamma \circ \bar{A}$ also satisfy the properties of (P1), (P2) and (P3) stated above. We note that $\bar{A}$ and $\gamma \circ \bar{A}$ have unique fixed points respectively. Let $\bar{w}$ be a path in N joining those fixed points and put $w = \pi' \circ \bar{w}$. Then the following diagram

$$
\begin{array}{ccc}
\pi_1(N/\Gamma, w(1)) & \xrightarrow{A_*} & \pi_1(N/\Gamma, w(1)) \\
{\scriptstyle w_*}\downarrow & & \downarrow{\scriptstyle w_*} \\
\pi_1(N/\Gamma, w(0)) & \xrightarrow{A_*} & \pi_1(N/\Gamma, w(0))
\end{array}
$$

commutes where w_* denotes the induced isomorphism. Apply the above discussion again, we can find a homeomorphism $T : N/\Gamma \to N/\Gamma$ such that $T(w(1)) = w(0)$, $T_* = w_*$ and $A \circ T = T \circ A$. See also Franks [Fr]. From the fact that N/Γ is of type $K(\Gamma, 1)$, it follows that T is homotopic to the identity map (cf. Spanier [Sp]).

Let $\bar{T} : N \to N$ be a lift of T such that $\bar{T}(\bar{w}(1)) = \bar{w}(0)$. Then

$$\bar{T}^{-1} \circ A \circ \bar{T} = \gamma^{-1} \circ A.$$

Combining this and (C) we have

$$\bar{A} \circ (\bar{T} \circ \bar{H}') = (\bar{T} \circ \bar{H}') \circ \bar{f}.$$

Since $(\bar{T} \circ \bar{H}') \circ \alpha = \phi(\alpha) \circ (\bar{T} \circ \bar{H}')$ for $\alpha \in F$, by Lemma 1.3 and (P2) we obtain that $\bar{T} \circ \bar{H}' = \bar{H}$. (2) was proved. $\qquad\square$

§4. Comments

In [Ha1] Handel showed that if M is a closed surface and $f : M \to M$ is a pseudo-Anosov homeomorphism then for $g \in \mathrm{Iso}(f)$ there is a g-invariant closed subset $X \subset M$ and a continuous surjection $h : X \to M$ homotopic to the inclusion such that $f \circ h = h \circ g$ on X. See also Handel [Ha2] and Fathi [Fa]. It will be possible to show that a pseudo-Anosov homeomorphism f of a closed surface is minimum in $\mathrm{Iso}(f)$ in the sense stated in §1. On the other hand, the author [Hi1] and Lewowicz [L] showed independently that an expansive homeomorphism of a closed surface is pseudo-Anosov. From these facts it will be natural to ask the following

Problem 4.1. *Is an expansive homeomorphism of a compact manifold minimum in its isotopy class ?*

Finally, we mention a following result to the above problem and the inverse of Theorem 0.2.

Theorem 4.2. *Let M be a closed topological manifold and $f : M \to M$ be a homeomorphism. Suppose M is of type $K(\pi, 1)$. If for $g \in \mathrm{Iso}(f)$ there is $h \in \mathrm{Hom}(id)$ such that $f \circ h = h \circ g$, then f is topologically conjugate to a hyperbolic infra-nilautomorphism.*

This will be obtained in showing that the homeomorphism f satisfies all the properties stated in Lemma 3.2. For the details the author hope to appear elsewhere.

References

[A]   L. Auslander, *Bieberbach's theorems on space groups and discrete uniform subgroups of Lie groups,* Ann. of Math. (2) **71**, (1960), 579-590.

[Ce]   A. Černavskii, *Local contractibility of the homeomorphism group of a manifold,* Math. U.S.S.R. Sbornik **8**, (1969), 287-333.

[D]   R. L. Devaney, *An Introduction to Chaotic Cynamical Systems,* Addison-Wesley, (1989).

[D-H]   P. Duvall and L. Husch, *Analysis on topological manifolds,* Fund. Math. **77**, (1972), 75-90.

[E-K]   R. Edwards and R. Kirby, *Deformations of spaces of imbeddings,* Ann. of Math. **93**, (1971), 63-88.

[Fa]   A. Fathi, *Homotopical stability of pseudo-Anosov diffeomorphisms,* Ergod. Th. Dynam. Sys. **10**, (1990), 287-294.

[Fr]   J. Franks, *Anosov diffeomorphisms,* Global Analysis, Proc. Sympos. Pure Math. 14, Amer. Math. Soc., (1970), 61-93.

[Ha1]   M. Handel, *Global shadowing of pseudo-Anosov,* Ergod. Th. Dynam. Sys. **5**, (1985), 373-377.

[Ha2]   ———— , *Entropy and semi-conjugacy in dimension two,* Ergod. Th. Dynam. Sys. **8**, (1988), 585-596.

[Hi1]   K. Hiraide, *Expansive homeomorphisms of compact surfaces are pseudo-Anosov,* Osaka J. Math. **27**, (1990), 117-162.

[Hi2]   ———— , *Positively expansive open maps of Peano spaces,* Topology and its Appl. **37**, (1990), 213-220.

[Ho]   G. Hochschild, *The Structure of Lie Groups,* Holden Day, San Fransisco, 1965.

[K]   C. Kuratowskii, *Topologie,* Monograf. Mat., Warszawa, I; 1948, II; 1965.

[L]   J. Lewowicz, *Expansive homeomorphisms of surfaces,* Bull. Brazil. Math. Soc. **20**, (1989), 113-133.

[P]   J. Palis, *The dynamics of a diffeomorphism and the rigidity of its centralizer,* Singularities and Dynamical Systems, North-Holland, 1985, 15-21.

[P-Y1] J. Palis and J. Yoccoz, *Rigidity of centralizers of diffeomorphisms,* Ann. Scient. Éc. Norm. Sup., 4e série **22**, (1989), 81-98

[P-Y2] ———— , *Centralizers of Anosov diffeomorphisms on tori,* Ann. Scient. Éc. Norm. Sup., 4e série **22**, (1989), 99-108.

[Sh]   M. Shub, *Structurally stable diffeomorphisms are dense,* Bull. Amer. Math. Soc. **78**, (1972), 817-818.

[Sp]   E. Spainer, *Algebraic Topology,* McGraw-Hill, New York, 1966.

POISSON LAW FOR AXIOM A SYSTEM

Masaki Hirata

Department of Mathematics
Faculty of Sciences
Tokyo Metropolitan University
Minami-Osawa, Hachioji-shi, Tokyo, Japan

Introduction

Recently, Ya.G. Sinai studied the distribution of spacings between nearest energy levels of a quantum particle on the two-dimensional compact Riemannian surfaces, and he shows the limiting Poisson distribution for spacings of quasi-classical eigenvalues for the quantum kicked rotator model ([S.I], [S.II]). The essential point of the proof is to reduce the problem to studying some ergodic transformation on $\mathbb{T}^2$. He considers the distribution of the visiting times of the trajectory to a certain horizontal strip, and obtained the limiting Poisson point process as the width of the strip tends to zero. And he points out that the way of appearance of the above Poisson point process is quite different from that in the usual situations in probability theory. This fact is very interesting from the ergodic theoretical view point. Inspired by it, we will consider the following problem.

Let X be a compact metric space, f a continuous map on X, and μ an f-invariant probability measure on X. Fix a point $z \in X$ and take its ϵ-neighborhoods $U_\epsilon(z)$. As a probability measure on $U_\epsilon(z)$, we will take the restriction of μ to $U_\epsilon(z)$, i.e.

$$\mu_\epsilon \equiv \frac{\mu\mid_{U_\epsilon(z)}}{\mu(U_\epsilon(z))}.$$

Denote the k-th return time of a point x from $U_\epsilon(z)$ to $U_\epsilon(z)$ by $T_{\epsilon,f}^{(k)}(x)$. Then, we want to know what is the limit distribution of the normalized k-th return times

$$\frac{T_{\epsilon,f}^{(k)}}{E_{\mu_\epsilon}(T_{\epsilon,f}^{(1)})}$$

as $\epsilon \to 0$.

Next let us introduce a counting measure ($\mathbb{N}^+$-valued Radon measure on $\mathbb{R}^+$), $Y_\epsilon(x)$, defined by

$$Y_\epsilon(x) = \sum_{k=1}^{\infty} \delta_{c_\epsilon \cdot T_{\epsilon,f}^{(k)}(x)},$$

where $c_\epsilon = 1/E_{\mu_\epsilon}(T_{\epsilon,f}^{(1)})$ and δ_p is the Dirac δ-measure at $p \in \mathbb{R}^+$. Then, $Y_\epsilon(\cdot)$ is a point process on $\mathbb{R}^+$. We will call it the normalized return time process. And the above problem can be considered as follows: what is the limit of the sequence of the normalized return time processes $\{Y_\epsilon\}_\epsilon$ as $\epsilon \to 0$?

It is expected that the limit distribution of the normalized first return time is the exponential distribution and that the limit distribution of the normalized return time process is the law of Poisson point process if the system (X, f, μ) is "chaotic" in some sense (for example, ergodic, mixing, etc.). Let us say that the Poisson law holds if it is true.

In this report the author considers the above problem for the typical "chaotic" system, namely, for the Axiom A system, and shows the Poisson law for it.

Let M be a compact C^∞ Riemannian manifold and $f : M \to M$ be an Axiom A diffeomorphism. We denote its non-wandering set by $\Omega = \Omega(f)$ and assume that $f|_\Omega$ is mixing. Take a Lipschitz continuous function $u : \Omega \to \mathbb{R}$ and denote the (unique) Gibbs measure (= the equilibrium state) for u by $\mu = \mu_u$. Fix a point $z \in \Omega$, and take its ϵ-neighborhoods $\{U_\epsilon(z)\}_\epsilon$. The main theorem is the following:

Theorem. *For $\mu - a.e.$ $z \in \Omega$, the sequence of the normalized return time processes converges to the Poisson point process in finite dimensional distribution: for any disjoint Borel sets $B_1, \cdots, B_n \in \mathcal{B}(\mathbb{R}^+)$, and any non-negative integers $k_1, \cdots, k_n$,*

$$\lim_{\epsilon \to 0} \mu_\epsilon(Y_\epsilon(B_1) = k_1, \cdots, Y_\epsilon(B_n) = k_n) = \prod_{i=1}^{n} \frac{\ell(B_i)^{k_i}}{k_i!} e^{-\ell(B_i)},$$

where ℓ is the Lebesgue measure.

It should be emphasized that the main theorem holds for $\mu - a.e.$ z, but not for every point.

Counter-example. *For a periodic point $z \in \Omega$ with period m, the limit distribution of the normalized first return time is the linear combination of the delta-distribution and the exponential distribution. Precisely,*

$$\lim_{\epsilon \to 0} \mu_\epsilon(c_\epsilon T_\epsilon^{(1)} < t) = 1 - \rho_z + \rho_z(1 - e^{-\rho_z t})$$

where $\rho_z = 1 - exp\{u(z) + u(f(z)) + \cdots + u(f^{m-1}(z))\}$.

The main theorem holds only if the eigenvalue of the operator $\tilde{\mathcal{L}}_N$ defined in Section 1 which goes to 1 as $N \to \infty$ is unique, or more precisely, if the number of the eigenvalues of $\tilde{\mathcal{L}}_N$ contained in a small neighborhood of 1 is only one for large N. Otherwise, the limit of the normalized return time process is expected to obey a compound Poisson law.

As there exists a Markov partition of the non-wandering set Ω, the essential part of a proof of the main theorem is to show the Poisson law for the (one-sided) symbolic dynamics. Then, in this note, we will sketch the outline of its proof.

The main theorem can be proved by approximating ϵ-neighborhood by a finite union of cylinder sets associated with a Markov partition. See [H] about the detail.

1. Set up

Let $J = \{1, \cdots, r\}$ be a finite set and $A = (A_{ij})_{i,j=1,\dots,r}$ be an irreducible $r \times r$ matrix with entries 0 or 1. Define the space Σ_A^+ by

$$\Sigma_A^+ = \{x = \{x_i\}_{i=0}^\infty \in J^{\mathbb{N}}; A_{x_i x_{i+1}} = 1 \quad \text{for all } i \in \mathbb{N}\}.$$

For a fixed $0 < \theta < 1$, we can define the metric $d = d_\theta$ on Σ_A^+ by

$$d_\theta(x,y) = \theta^n \quad \text{if } x_i = y_i \text{ for } i = 0, \cdots, n-1 \quad \text{and } x_n \neq y_n.$$

We denote the shift on Σ_A^+ by σ :

$$(\sigma x)_i = x_{i+1}.$$

Let $\mathcal{F}_\theta(\Sigma_A^+)$ be the totality of real valued Lipschitz continuous functions on Σ_A^+ (with respect to d_θ) and define the norm on $\mathcal{F}_\theta(\Sigma_A^+)$ by

$$\|\|g\|\|_\theta = \|g\|_\infty + \|g\|_\theta$$

where $\|g\|_\infty$ is the supremum norm and $\|g\|_\theta$ is the Lipschitz constant for g:

$$\|g\|_\theta = \sup\left\{\frac{|g(x) - g(y)|}{d_\theta(x,y)}; x \neq y\right\}.$$

For $u \in \mathcal{F}_\theta(\Sigma_A^+)$, we define the Ruelle-Perron-Frobenius operator $\mathcal{L} = \mathcal{L}_u : \mathcal{F}_\theta(\Sigma_A^+) \to \mathcal{F}_\theta(\Sigma_A^+)$ by

(1.1)
$$\mathcal{L}_u f(x) = \sum_{\sigma y = x} e^{u(y)} f(y).$$

We assume that

(1.2)
$$\mathcal{L}_u 1 = 1.$$

If not, we can obtain (1.2) by replacing u by $u' = u + \log h - \log(h \circ \sigma) - P(u)$ where $P(u)$ is the topological pressure for u and h is the eigenfunction of $\mathcal{L}_u$ corresponding to the maximal eigenvalue $e^{P(u)}$. Hence we may assume (1.2) without loss of generality. So we make this assumption throughout this note.

Let $\mu = \mu_u$ be the Gibbs measure for u. In our situation, the Gibbs measure coincides with the equilibrium state. Hence μ satisfies the following equality

(1.3)
$$P(u) = h_\mu(\sigma) + \int u d\mu = 0$$

where $h_\mu(\sigma)$ is the metrical entropy. We remark that $P(u) = 0$ follows from our assumption (1.2) and that $h_\mu(\sigma) > 0$.

Now we fix a point $z \in \Sigma_A^+$, and denote cylinder sets by

$$[z]_N \equiv [z_0 z_1 \ldots z_{N-1}] = \{y \in \Sigma_A^+; y_i = z_i, i = 0, \ldots N-1\}, \qquad N = 1, 2, \cdots.$$

Since the measure μ is σ-invariant, we can define a first return time from $[z]_N$ to $[z]_N$, denoted by $T_N(x)$, for $\mu - a.e.$ $x \in [z]_N$ for each N:

$$T_N(x) = \inf\{i \in \mathbb{N}^+; \sigma^i x \in [z]_N\}.$$

We introduce the following singularly perturbed Ruelle-Perron-Frobenius operator $\tilde{\mathcal{L}}_N : \mathcal{F}_\theta(\Sigma_A^+) \to \mathcal{F}_\theta(\Sigma_A^+)$:

$$
\begin{aligned}
(1.4) \qquad \tilde{\mathcal{L}}_N f(x) &= \mathcal{L}(1_{[z]_N^c} \cdot f)(x) \\
&= \sum_{\sigma y = x} e^{u(y)} 1_{[z]_N^c}(y) f(y),
\end{aligned}
$$

where $[z]_N^c$ denotes the complement of the set $[z]_N$ and $1_{[z]_N^c}$ is its indicator function.

Then, we can obtain the following lemma immediately by using the equality:

$$(1.5) \qquad \int \mathcal{L}f \cdot g \, d\mu = \int f \cdot (g \circ \sigma) d\mu$$

for $f, g \in \mathcal{F}_\theta(\Sigma_A^+)$.

Lemma 1.1.

$$\mu(\{x \in [z]_N; T_N(x) = i\}) = \int \tilde{\mathcal{L}}_N^{i-1}(\mathcal{L}(1_{[z]_N}))(x) 1_{[z]_N}(x) \mu(dx).$$

2. Basic properties of $\tilde{\mathcal{L}}_N$

The properties of spectrum of analytically perturbed Ruelle operator is well known by the results of Ruelle and Pollicott ([R.],[P.]). But we can not apply their results directly to $\tilde{\mathcal{L}}_N$ because it is a singularly perturbed one. So, in this section, we will study some basic properties of $\tilde{\mathcal{L}}_N$.

Lemma 2.1(Lasota-Yorke type inequality). *For each $N \in \mathbb{N}^+$, there exists a constant c_N, which depends only on N, such that the following inequality holds for any $h \in \mathcal{F}_\theta(\Sigma_A^+)$ and any $p \in \mathbb{N}$:*

$$(2.1) \qquad \|\tilde{\mathcal{L}}_N^p h\|_\theta \leq \theta^p \|h\|_\theta + c_N \|h\|_\infty.$$

By Lemma 2.1, we can estimate the upper bound of the essential spectral radius of $\tilde{\mathcal{L}}_N$ by almost the same technique used in [P].

Lemma 2.2. *The essential spectral radius of $\tilde{\mathcal{L}}_N$ is not greater than θ.*

Remark From the definition of $\tilde{\mathcal{L}}_N$ and the assumption (1.2): $\mathcal{L}_u 1 = 1$, it follows that the spectral radius of $\tilde{\mathcal{L}}_N$ is not greater than that of $\mathcal{L}_u$ which is equal to 1. So, by Lemma 2.2, the spectra of $\tilde{\mathcal{L}}_N$ in the annulus $\{t \in \mathbb{C}; \theta < |t| \leq 1\}$ consists only of isolated eigenvalues of finite mutiplicity. We will denote them by $\{\lambda_N^{(j)}\}_j$. Similarly, we will denote by $\{\lambda^{(j)}\}_j$ the isolated eigenvalues of $\mathcal{L}_u$ in the annulus $\{t \in \mathbb{C}; \theta < |t| \leq 1\}$.

Now, we are especially interested in the asymptotic behavior of the eigenvalue of $\tilde{\mathcal{L}}_N$ of the maximal modulus. In order to know it, we will consider the zeta function associated with $\tilde{\mathcal{L}}_N$.

3. The Zeta-function associated with $\tilde{\mathcal{L}}_N$

The Ruelle-Artin-Mazur zeta function $\zeta(t)$ is defined as follows:

$$
\begin{aligned}
\zeta(t) &= \exp\{\sum_{p=1}^{\infty} \frac{t^p}{p} \sum_{x \in Fix_p\sigma} e^{S_p u(x)}\} \\
&= \exp\{\sum_{p=1}^{\infty} \frac{t^p}{p} \sum_{\substack{a_1 \cdots a_p \\ A_{a_p a_1} = 1}} \mathcal{L}^p 1_{[a_1 \cdots a_p]}(\dot{a}_1 \cdots \dot{a}_p)\}
\end{aligned}
$$

(3.1)

where $\dot{a}_1 \cdots \dot{a}_p$ is a periodic point $a \in \Sigma_A^+$ such that $a_{kp+i} = a_i$ for any $k \in \mathbb{N}$. It is well known that the poles of $\zeta(t)$ are corresponding to the eigenvalues of $\mathcal{L}$.

Proposition 3.1 (Ruelle [R.]). *Let $\lambda^{(j)}$ be the eigenvalue of $\mathcal{L}$ in the annulus $\{t \in \mathbb{C}; \theta < |t| \le 1\}$ of multiplicity m_j. Then $\frac{1}{\lambda^{(j)}}$ is the pole of $\zeta(t)$ in $\{t \in \mathbb{C}; 1 \le |t| < \theta^{-1}\}$ of the same multiplicity m_j, and vice versa.*

Now, we define a formal power series $\tilde{\zeta}_N(t)$ as follows:

$$
\begin{aligned}
\tilde{\zeta}_N(t) &= \exp\{\sum_{p=1}^{\infty} \frac{t^p}{p} \sum_{x \in Fix_p\sigma} e^{S_p u(x)} \prod_{j=0}^{p-1} 1_{[z]_N^c}(\sigma^j x)\} \\
&= \exp\{\sum_{p=1}^{\infty} \frac{t^p}{p} \sum_{\substack{a_1 \cdots a_p \\ A_{a_p a_1} = 1}} \tilde{\mathcal{L}}_N^p 1_{[a_1 \cdots a_p]}(\dot{a}_1 \cdots \dot{a}_p)\}.
\end{aligned}
$$

(3.2)

We call it the zeta function associated with $\tilde{\mathcal{L}}_N$. Then, we can show the same correspondence as in Proposition 3.1 between the eigenvalues of $\tilde{\mathcal{L}}_N$ and the poles of $\tilde{\zeta}_N(t)$.

Proposition 3.2. *Let $\lambda_N^{(j)}$ be the eigenvalue of $\tilde{\mathcal{L}}_N$ in $\{t \in \mathbb{C}; 0 < |t| \le 1\}$ of multiplicity m_j. Then, $\frac{1}{\lambda_N^{(j)}}$ is the pole of $\tilde{\zeta}_N(t)$ in $\{t \in \mathbb{C}; 1 \le |t| < \theta^{-1}\}$ of the same multiplicity m_j, and vice versa.*

Proposition 3.3. *Let us denote the convergence radius of $\tilde{\zeta}_N(t)$ by $\tilde{t}_N$. Then,*

$$
\lim_{N \to \infty} \tilde{t}_N = 1 \qquad \text{for} \quad \mu - a.e. \ z.
$$

Corollary 3.4. *Denote by $\tilde{\lambda}_N$ the eigenvalue of $\tilde{\mathcal{L}}_N$ of maximal modulus. Then,*

$$
\lim_{N \to \infty} \tilde{\lambda}_N = 1 \qquad \text{for} \quad \mu - a.e. \ z.
$$

Remark We can see that the convergence radius of $(1 - \tilde{\lambda}_N t) \cdot \tilde{\zeta}_N^{(E)}(t)$ goes to that of $(1-t) \cdot \zeta(t)$ as $N \to \infty$. Since 1 is the simple pole of $\zeta(t)$, the modulus of the eigenvalues of $\tilde{\mathcal{L}}_N$ execpt $\tilde{\lambda}_N$ do not go to 1 as $N \to \infty$. Precisely, there exists a number $0 < q < 1$ such that for any $N \in \mathbb{N}$,

$$
\sup\{|\lambda|; \ \lambda \in \text{Spec}(\tilde{\mathcal{L}}_N) \setminus \tilde{\lambda}_N\} < q
$$

where $\text{Spec}(\tilde{\mathcal{L}}_N)$ is the spectrum of $\tilde{\mathcal{L}}_N$.

4. Poisson law for Symbolic Dynamics

Now, we will show the Poisson law for symbolic dynamics $(\Sigma_A^+, \sigma, \mu)$.

We fix a point $z \in \Sigma_A^+$, and take a cylinder set $[z]_N$ as a neighborhood of z. On $[z]_N$, we define a probability measure μ_N as the restriction of the equilibrium state μ to $[z]_N$, i.e.,

$$\mu_N = \frac{\mu|_{[z]_N}}{\mu([z]_N)}.$$

In order to study the limit distribution of the normalized first return time $\epsilon_N T_N$, where $\epsilon_N = 1/E_{\mu_N}(T_N)$, as $N \to \infty$, we consider its Laplace transform $\phi_N(\alpha)$:

$$\begin{aligned}
\phi_N(\alpha) &\equiv \mu_N(e^{-\alpha \epsilon_N T_N}) \\
&= \int e^{-\alpha \epsilon_N T_N(x)} \mu_N(dx).
\end{aligned}$$

Before we compute the limit of $\phi_N(\alpha)$ as $N \to \infty$, we prepare several lemmas.

Lemma 4.1. *The operator $\tilde{\mathcal{L}}_N : \mathcal{F}_\theta(\Sigma_A^+) \to \mathcal{F}_\theta(\Sigma_A^+)$ can be decomposed as follows:*

$$(4.1) \qquad \tilde{\mathcal{L}}_N = \tilde{\lambda}_N \tilde{E}_N + \tilde{\Psi}_N$$

where $\tilde{E}_N$ is the projection to the eigenspace corresponding to the eigenvalue $\tilde{\lambda}_N$ of maximal modulus, and $\tilde{\Psi}_N$ is a bounded linear operator such that

$$\tilde{E}_N \tilde{\Psi}_N = \tilde{\Psi}_N \tilde{E}_N = 0.$$

proof By Lemma 2.1, we can apply the Ionescu-Turcia-Marinescu theorem ([I.T.M.]) to $\tilde{\lambda}_N^{-1} \cdot \tilde{\mathcal{L}}_N$. □

From Lemma 4.1, we obtain the following decomposition:

$$(4.2) \qquad \tilde{\mathcal{L}}_N^i = \tilde{\lambda}_N^i \tilde{E}_N + \tilde{\Psi}_N^i.$$

Then, by Lemma 1.1, we get for $i \geq 2$,

$$(4.3) \qquad \begin{aligned}
\mu_N(T_N = i) &= \int \tilde{\mathcal{L}}_N^{i-1}(\mathcal{L}1_{[z]_N}) d\mu_N \\
&= \tilde{\lambda}_N^{i-1} \int \tilde{E}_N(\mathcal{L}1_{[z]_N}) d\mu_N + \int \tilde{\Psi}_N^{i-1}(\mathcal{L}1_{[z]_N}) d\mu_N.
\end{aligned}$$

The following lemma is well-known as a part of Ambrose-Kakutani's theorem ([PE.]).

Lemma 4.2.

$$\epsilon_N = \frac{1}{E_{\mu_N}(T_N)} = \mu([z]_N)$$

Lemma 4.3. *For $\mu - a.e.$ z, the following equalities hold.*

$$(4.4) \qquad \lim_{N \to \infty} \frac{\int \tilde{E}_N(\mathcal{L}1_{[z]_N}) d\mu_N}{1 - \tilde{\lambda}_N} = \lim_{N \to \infty} \int \tilde{E}_N 1 d\mu_N = 1.$$

$$(4.5) \qquad \lim_{N \to \infty} \frac{\int \tilde{E}_N(\mathcal{L}1_{[z]_N})d\mu_N}{\epsilon_N} = 1.$$

$$(4.6) \qquad \lim_{N \to \infty} \epsilon_N \sum_{i=1}^{\infty} \int \check{\Psi}_N^i 1 d\mu_N = 0.$$

Now, we will consider the limit of $\phi_N(\alpha)$.

Put $[\tilde{E}_N] = \int \tilde{E}_N(\mathcal{L}1_{[z]_N})d\mu_N$, $[\check{\Psi}_N^i] = \int \check{\Psi}_N^i(\mathcal{L}1_{[z]_N})d\mu_N$.

$$\begin{aligned}
\phi_N(\alpha) &= \int e^{-\alpha \epsilon_N T_N} d\mu_N \\
&= \sum_{i=1}^{\infty} e^{-\alpha \epsilon_N i} \mu_N(T_N = i) \\
&= e^{-\alpha \epsilon_N}\{\int \mathcal{L}\infty_{[t]_N}[\mu_N + \sum^{\infty} 1^{-\alpha \epsilon_N)}\hat{\mathcal{R}}_N[\hat{\mathcal{E}}_N] + \sum^{\infty} 1^{-\alpha \epsilon_N)}[\hat{\mathcal{B}}_N]\} \\
&= e^{-\alpha \epsilon_N}\{1 - \tilde{\lambda}_N \int \tilde{E}_N 1 d\mu_N + \frac{\tilde{\lambda}_N}{e^{\alpha \epsilon_N} - \tilde{\lambda}_N} \cdot [\tilde{E}_N] + \sum_{i=1}^{\infty}(e^{-\alpha \epsilon_N i} - 1)[\check{\Psi}_N^i]\}.
\end{aligned}$$

where we used (4.3).

By (4.4), we can see that for $\mu - $ a.e. z,

$$\lim_{N \to \infty}(1 - \tilde{\lambda}_N \int \tilde{E}_N 1 d\mu_N) = 0.$$

The following equality is obtained by (4.4) and (4.5),

$$\lim_{N \to \infty} \frac{\tilde{\lambda}_N}{e^{\alpha \epsilon_N} - \tilde{\lambda}_N} \cdot [\tilde{E}_N] = \frac{1}{1 + \alpha}.$$

And we can see by (4.6),

$$\begin{aligned}
|\sum_{i=1}^{\infty}(e^{-\alpha \epsilon_N i} - 1)[\check{\Psi}_N^i]| &\leq \alpha \epsilon_N \sum_{i=1}^{\infty} i[\check{\Psi}_N^i] \\
&= \alpha \epsilon_N \sum_{i=1}^{\infty} \int \check{\Psi}_N^i 1 d\mu_N \\
&\to 0.
\end{aligned}$$

Therefore, for $\mu - $ a.e. z,

$$\lim_{N \to \infty} \phi_N(\alpha) = \frac{1}{1 + \alpha}.$$

This implies that the following theorem.

Theorem 4.4. *For $\mu - $ a.e. z, the limit distribution of $\epsilon_N T_N$ as $N \to \infty$ exists and it is the exponential distribution with parameter 1, where $\epsilon_N = 1/E_{\mu_N}(T_N)$.*

Next we will study the k-th return times $T_N^{(k)}$:

$$T_N^{(k)}(x) \equiv \sum_{j=0}^{k-1} T_N(\sigma^{T_N^{(j)}(x)}x), \qquad k = 1, 2, \cdots,$$

where $T_N^{(0)}(x) \equiv 0$.

The measure μ_N on $[z]_N$ is an invariant measure of the induced transformation of the shift σ to $[z]_N$, i.e. $\sigma^{T_N(\cdot)}(\cdot) : [z]_N \to [z]_N$. Threrefore the following lemma holds.

Lemma 4.5. *For each $k \geq 1$, $\epsilon_N(T_N^{(k+1)} - T_N^{(k)})$ has the same distribution as $\epsilon_N T_N$. Therefore, for $\mu - a.e.$ z, the limit distribution of $\epsilon_N(T_N^{(k+1)} - T_N^{(k)})$ is the exponential distribution.*

We remark that the limit distributions of $\epsilon_N(T_N^{(k+1)} - T_N^{(k)})$ are mutualy independent because $(\Sigma_A^+, \sigma, \mu)$ is weakly Bernoulli. (See [B.])

Hence we obtain the following proposition.

Proposition 4.7. *For $\mu - a.e.$ z,*

$$\lim_{N \to \infty} \mu_N(\epsilon_N T_N^{(k)} \leq t \ \text{ and } \ \epsilon_N T_N^{(k+1)} > t) = \frac{t^k}{k!} e^{-t}.$$

Here, let us define a point process on $\mathbb{R}^+$, say $Y_N(\cdot)$, as follows:

$$Y_N(\cdot) = \sum_{k=1}^{\infty} \delta_{\epsilon_N T_N^{(k)}}(\cdot)$$

where δ_p is the Dirac δ measure at $p \in \mathbb{R}^+$. We will call it the normalized return time process.

Then, the above proposition implies the following .

Theorem 4.7. *For $\mu - a.e.$ $z \in \Sigma_A^+$, the sequence of the normalized return time processes $\{Y_N\}_N$ converges to the Poisson point process as $N \to \infty$ in finite dimensional distribution, i.e. for any disjoint Borel sets $B_1, \cdots, B_n \in \mathcal{B}(\mathbb{R}^+)$ and any non-negative integers $k_1, \cdots, k_n$,*

$$\lim_{N \to \infty} \mu_N(Y_N(B_1) = k_1, \cdots, Y_N(B_n) = k_n) = \prod_{i=1}^{n} \frac{\ell(B_i)^{k_i}}{k_i!} e^{-\ell(B_i)},$$

where ℓ is the Lebesgue measure.

We remark that we can easily extend the above theorem to the two-sided symbolic dynamics. As we have mentioned in Introdution , the main theorem can be proved by approximating ϵ-neighborhood by a finite union of cylinder sets. See [H] about the way of approximating $U_\epsilon(z)$.

References

[B.]   R. Bowen, *Equilibrium states and the ergodic theory of Anosov diffeomorphisms* Lec. Note. in Math. vol.470, Springer (1975).

[H.]   M. Hirata, *Poisson law for Axiom A diffeomorphisms* To appear in Ergodic th. and Dynam. sys.

[I.T.M.]   C. T. Ionescu-Tulcea, G. Marinescu, *Théorie ergodique pour des classes d'operations non complètement continues* Ann. of Math. **52**, (1950), 140–147

[PE.]   K. Petersen, *Ergodic theory* Cambridge University Press, (1983).

[P.]   M. Pollicott, Meromorphic extensions of generalized zeta function Invent. math. **85**, (1986), 147–164

[R]   D. Ruelle, *Thermodynamic formalism* Encyclopedia of Mathematics and its Applications, vol.5 Addison-Wesley, (1978).

[S.I]   Ya. G. Sinai, *Some mathematical problems in the theory of quantum chaos* Physica A, **163**, (1990), 197–204.

[S.II]   Ya. G. Sinai, *Mathematical problems in the theory of quantum chaos* Lec. Note. in Math. vol.1469, 41–59. Springer

QUADRATIC MAPS WITH MAXIMAL OSCILLATION

Franz Hofbauer[1] and Gerhard Keller[2]

[1] Universität Wien, Austria
[2] Universität Erlangen, Germany

INTRODUCTION

Let $(f_t)_{0 \leq t \leq 1}$ denote the family of quadratic maps $f_t(x) = 2t(1 - x^2) - 1$ on $[-1, 1]$. An important aspect of the asymptotics of interates of a map f_t is the behaviour of mass distributions along individual orbits. If δ_x denotes the point mass in x, one can study the uniform distributions

$$\nu_{x,t,n} = \frac{1}{n} \sum_{k=1}^{n} \delta_{f_t^k x}$$

along orbit pieces of length n. Denote by $\bar{\omega}_t(\delta_x)$ the set of all weak accumulation points of the sequence $(\nu_{x,t,n})_{n>0}$, and let λ be the normalized Lebesgue measure on $[-1, 1]$. The following facts are well known:

- If f_t has a stable periodic point p of period m, then $\bar{\omega}_t(\delta_x) = \{\nu_{p,t,m}\}$ for λ-a.e. x.

- If f_t has an invariant probability measure $\mu \ll \lambda$, then $\bar{\omega}_t(\delta_x) = \{\mu\}$ for λ-a.e. x.

- If f_t has a solenoidal attractor with unique invariant probability measure ρ, then $\bar{\omega}_t(\delta_x) = \{\rho\}$ for λ-a.e. x.

In all these cases we say that f_t has an asymptotic measure. On the other hand, Hofbauer and Keller (1990 a,b) constructed examples of quadratic maps for which $\bar{\omega}_t(\delta_x)$ is a large set with prescribed properties. For example, given $0 \leq h_0 < h_1 < \log \frac{1+\sqrt{5}}{2}$, there are uncountably many parameters t such that

$$\{h_\nu(f_t) : \nu \in \bar{\omega}_t(\delta_x), \nu \ \text{ergodic}\} = [h_0, h_1]$$

for λ-a.e. x. Based on similar ideas we will prove in this note that there are quadratic maps f_t for which Lebesgue-a.e. x has maximal oscillation, i.e.

Theorem 1. *Let $\mathcal{M}_1(f_t)$ be the set of all f_t-invariant Borel probability measures on $[-1, 1]$. There are uncountably many parameters t such that $\bar{\omega}_t(\delta_x) = \mathcal{M}_1(f_t)$ for λ-a.e. x.*

Algorithms, Fractals, and Dynamics
Edited by Y. Takahashi, Plenum Press, New York, 1995

This result is not only true for the quadratic family, but for each full continuous family of unimodal maps with negative Schwarzian derivative, see Hofbauer and Keller (1990a) for details.

Acknowledgement: The question, whether maps with maximal oscillation exist, was asked to one of us (G.K.) by A. Blokh.

KNEADING MAPS

We recall some concepts and results from the theory of interval maps in the form they have been used in Hofbauer and Keller (1990a).

The itinerary of a point $x \in [-1,1]$ is the $0, 1$-sequence $\varphi(x) = \omega_1 \omega_2 \omega_3 \cdots$ where $\omega_i = 0$ if $f_t^{i-1}(x) < 0$ and $\omega_i = 1$ otherwise. (As we do not make use of itineraries of preimages of the critical point 0, we can use this simplified definition.)

We suppose throughout that f_t has no stable periodic orbit. In this case, the kneading sequence $\underline{e}$ of f_t, which is the itinerary of $f_t(0)$, is not periodic.

$Q : \mathbb{N} \to \mathbb{N} \cup \{0\}$ is a *kneading map*, if $Q(k) < k$ for all $k \in \mathbb{N}$ and if

$$(Q(k+j))_{j \geq 1} \geq (Q(Q^2(k) + j))_{j \geq 1}$$

where $\geq$ denotes the lexicographic ordering between sequences. Each nonperiodic kneading sequence $\underline{e}$ determines a kneading map $Q : \mathbb{N} \to \mathbb{N} \cup \{0\}$ in the following way:

Define positive integers $1 = r_1, r_2, r_3, \cdots$ and $S_i = 1 + r_1 + \cdots r_i$ inductively by

$$e_{S_i+1} \cdots e_{S_{i+1}} = e_1 \cdots e_{r_{i+1}} e'_{r_{i+1}} \quad \text{for } i \geq 0. \tag{1}$$

Hofbauer (1980) showed that there is a unique kneading map $Q : \mathbb{N} \to \mathbb{N} \cup \{0\}$ such that

$$r_k = S_{Q(k)} \quad \text{for all } k \geq 1. \tag{2}$$

Vice versa, each kneading map $Q : \mathbb{N} \to \mathbb{N} \cup \{0\}$ defines via (2) and (1) a unique $0, 1$-sequence $\underline{e}$ with $e_1 = 1$, and Theorem 4 of Hofbauer and Keller (1990a) asserts that each sequence $\underline{e}$ arising from a kneading map which is not eventually periodic is the kneading sequence of at least one map f_t. Denote by $J_\infty(Q)$ the set of all parameters t for which kneading sequence of f_t is determined by Q. (Indeed, the results from that reference are more complete in that they include the case of periodic kneading sequence.) We also note that

$$S_k \leq 2^k \tag{3}$$

by (2) because $Q(k) < k$.

In order to construct kneading sequences with special properties we introduced the notions of frame and skeleton. A sequence $\mathcal{F} = (0 = V_0 < U_1 < V_1 < U_2 < V_2 < \cdots)$ of integers is a *frame*, if

$$U_{k+1} \geq k \cdot 2^{k+V_k} \quad \text{for all } k \geq 0 \tag{4}$$

and

$$V_k \geq k \cdot 2^{U_k} \quad \text{for all } k \geq 1. \tag{5}$$

(At present it is only important to notice that a frame is a very sparse increasing sequence of integers.) Given such a frame $\mathcal{F}$ we define the *skeleton* $\mathcal{S}(\mathcal{F})$ as the set of all kneading maps $Q : \mathbb{N} \to \mathbb{N} \cup \{0\}$ satisfying

$$U_k < i \leq V_k \Rightarrow Q(i) = U_k$$

and

$$Q(U_{k+1}) < U_k$$

for all $k \geq 1$.

Let be $b := f_t(0)$. Important for us is

Proposition 1 [Proposition 1 of Hofbauer and Keller (1990a)] *There are uncountably many different frames $\mathcal{F}$ such that for each $Q \in \mathcal{S}(\mathcal{F})$ and each $t \in J_\infty(Q)$ holds:*
For λ-a.e.x and each $\psi \in C([-1,1])$

$$\lim_{n \to \infty} (\nu_{x,t,S(V_n)}(\psi) - \nu_{b,t,S(U_n)}(\psi)) = 0.$$

As our construction yields indecomposable kneading sequences, the corresponding maps f_t are topologically transitive on $I_t := [f_t^2(0), f_t(0)]$ such that the set of uniform distributions on periodic orbits is dense in the space of all f_t-invariant probability measures on I_t, see Hofbauer (1988) or Theorem 10.3 of Blokh (1991). Since the point mass in $z = -1$ is the only f_t-invariant measure which is not concentrated on I_t, the main theorem follows from Proposition 1 together with

Proposition 2 *For each frame $\mathcal{F}$ there is a $Q \in \mathcal{S}(\mathcal{F})$ such that for all $t \in J_\infty(Q)$ and each periodic point $z = f_t^p(z) \in I_t$ holds*
$$\nu_{z,t,p} \text{ is an accumulation point of the sequence } (\nu_{b,t,S(U_n)})_{n \geq 1}.$$

The proof is give in the next section.

Indeed, these two propositions allow a slightly stronger conclusion:

Corollary 1 *There are uncountably many parameters t for which $\bar{\omega}_t(\lambda) = \mathcal{M}_1(f_t)$.*

PROOF OF PROPOSITION 2

As in Hofbauer (1980) we use the following order relation between $0, 1$-sequences $u_1 u_2 \cdots$ and $v_1 v_2 \cdots$:

$u_1 u_2 \cdots \trianglelefteq v_1 v_2 \cdots$ if the sequence are identical or if the following holds: Let $n = \min\{j : u_j \neq v_j\}$ and denote by $p(n)$ the number of ones among $u_1 \cdots u_{n-1}$. Now $u_n = 0, v_n = 1$ if $p(n)$ is even and $u_n = 1, v_n = 0$ if $p(n)$ is odd.
The symbol $\triangleleft$ denotes the relation "$\trianglelefteq$ but not $=$".

We start with a lemma which describes how an arbitrary itinerary can be decomposed into initial segments of the kneading sequence.

Lemma 1 *Let $u_1 u_2 \cdots$ be a $0, 1$-sequence with $u_1 = 1$. Fix m and suppose that*

$$u_{l+1} \cdots u_{l+S_m} \triangleleft e_1 \cdots e_{S_m} \quad \text{for } l \geq 1. \tag{6}$$

Define numbers $(\tilde{r}_i)_{i \geq 1}$ by

$$u_{\tilde{R}_{i-1}+j} = e_j \quad \text{for } 1 \leq j < \tilde{r}_i \quad \text{and} \quad u_{\tilde{R}_i} \neq e_{\tilde{r}_i} \quad \text{for } i \geq 1 \tag{7}$$

where $R_1 = \tilde{r}_1 + \cdots + \tilde{r}_i$ and $R_0 = 0$.

Then there is for each $i \geq 1$ a number $\tilde{Q}(i) \in \{0, 1, \cdots, m\}$ such that $\tilde{r}_i = S_{\tilde{Q}(i)}$. We have $\tilde{Q}(1) \geq 1$ and

$$(Q(t+j))_{1 \leq j \leq m-t} < (\tilde{Q}(i+j))_{1 \leq j \leq m-t} \quad \text{for each } i \geq 1 \text{ with } \tilde{Q}(i) \geq 1 \qquad (8)$$

where $t = Q(\tilde{Q}(i))$. ($<$ means strictly smaller in the lexicographic ordering.)

Proof. As $e_1 \cdots e_{S_m}$ is not contained as a substring in $u_1 u_2 \cdots$ by (6), we have $\tilde{r}_i \leq S_m$ for $i \geq 1$. By Lemma 1(iii) of Hofbauer (1980) we have $\tilde{r}_i = S_{\tilde{Q}(i)}$ for some $\tilde{Q}(i) \in \{0, 1, \cdots, m\}$. As $u_1 = 1$, we have $\tilde{Q}(1) \geq 1$. It remains to show (8).

Fix $i \geq 1$ with $\tilde{Q}(i) \geq 1$ and set $t = Q(\tilde{Q}(i))$. Choose $k \geq 1$ such that

$$\tilde{r}_{i+j} = r_{t+j} \text{ for } 1 \leq j < k \quad \text{and} \quad \tilde{r}_{i+k} \neq r_{t+k}. \qquad (9)$$

By (7) and (1) we get

$$u_{\tilde{R}_{i-1}+S_{\tilde{Q}(i)-1}+1} \cdots u_{\tilde{R}_i} = e_{S_{\tilde{Q}(i)-1}+1} \cdots e'_{\tilde{r}_i} = e_1 \cdots e_{S_t}$$

because $\tilde{R}_i - \tilde{R}_{i-1} = \tilde{r}_i = S_{\tilde{Q}(i)}$ and $r_{\tilde{Q}(i)} = S_t$.

For $1 \leq j < k$ we get in the some way that

$$u_{\tilde{R}_{i+j-1}+1} \cdots u_{\tilde{R}_{i+j}} = e_1 \cdots e'_{\tilde{r}_{i+j}} = e_{S_{t+j+1}+1} \cdots e_{S_{t+j}}$$

because $\tilde{r}_{i+j} = r_{t+j}$. Setting $l = \tilde{R}_{i-1} + S_{\tilde{Q}(i)-1}$ this together gives

$$u_{l+1} \cdots u_{\tilde{R}_{i+k-1}} = e_1 \cdots e_{S_{t+k-1}}. \qquad (10)$$

Since $u_1 u_2 \cdots$ does not contain $e_1 \cdots e_{S_m}$ by (6), we get $S_{t+k-1} < S_m$, i.e. $k \leq m - t$, and we can choose k maximal with respect to (9) such that $\tilde{r}_{i+k} \neq r_{t+k}$.

Suppose that $\tilde{r}_{i+k} < r_{t+k}$ and let $n = S_{t+k-1} + \tilde{r}_{i+k}$. By (7) and (1) we get

$$u_{\tilde{R}_{i+k-1}+1} \cdots u_{\tilde{R}_{i+k}} = e_1 \cdots e'_{\tilde{r}_{i+k}} = e_{S_{t+k-1}+1} \cdots e'_n.$$

Together with (10) this gives

$$u_{l+1} \cdots u_{\tilde{R}_{i+k}} = e_1 \cdots e'_n.$$

Applying (1) once more we get, as $r_{t+k} > \tilde{r}_{i+k} = S_{\tilde{Q}(i+k)}$

$$e_1 \cdots e_n = e_1 \cdots e_{S_{t+k-1}} e_1 \cdots e_{S_{\tilde{Q}(i+k)}},$$

and Lemma 1(ii) of Hofbauer (1980) implies that the number of ones in $e_1 \cdots e_n$ is even. Considering the two cases $e_n = 0$ ane $e_n = 1$ one shows that $e_1 \cdots e_n \triangleleft e_1 \cdots e_{n-1} e'_n$. Together with $u_{l+1} \cdots u_{\tilde{R}_{i+k}} = e_1 \cdots e'_n$ this contradicts to (6) because $n = S_{t+k-1} + \tilde{r}_{i+k} < S_{t+k-1} + r_{t+k} = S_{t+k} \leq S_m$. Hence $\tilde{r}_{i+k} > r_{t+k}$. So we have

$$(r_{t+k})_{1 \leq j \leq k} < (\tilde{r}_{i+j})_{1 \leq j \leq k},$$

which implies (8) since $k \leq m - t$. $\qquad \square$

Lemma 2 *Let $\mathcal{F} = (V_0, U_1, V_1, \cdots)$ be a frame with $U_1 > 1$. Then there is $Q \in \mathcal{S}(\mathcal{F})$ such that for all $t \in J_\infty(Q)$ the following holds: There is an enumeration $(p_1, p_2 \cdots)$ of the periodic points of f_t in $(0,1)$, which contains each of these periodic points infinitely often. If $a_k = S_{V_k + k}$ and $b_k = S_{U_{k+1}}$, then $e_{a_k + 1} \cdots e_{b_k}$ is an initial segment of the itinerary of p_k for each $k \geq 1$.*

Proof: First fix a bijective map $\gamma : \mathbb{N} \to \mathbb{N} \times \mathbb{N}$ such that $i_1 \leq i$ and $i_2 \leq i$ if $(i_1, i_2) = \gamma(i)$.

We define the required kneading map $Q \in \mathcal{S}(\mathcal{F})$ by induction. After the n-th step, $Q(i)$ will be defined for $1 \leq i \leq U_n$. This determines the initial segment $e_1 \cdots e_{S_{U_n}}$ of the corresponding kneading sequence using (2) and (1). After the n-th step we fix an enumeration $(v_{(i,n)})_{i \geq 1}$ of all periodic $0,1$-sequences $u_1 u_2 \cdots$ with $u_1 = 1$ satisfying

$$u_{l+1} \cdots u_{l + S_{U_n} - 1} \triangleleft e_1 \cdots e_{S_{u_n} - 1} \text{ for all } l \geq 0. \tag{11}$$

In order to start this induction set $Q(i) = 0$ for $1 \leq i \leq U_1$. Suppose that $k \geq 1$ and that $Q(i) \in \{0, 1, \cdots i - 1\}$ is defined for $1 \leq i \leq U_k$ in such a way that

$$(Q(Q^2(i) + j))_{1 \leq j \leq U_k - i} \leq (Q(i+j))_{1 \leq j \leq U_k - i} \text{ for } 1 \leq i < U_k \text{ with } Q(i) \geq 1. \tag{12}$$

and such that

$$Q(1) = Q(U_1) = 0 \text{ and } Q(U_l) < U_{l-1} \text{ for } 2 \leq l \leq k. \tag{13}$$

(It is easy to see that (12) and (13) are satisfied for $k = 1$.) Furthermore suppose that $(v_{(i,j)})_{i \geq 1}$ is fixed for $1 \leq j \leq k$.

In order to define $Q(i)$ for $U_k < i \leq U_{k+1}$, denote the $0,1$-sequence $v_{\gamma(k)}$ by $u_1 u_2 \cdots$, which is well defined, because $\gamma(k) \in \mathbb{N} \times \{1, 2, \cdots, k\}$. Furthermore, (11) holds for some $n \leq k$. We apply Lemma 1 with $m = U_n - 1$. Let $\tilde{Q} : \mathbb{N} \to \{0, 1, \cdots, U_n - 1\}$ be as in Lemma 1. We define $Q(i)$ for $U_k < i \leq U_{k+1}$ as follows:

$$\begin{aligned}
Q(U_k + s) &= U_k \quad \text{for } 1 \leq s \leq V_k - U_k \\
Q(V_k + s) &= U_{k-s} \quad \text{for } 1 \leq s \leq k - 1 \\
Q(V_k + k) &= 1 \\
Q(V_k + k + s) &= \tilde{Q}(s) \quad \text{for } 1 \leq s \leq U_{k+1} - V_k - k.
\end{aligned}$$

One easily sees that $Q(i) \leq i - 1$ for all i. We show (12) for $k + 1$: Since $Q(U_k + 1) = U_k > Q(i)$ for all $i \leq U_k$, we get from (12) that

$$(Q(Q^2(i) + j))_{1 \leq j \leq U_k - i + 1} < (Q(i+j))_{1 \leq j \leq U_k - i + 1}$$

for all $2 \leq i \leq U_k$ with $Q(i) \geq 1$. For $U_k < i \leq V_k + k$ one easily shows that $Q(Q^2(i) + 1) < Q(i+1)$ using (13). For $i = V_k + k$ one needs also that $\tilde{Q}(1) \geq 1$, which follows from Lemma 1. If $V_k + k < i < U_{k+1}$ and $Q(i) \geq 1$ it follows from (8) that

$$(Q(Q^2(i) + j))_{1 \leq j \leq U_{k+1} - i} < (Q(i+j))_{1 \leq j \leq U_{k+1} - i}.$$

We get only $\leq$ and not $<$, since the number $m - t$ in (8) may be larger than $U_{k+1} - i$ for certain i.

This finishes the proof of (12) for $k + 1$. Since $\tilde{Q}(i) \leq U_n - 1 \leq U_k - 1$ by Lemma 1, we get $Q(U_{k+1}) < U_k$, which proves (13) for $k + 1$. This ends the definition of Q and shows that $Q \in \mathcal{S}(\mathcal{F})$.

By Theorem 4 of Hofbauer and Keller (1990a) the kneading sequence $e_1 e_2 \cdots$ of f_t for $t \in J_\infty(Q)$ is determined by Q via (1) and (2). Recall that $\varphi(x)$ denotes the itinerary of x under f_t. Let $\mathcal{P}$ be the set of all periodic $0, 1$-sequences $u_1 u_2 \cdots$ with $u_1 = 1$ which satisfy $u_{l+1} u_{l+2} \cdots \lhd e_1 e_2 \cdots$ for all $l \geq 0$. Since $e_1 e_2 \cdots$ is not periodic by (2.10) of that reference, φ is a bijection from the set of periodic points of f_t in $(0, 1]$ to $\mathcal{P}$.

We show that each $u_1 u_2 \cdots \in \mathcal{P}$ satisfies (11) for infinitely many n. There is $k \geq 1$ such that $u_1 u_2 \cdots$ does not contain $e_1 \cdots e_k$. Otherwise $u_1 u_2 \cdots$ would contain arbitrarily long initial segment of $e_1 e_2 \cdots$ and, as $u_1 u_2 \cdots$ is of some period q, all these initial setments of $e_1 e_2 \cdots$ start at one of the first q coordinates of $u_1 u_2 \cdots$, which implies $u_{l+1} u_{l+2} \cdots = e_1 e_2 \cdots$ for some l, an contradiction. Hence, $u_1 u_2 \cdots$ does not contain $e_1 e_2 \cdots e_k$, such that (11) holds for all n with $S_{U_n - 1} \geq k$. This implies that each elment of $\mathcal{P}$ occurs infinitely often in $(v_{(i,n)})_{i \geq 1, n \geq 1}$.

Set $v_i := v_{\gamma(i)}$ and $p_i := \varphi^{-1}(v_i)$. Then each periodic point in $(0, 1]$ occurs infinitely often in $(p_i)_{i \geq 1}$. The definition of $Q(i)$ for $V_k + k < i \leq U_{k+1}$ together with (1) and (7) implies that $e_{a_k+1} \cdots e_{b_k}$ is an initial segment of $\varphi(p_k) = v_{\gamma(k)}$ for $k \geq 1$, where $a_k = S_{V_k + k}$ and $b_k = S_{U_{k+1}}$. $\qquad \square$

Proof of Proposition 2: Fix a periodic point $z = f_t^p(z)$, and let $d_k := b_k - a_k$. As $\lim_{k \to \infty} d_k / b_k = 1$ by (4) and (3) and as $\varphi^{-1} : \varphi([-1, 1]) \to [-1, 1]$ is continuous, it follows from Lemma 2 that for any $\psi \in C([-1, 1])$

$$\lim_{k \to \infty} \left(\nu_{z, t, d_k}(\psi) - \nu_{b, t, b_k}(\psi) \right) = 0.$$

Now the proposition follows from the obvious fact that

$$\lim_{k \to \infty} \left(\nu_{z, t, d_k}(\psi) - \nu_{z, t, p}(\psi) \right) = 0.$$

$\qquad \square$

References

1. Blokh, A., *The "spectral" decomposition for one-dimensional maps*, Preprint, (1991).
2. Hofbauer, F., *The topological entropy of the transformation $x \to ax(1-x)$*, Monatshefte Math. **90**, (1980), 117-141.
3. Hofbauer, F., *Generic properties of invariant measures for continuous piecewise monotonic transformations*, Monatshefte Math. **106**, (1988), 301-312.
4. Hofbauer, F., and Keller, G., *Quadratic maps without asymptotic measure*, Commun. Math. Phys. **127**, (1990a), 319-337.
5. Hofbauer, F., and Keller, G., *Some remarks about recent results on S-unimodal maps*, Annales de l'Institut Henri Poincaré, Physique Théorique **53**,4 (1990b), 13-425.

FRACTAL DOMAINS OF QUASI–PERIODIC MOTIONS ON T^2

Shunji Ito

Tsuda College
Kodaira-shi, Tokyo 187 Japan

Introduction

We mention the following theorem in [1] without definitions or explanations.

Theorem A *(Ito–Ohtsuki) Let $(\alpha, \beta, \gamma, \delta) \in [0,1]^4$ be a periodic point of natural extension of Modified Jacobi–Perron algorithm with period $\begin{pmatrix} a_1 & \cdots & a_k \\ \varepsilon_1 & \cdots & \varepsilon_k \end{pmatrix}$, and P be the contractive invariant surface of linear map ϕ:*

$$\phi \begin{pmatrix} x \\ y \\ z \end{pmatrix} := A_{\begin{pmatrix} a_1 \\ \varepsilon_1 \end{pmatrix}} \cdots A_{\begin{pmatrix} a_k \\ \varepsilon_k \end{pmatrix}} \begin{pmatrix} x \\ y \\ z \end{pmatrix},$$

where

$$A_{\begin{pmatrix} a \\ \varepsilon \end{pmatrix}} = \begin{cases} \begin{pmatrix} a & 0 & 1 \\ 1 & 0 & 0 \\ 0 & 1 & 0 \end{pmatrix} & \text{if } \varepsilon = 0, \\ \begin{pmatrix} a & 1 & 0 \\ 0 & 0 & 1 \\ 1 & 0 & 0 \end{pmatrix} & \text{if } \varepsilon = 1. \end{cases}$$

Let $\pi : R^3 \rightarrow P$ be the projection to P along ${}^t(1, \gamma, \delta)$. Then there exists a set X and a partition $\{X_i : i = 1, 2, 3\}$ of X on P satisfying the following property: (1)(*tiling*)

$$\bigcup_{x \in L} (X + x) = P,$$
$$\text{int}(X + x) \cap \text{int}(X + x') = \emptyset \quad (x \neq x' \in L),$$

where $L := \{m\pi(e_2 - e_1) + n\pi(e_3 - e_1) : m, n \in Z\}$.

(2) *The domain exchange transformation W on X, which is isomorphic to a quasi-periodic motion on 2–dimensional torus T^2, is well-defined:*

$$W x = x - \pi e_i \quad \text{if } x \in X_i.$$

(3) Let $X^{(1)} = \phi X$, $X_i^{(1)} = \phi X_i$. Then the induced transformation $W|_{X^{(1)}}$ of W to the set $X^{(1)}$ is isomorphic to W with the isomorphism ϕ, that is, the following commutative relation holds:

$$
\begin{array}{ccc}
 & W & \\
X & \longrightarrow & X \\
\phi \downarrow & & \downarrow \phi \, , \\
X^{(1)} & \longrightarrow & X^{(1)} \\
 & W|_{X^{(1)}} &
\end{array}
$$

moreover, the induced automorphism has $\sigma\begin{pmatrix} a_1 \\ \varepsilon_1 \end{pmatrix} \cdots \sigma\begin{pmatrix} a_k \\ \varepsilon_k \end{pmatrix}$ -structure, where

$$
\sigma\begin{pmatrix} a \\ 0 \end{pmatrix} : \begin{array}{l} 1 \to \cdots 1 \cdots 12 \\ 2 \to 3 \\ 3 \to 1 \end{array}
\qquad
\sigma\begin{pmatrix} a \\ 1 \end{pmatrix} : \begin{array}{l} 1 \to \cdots 1 \cdots 13 \\ 2 \to 1 \\ 3 \to 2 \end{array}
$$

(The definition of σ-structure will be found in Theorem D.)

The purpose of this paper is to show that the statement of the theorem is correct for not only periodic points but also for almost all $(\alpha, \beta) \in [0,1]^2$ as Theorem C and D. To say this assertion, we use the following simultaneous approximation theorem:

Theorem B (Ito, Keane, Ohtsuki) Let $\begin{pmatrix} a_1 & a_2 & \cdots \\ \varepsilon_1 & \varepsilon_2 & \cdots \end{pmatrix}$ be the sequence of partial quotients of (α, β) by Modified Jacobi Perron algorithm. Let $\left(\frac{p_n^{(1)}}{q_n^{(1)}}, \frac{r_n^{(1)}}{q_n^{(1)}} \right)$ be simultaneous approximation of (α, β) given by

$$
\begin{pmatrix} q_n^{(1)} & q_n^{(2)} & q_n^{(3)} \\ p_n^{(1)} & p_n^{(2)} & p_n^{(3)} \\ r_n^{(1)} & r_n^{(2)} & r_n^{(3)} \end{pmatrix} := A\begin{pmatrix} a_1 \\ \varepsilon_1 \end{pmatrix} \cdots A\begin{pmatrix} a_n \\ \varepsilon_n \end{pmatrix}.
$$

Then there exists a positive number ε such that for all most all $(\alpha, \beta) \in [0,1]^2$

$$
\left| \alpha - \frac{p_n^{(1)}}{q_n^{(1)}} \right| < \frac{1}{(q_n^{(1)})^{1+\varepsilon}}, \qquad \left| \beta - \frac{r_n^{(1)}}{q_n^{(1)}} \right| < \frac{1}{(q_n^{(1)})^{1+\varepsilon}}
$$

for large n.

The proof of the theorem is found in article [2].

1. Proofs of the theorems C and D

To obtain the theorems C and D mentioned later, it is sufficient to show that for all most all $(\alpha, \beta) \in [0,1]^2$ there exists a limit set X such that

$$
X = \lim_{n \to \infty} \phi\begin{pmatrix} a_1 \\ \varepsilon_1 \end{pmatrix} \cdots \phi\begin{pmatrix} a_n \\ \varepsilon_n \end{pmatrix} \pi_n K \Sigma\begin{pmatrix} a_n \\ \varepsilon_n \end{pmatrix} \cdots \Sigma\begin{pmatrix} a_1 \\ \varepsilon_1 \end{pmatrix}(u).
$$

(Definitions u and K are found in [1].) In other words, we only need to show the existence of the boundary as a closed curve of the domain X such that

$$
\partial X = \lim_{n \to \infty} \partial\phi\begin{pmatrix} a_1 \\ \varepsilon_1 \end{pmatrix} \cdots \phi\begin{pmatrix} a_n \\ \varepsilon_n \end{pmatrix} \pi K \Sigma\begin{pmatrix} a_n \\ \varepsilon_n \end{pmatrix} \cdots \Sigma\begin{pmatrix} a_1 \\ \varepsilon_1 \end{pmatrix}(u).
$$

96

LEMMA 1 *For each* n

$$d(\partial\pi_n K\Sigma{\binom{a_n}{\varepsilon_n}} \cdots \Sigma{\binom{a_1}{\varepsilon_1}}(u), \partial\phi{\binom{a_n}{\varepsilon_n}}\pi_{n-1}K\sigma{\binom{a_n-1}{\varepsilon_n-1}} \cdots \Sigma{\binom{a_1}{\varepsilon_1}}(u))$$
$$\leq \max_{i=1,2,3}\|\pi e_i\|$$

Proof: From the definition of Σ, we know that the boundary of the set

$$\pi_n K\Sigma{\binom{a_n}{\varepsilon_n}} \cdots \Sigma{\binom{a_1}{\varepsilon_1}}(u)$$

is constructed by the segments $K(\pi e_i)$, $i = 1, 2, 3$, and from the definition of the lifting endomorphisms Θ we also know that the distance of both sets is less than $\max_{i=1,2,3}\|\pi e_i\|$.

COROLLARY 2 *For each* n, *there exists* $C > 0$ *such that*

$$d(\partial D_n, \partial D_{n-1}) \leq C \max_{i=1,2,3}(|q_n^{(i)}\alpha - p_n^{(i)}|, |q_n^{(i)}\beta - r_n^{(i)}|),$$

where $\partial D_n = \phi^{-1}{\binom{a_1}{\varepsilon_1}} \cdots \phi^{-1}{\binom{a_n}{\varepsilon_n}}(\partial\pi_n K\Sigma{\binom{a_n}{\varepsilon_n}} \cdots \Sigma{\binom{a_1}{\varepsilon_1}}u).$

Proof: From lemma 1, the distance $d(\partial D_n, \partial D_{n-1})$ is estimated by $\pi{\begin{pmatrix} q_n^{(i)} \\ p_n^{(i)} \\ r_n^{(i)} \end{pmatrix}}$, $i = 1, 2, 3$.

By the way, using the basis ${\begin{pmatrix} 1 \\ \alpha \\ \beta \end{pmatrix}}$, ${\begin{pmatrix} \alpha \\ -1 \\ 0 \end{pmatrix}}$, ${\begin{pmatrix} \beta \\ 0 \\ -1 \end{pmatrix}}$, any element ${\begin{pmatrix} x \\ y \\ z \end{pmatrix}} \in R^3$ is denoted

by

$$x = \lambda{\begin{pmatrix} 1 \\ \alpha \\ \beta \end{pmatrix}} + \mu{\begin{pmatrix} \alpha \\ -1 \\ 0 \end{pmatrix}} + \nu{\begin{pmatrix} \beta \\ 0 \\ -1 \end{pmatrix}},$$

and ν, μ are given by the linear combination of inner products $\left({\begin{pmatrix} x \\ y \\ z \end{pmatrix}}, {\begin{pmatrix} \alpha \\ -1 \\ 0 \end{pmatrix}}\right)$ and

$\left({\begin{pmatrix} x \\ y \\ z \end{pmatrix}}, {\begin{pmatrix} \beta \\ 0 \\ -1 \end{pmatrix}}\right)$. Therefore, the values $\pi{\begin{pmatrix} q_n^{(i)} \\ p_n^{(i)} \\ r_n^{(i)} \end{pmatrix}}$ are estimated by

$\left({\begin{pmatrix} q_n^{(i)} \\ p_n^{(i)} \\ r_n^{(i)} \end{pmatrix}}, {\begin{pmatrix} \alpha \\ -1 \\ 0 \end{pmatrix}}\right)$ and $\left({\begin{pmatrix} q_n^{(i)} \\ p_n^{(i)} \\ r_n^{(i)} \end{pmatrix}}, {\begin{pmatrix} \beta \\ 0 \\ -1 \end{pmatrix}}\right)$.

By the simultaneous approximation Theorem B and corollary 2, we have

COROLLARY 3 *For almost all* $(\alpha, \beta) \in [0,1]^2$ *there exists a limit set* X *in the sence of Hausdorff metric such that*

$$X = \lim_{n\to\infty} \phi{\binom{a_1}{\varepsilon_1}} \cdots \phi{\binom{a_n}{\varepsilon_n}}\pi_n K(\Sigma{\binom{a_n}{\varepsilon_n}} \cdots \Sigma{\binom{a_1}{\varepsilon_1}}(u)).$$

Now, the statement of Theorem A in the introduction is slightly changed as follows.

Theorem C *Let $P^{(\alpha,\beta)}$ be the orthogonal plane of $^t(1,\alpha,\beta)$ and let $\pi^{(\alpha,\beta)}$ be the projection to P along $^t(1,\gamma,\delta)$. For almost all $(\alpha,\beta,\gamma,\delta) \in [0,1]^4$, there exists a limit set $X^{(\alpha,\beta)}$ and a partition $\{X_i^{(\alpha,\beta)} : i = 1,2,3\}$ on $P^{(\alpha,\beta)}$ satisfying the following properties.*
  (1)*(tiling)*

$$\bigcup(X^{(\alpha,\beta)} + x) = P^{(\alpha,\beta)},$$
$$\text{int}(X^{(\alpha,\beta)} + x) \cap \text{int}(X^{(\alpha,\beta)} + x') = \emptyset \quad (x \neq x' \in L),$$

where $L = \{m\pi^{(\alpha,\beta)}(e_2 - e_1) + n\pi^{(\alpha,\beta)}(e_3 - e_1) : m,n \in Z\}$.
  (2) *The domain exchange transformation $W^{(\alpha,\beta)}$ on $X^{(\alpha,\beta)}$ is well–defined, which is isomorphic to a quasi–periodic motion on T^2,*

$$W^{(\alpha,\beta)}x = x - \pi^{(\alpha,\beta)}e_i \quad \text{if } x \in X_i^{(\alpha,\beta)}.$$

  (3) *Let $X^{(1)} := \phi_{\binom{a_1}{\varepsilon_1}} X^{(\alpha_1,\beta_1)}$, then the induced automorphism of $W^{(\alpha,\beta)}$ to the set $X^{(1)}$ is isomorphic to $W^{(\alpha_1,\beta_1)}$ with the isomorphim $\phi_{\binom{a_1}{\varepsilon_1}}$. Moreover, the induced automorphism has $\sigma_{\binom{a_1}{\varepsilon_1}}$–structure.*

As a corollary of the theorem, we have a following theorem related to the recurrence rule of the origin point of quasi–periodic motions.

Theorem D *For almost all $(\alpha,\beta,\gamma,\delta) \in [0,1]^4$, let us define the decreasing sequence $X^{(n)}$ of neighborhoods of origin point and its partition $X_i^{(n)}$, $i = 1,2,3$ as follows:*

$$X^{(n)} = \phi_{\binom{a_1}{\varepsilon_1}} \cdots \phi_{\binom{a_n}{\varepsilon_n}} \left(X^{(\alpha_n,\beta_n)}\right),$$
$$X_i^{(n)} = \phi_{\binom{a_1}{\varepsilon_1}} \cdots \phi_{\binom{a_n}{\varepsilon_n}} \left(X_i^{(\alpha_n,\beta_n)}\right)$$

and put

$$l_i^{(n)} := q_n^{(i)} + p_n^{(i)} + r_n^{(i)}.$$

Then for the domain exchange $W(= W^{(\alpha,\beta)})$, we have the following property called $\sigma_{\binom{a_1}{\varepsilon_1}} \cdots \sigma_{\binom{a_n}{\varepsilon_n}}$–structure:

$$W^k X_i^{(n)} \cap X_i^{(n)} = \emptyset \quad 0 \le k < l_i^{(n)} - 1,$$
$$W^k X_i^{(n)} \subset X_{s_k^{(i)}} \quad 0 \le k < l_i^{(n)} - 1,$$
$$W^{l_i^{(n)}} X_i^{(n)} = W_{X^{(n)}} X_i^{(n)},$$
$$\bigcup_{i=1,2,3} \bigcup_{k=0,1,\dots,l_i^{(n)}-1} W^k X_i^{(n)} = X,$$

98

where $W_{X^{(n)}}$ is the induced automorphism of $W(= W^{(\alpha,\beta)})$ to the set $X^{(n)}$ and

$$\sigma\begin{pmatrix} a_1 \\ \varepsilon_1 \end{pmatrix} \cdots \sigma\begin{pmatrix} a_n \\ \varepsilon_n \end{pmatrix} : \begin{array}{l} 1 \to s_1^{(1)} s_2^{(1)} \cdots s_{l_1^{(n)}}^{(1)} \\[4pt] 2 \to s_1^{(2)} s_2^{(2)} \cdots s_{l_2^{(n)}}^{(2)} \\[4pt] 3 \to s_1^{(3)} s_2^{(3)} \cdots s_{l_3^{(n)}}^{(3)} \end{array} .$$

References

1. ITO,S. and OHTSUKI.M., *Modified Jacobi–Perron algorithm and generating Markov partitions for special hyperbolic toral automorphisms,* Tokyo J. Math. **16**, No.2 (1993), 441-472.
2. ITO,S., KEANE,M. and OHTSUKI,M., *Almost everywhere exponential convergence of modified Jacobi-Perron algorithm.* Ergodic Theory and Dynam. Sys. **13**, (1993), 319-334.

PRIME TYPE III_λ AUTOMORPHISMS: AN INSTANCE OF CODING TECHNIQUES APPLIED TO NON-SINGULAR MAPS

Andrés del Junco[1] and César Silva[2]

[1] Department of Mathematics
University of Tronto, Tronto M5S 1A1, CANADA
[2] Department of Mathematics
Williams College, Williamstown Mass. 01267, U.S.A.

Rudolph and Silva introduced a notion of minimal self-joinings for non-singular automorphisms and used it to construct an example T of each Krieger type which commutes only with its powers and has only trivial invariant σ-algebras. Here we show that such examples can be obtained more directly using coding ideas. In fact, coding techniques yield results which do not seem obtainable via joinings, e.g. a complete classification of the factor algebras of $T \times T$. Such coding techniques are quite standard in the context of finite measure-preserving maps but have not, as far as we know, been applied to non-singular maps. In the classical setting the $\bar{d}$-metric on sequences plays a central role and here we use a weighted version of it. The example T which we work with is a type III_λ version of Chacón's map.

Introduction

We consider non-singular systems $\mathbf{X} = (X, \mathcal{B}, \mu, T)$, by which we mean that $(X, \mathcal{B}, \mu)$ is a Lebesgue probability space and $T : (X, \mathcal{B}, \mu) \to (X, \mathcal{B}, \mu)$ is a non-singular automorphism. We denote by $C(\mathbf{X})$ or $C(T)$ the centralizer of T, that is, all non-singular endomorphisms S of $(X, \mathcal{B}, \mu)$ such that $ST = TS$ a.e. We denote by $F(\mathbf{X})$ the class of factor algebras $\mathcal{F}$ of T, that is $\mathcal{F}$ is sub-σ-algebra of $\mathcal{B}$ invariant under T.

We are interested in $\mathbf{X}$'s for which $C(\mathbf{X})$ and $F(\mathbf{X})$ are small. The extreme cases are when $C(\mathbf{X}) = \{T^n : n \in \mathbb{Z}\}$, in which case we say $\mathbf{X}$ has trivial centralizer, and when $F(\mathbf{X}) = \{\{\phi, X\}, \mathcal{B}\}$, in which case we say $\mathbf{X}$ is prime. Both these equations are of course to be constructed in the a.e sense, for example the second one means that every complete factor algebra is either $\mathcal{B}$ or the algebra of null or co-null sets.

The first example of a finite measure-preserving prime map with trivial centralizer was Ornstein's rank one mixing map [O]. (Adams and Friedman [A,F] have recently given an explicit description of a rank one construction which is mixing.) Then it was shown in [J1] that a very simple (weakly mixing but not mixing) rank one map due to Chacón is prime and has trivial centralizer. [J2] implies that a certain one-parameter family of interval exchanges also have this property.

The original proofs of primality and trivial centralizer for the rank one mixing map and Chacón's map used so-called coding arguments. In 1978 Rudolph [R1] introduced the notion of minimal self-joinings (MSJ), a powerful tool which simplified and systematized these kinds of arguments as well as giving much more: for example if T has MSJ one has a complete description of $C(S)$ and $F(S)$ for any S which is a cartesian product of non-zero powers of T.

So far we have discussed only type II_1 (i.e. finite measure preserving) maps. Hamachi [H] and later and independently Aaronson [A] have both observed that there are type III_λ odometers with trivial centralizer, unlike II_1 case where an odometer always has uncountable centralizer. Prime non-singular maps are less easy to come by. Aaronson and Nadkarni [A,N] have constructed type II_∞ map which is prime in weak sense: with respect to the infinite invariant measure each proper factor algebra is purely infinite.

The breakthrough was Rudolph and Silva's introduction of a notion of MSJ for non-singular maps ([R,S]). They proved that it implies primality and trivial centralizer and constructed an example of each Krieger type (II_∞ and III_λ, $0 \leq \lambda \leq 1$). Their notion of MSJ is only a partial generalization of the II_1 case, in that it only deals with joinings of two copies of T. For this reason it gives no information about centralizer and factors for $T \times T$, or higher cartesian powers. Indeed, there are some serious obstacles to a definition of higher order MSJ which we have not been able to surmount.

Our aim here is to show that the older coding arguments can be applied in the non-singular setting to obtain at least some of the desired results on Cartesian powers. The motivation is two-fold: firstly to circumvent the above-mentioned obstacles but secondly and more importantly to demonstrate that coding arguments can be made to work in the non-singular setting.

Indeed, we believe that there is a much broader scope for applications of such ideas. We mention just a couple of possibilities: factors and isomorphisms of non-singular odometers, Kakutani equivalence for non-singular maps and generalization of King's weak closure theorem [K] to non-singular rank one maps. This last will be the subject of a future paper.

What follows is a preliminary report on our work on type $III_\lambda(0 < \lambda < 1)$ version $T = T_\lambda$ of Chacón's map, which is described in §2. (When $\lambda = 1$, T_λ reduces to the classical II_1 Chacón's map). $T^{\otimes n}$ denotes the n-fold Cartesian power of T acting on $(X^n, \mathcal{B}^n, \mu^n)$.

Theorem 1: For $k \neq 0$, $C(T_\lambda^k) = \{T_\lambda^n : n \in \mathbb{Z}\}$ and T_λ^k is prime.

Theorem 2: If $0 < \lambda_1 < \lambda_2 < \cdots < \lambda_k \leq 1$ and $n_1, \cdots, n_k \in \mathbb{Z}$ then $C(T_{\lambda_1}^{\otimes n_1} \times \cdots \times T_{\lambda_k}^{\otimes n_k})$ is generated by maps of the form $U_1 \times \cdots \times U_k$, where each U_i acting on X^{n_i} is a Cartesian product of powers of T_{λ_i}, or a co-ordinate permutation on X^{n_i}.

Theorem 3: $F(T_{\lambda_1} \times T_{\lambda_2})$ is minimal, namely if $\lambda_1 \neq \lambda_2$ then

$$F(T_{\lambda_1} \times T_{\lambda_2}) = \{\{\phi, X \times X\}, \mathcal{B} \times X, X \times \mathcal{B}, \mathcal{B} \times \mathcal{B}\},$$

while if $\lambda_1 = \lambda_2$ then

$$F(T_{\lambda_1} \times T_{\lambda_2}) = \{\{\phi, X \times X\}, \mathcal{B} \times X, X \times \mathcal{B}, \mathcal{B} \times \mathcal{B}\} \bigcup \{(id \times T^m)\mathcal{B}^{2\odot} : m \in \mathbb{Z}\}$$

where $\mathcal{B}^{2\odot}$ denotes the symmetric σ-algebra in $\mathcal{B} \times \mathcal{B}$, that is, those sets invariant under the co-ordinate interchange.

Theorem 4: *There is a special flow $\{F_t\}$ with base T_λ and two-step ceiling function such that F_1 is prime and $C(F_1) = \{F_t\}$.*

Of course, our proofs of the above theorems are valid in the measure-preserving case $\lambda = 1$, and thus provide in some cases, notably Theorem 3, new proofs, via coding, of results previously obtained only via *MSJ*.

Complete proof of these results will appear elsewhere. What we will do here, to illustrate the coding techniques, is to prove the following special case of Theorem 1.

Theorem 0: *T_λ is prime.*

Before proceeding with the proof of Theorem 0 we discuss briefly the essential difficulties which arise in generalizing coding arguments to the non-singular setting. In a nutshell, the proof of primality for the classical Chacón map goes as follows. It is enough to prove that every homeomorphism $\phi : \mathbf{X} \to \mathbf{Y}$, where $\mathbf{Y}$ is a system with phase space $Y = \{0,1\}^{\mathbb{Z}}$, is a.e. one-to-one or a.e. constant. We approximate ϕ in $\bar{d}$ by finite codes ϕ_m, say $\bar{d}(\phi(x), \phi_m(x)) < \frac{1}{m}$. If ϕ is not a.e one-to-one we can find generic points $x, y \in X$ such that $x \notin \{T^n y\}$ and $\phi(x) = \phi(y)$. Since $\bar{d}(\phi_m(x), \phi_m(y)) < 2/m.$, using the fact that x and y are on different orbits, we can argue that for large n the rank one n-block B_n is coded by ϕ_m $\bar{d}$-close to its shift. This approximate invariance easily gives exact invariance of ϕ, $\phi = \phi \circ T$ so ϕ is a.e. constant by ergodicity.

The difficulty in the non-singular case is that using the Hurewicz ergodic theorem instead of Birkhoff theorem for the $\bar{d}$-approximation one is naturally led to $\bar{d}_x(\phi(x), \phi_m(x)) < 1/m$ where $\bar{d}_x$ now refers to a weighted $\bar{d}$-metric with weights given by the Radon-Nikodym derivatives $\frac{d\mu \circ T^i}{d\mu}(x)$. Then when working with a pair (x,y) we have to deal with both $\bar{d}_x$ and $\bar{d}_y$ which may be quite different. The natural way to get around this is to work with the relative product measure over ϕ getting just one metric $\bar{d}_{(x,y)}$ with weights determined by the pair (x,y). In order to apply the Hurewicz theorem to the relative product we need to know that it is conservative, which can be shown using the notion of rationality from [R,S]. However the relative product need not be ergodic, so that the $\bar{d}$-approximation we get will now be only for ε-a.a. (x,y), not a.a. (x,y). In addition the weighted $\bar{d}$ means that we require some estimates comparing the weights at different parts of a sequence.

The rest of the paper is organized as follows. In §1 we develop the basic ideas and tools which we will need. We recall the key idea of a rational joining of non-singular systems from [R.S]. It is also important in our approach and we highlight one consequence of it (Proposition 1.2) which we shall use heavily. We also introduce the weighted $\bar{d}$ metric upon which our argument is based and prove that any homomorphism can be approximated in $\bar{d}$ by a finite code.

In §2 we describe a type III_λ version T_λ of Chacón's map for each $\lambda \in (0,1)$, first as a cutting and stacking construction and then as a non-singular measure for a shift map. We establish some of its basic properties.

In §3 we consider the relative product of T_λ with itself, relative to a given factor. We establish some properties of it which we will need, in particular conservativity, which follow from rationality of the relative product. Most of this is more or less explicitly contained in [R,S]. Finally §4 contains the proof of Theorem 0.

§1 Preliminaries

If $\mathbf{X} = (X, \mathcal{B}, \mu, T)$ is a non-singular system we will denote by $\omega = \omega_{\mathbf{X}}$ the Jacobian (or Radon-Nikodym derivative) of T:

$$\omega = \frac{d\mu \circ T^{-1}}{d\mu}.$$

When μ or T needs to be emphasized we sometimes write ω_μ or ω_T. We will also write ω_i (sometimes $\omega_{X,i}$ for ω_{T^i}). One has the cocycle relation

$$\omega_i(x)\omega_j(T^i x) = \omega_{i+j}(x).$$

If $\mathbf{X} = (X, \mathcal{B}, \mu, T)$ and $\mathbf{Y} = (Y, \mathcal{C}, \nu, S)$ are non-singular, a (measure-preserving) homomorphism $\phi : \mathbf{X} \to \mathbf{Y}$ is a measurable $\phi : X \to Y$ s.t. $\mu \circ \phi^{-1} = \nu$ and $\phi T = S\phi$ μ-a.e. One can, of course, consider homomorphisms which are merely non-singular ($\mu \circ \phi^{-1} \sim \nu$) but one can always replace ν by $\mu \circ \phi^{-1}$ to make ϕ measure-preserving, so we work only with measure-preserving homomorphisms. Clearly if ϕ is a homomorphism $\phi^{-1}(\mathcal{C})$ is a factor algebra.

$\mathbf{Y} = (Y, \mathcal{C}, \nu, S)$ is called a symbolic system, with alphabet A, if $Y = A^{\mathbb{Z}}$, A is finite, $\mathcal{C}$ is a Borel σ-algebra on Y, S is the left shift on Y and ν is any measure which is non-singular for the shift map. We denote intervals in $\mathbb{Z}$ by $[m, n] = \{m, m+1, \cdots, n\}$. If $x \in A^{\mathbb{Z}}$ (i.e. $x : \mathbb{Z} \to A$) and $I = [m, n]$ we denote the restriction $x|_I$ by $x[m, n] \in A^I$. When it is clear from the context we will sometimes identify A^I with A^{m-n+1}, i.e. we think of $x[m, n]$ as a word of length $m - n + 1$ with no particular indexing.

Lemma 1.1: *To prove that a system $\mathbf{X}$ is prime it is enough to show that any homomorphism $\phi : \mathbf{X} \to \mathbf{Y}$ onto a symbolic system $\mathbf{Y}$ with alphabet $\{0, 1\}$ is either a.e. 1-1 or a.e. constant. (a.e. 1-1 means there is a subset of full measure on which ϕ is 1-1.)*

Proof: If $\mathcal{F}$ is a factor algebra of $\mathbf{X}$ then any tow-set $\mathcal{F}$ measurable partition, i.e. an $\mathcal{F}$-measurable $P : X \to \{0, 1\}$, defines a homomorphism onto $(\{0, 1\}^{\mathbb{Z}}, \mathcal{C}, \nu, S)$ via

$$\phi(x)(i) = P(T^i x),$$

where we take $\nu = \mu \circ \phi^{-1}$. If, for more P, ϕ is a.e. 1-1 this means the sequence $\{T^i P\}$ separates points, so by the theory of Lebesgue spaces, e.g. [R2,Chapt. 2], $\mathcal{F} = \mathcal{B}$. On the other hand, if for every $P\phi$ is a.e. constant, $\mathcal{F}$ is trivial. □

If $\phi : \mathbf{X} \to \mathbf{Y}$ is a factor map μ may be disintegrated with respect to ϕ:

$$\mu = \int \mu_y d\nu(y),$$

where, for ν-a.a. y, μ_y is a probability measure on X, supported on $\phi^{-1}\{y\}$ (see[F]). (Here, the assumption that ϕ is measure-preserving plays a role!) One easily checks that

$$w_Y(y) = \int_X w_X(x) d\mu_y(x), \quad \text{for } \nu \text{ a.a. } y. \tag{1}$$

If $X_j = (X_j, \mathcal{B}_j, \mu_j, T_j), j = 1, 2$ are systems a (non-singular) *joining* of X_1 and X_2 is measure $\hat{\mu}$ on $\mathcal{B}_1 \times \mathcal{B}_2$ projecting onto μ_1 and μ_2 and non-singular for $T_1 \times T_2$. In other words, the projections $X_1 \times X_2 \to X_j, j = 1, 2$ are homomorphisms. We write

$$\omega_i^1 = \omega_{\mu_1, T_1^i}, \quad \omega_i^2 = \omega_{\mu_2, T_2^i}, \quad \hat{\omega}_i = \omega_{\hat{\mu}, T_1^i \times T_2^i},$$

for $i \in \mathbb{Z}$. As a special case of (1) one has

$$\omega_i^1 = \int_{X_2} \hat{\omega}_i(x_1, x_2) d\hat{\mu}_{x_1}(x_2)$$

where $\hat{\mu} = \int_{X_1} \hat{\mu}_{x_1} d\mu_1(x_1)$ is the disintegration of $\hat{\mu}$ over the first co-ordinate in $X \times X$. As in [R,S] $\hat{\mu}$ is called *rational* if there are measurable functions $c_1(x_1)$ and $c_2(x_2)$ such that

$$\hat{\omega}(x_1, x_2) = \omega^1(x_1)\omega^2(x_2)c_1(x_1) = \omega^1(x_1)\omega^2(x_2)c_2(x_2) \quad \hat{\mu} \text{ a.e.}$$

If $\phi : X \to Y$ is a homomorphism the relative product measure $\hat{\mu} = \mu \times_\phi \mu$ on $X \times X$ is defined by

$$\hat{\mu} = \int \mu_y \times \mu_y d\nu(y)$$

where $\mu = \int \mu_y d\nu(y)$ is the disintegration of μ over Y. It is straight forward to check that $\hat{\mu}$ is non-singular joining of X with X and that

$$\hat{\omega}(x_1, x_2) = \frac{\omega^1(x_1)\omega^2(x_2)}{\omega_Y(\phi(x_1))} = \frac{\omega^1(x_1)\omega^2(x_2)}{\omega_Y(\phi(x_2))}$$

(Here $\omega^1 = \omega^2 = \omega_X$. Note that $\phi(x_1) = \phi(x_2)$ μ-a.e. since $\hat{\mu}$ is, by its definition, supported on $\{(x_1, x_2) : \phi(x_1) = \phi(x_2)\}$.) This formula in turn makes it clear that $\hat{\mu}$ is a rational joining.

Now we state a simple consequence of rationality which is implicit in [R,S](Proposition 4.3.2) and which we feel deserves more prominent billing. It is a key point in [R,S] and in the present paper it contains all the use which we shall make of rationality. For completeness we include the proof, which is the same as in [R,S].

Lemma 1.2 *Suppose that $\hat{\mu}$ is a rational joining of X_1 and X_2, $i \in \mathbb{Z}$ and $\exists C \in (0, \infty)$ such that*

$$C^{-1} < \omega_i^j < C \quad \mu_j\text{-a.e.} \quad j = 1, 2.$$

Then

$$C^{-3} < \hat{\omega}_i < C^3 \quad \mu_j\text{-a.e..}$$

Proof: Note that by the cocycle relation we can write

$$\hat{\omega}_i(x_1, x_2) = \omega_i^1(x_1)\omega_i^2(x_2)c_i(x_1) \quad \hat{\mu}\text{-a.e..} \tag{2}$$

Let $\hat{\mu} = \int \hat{\mu}_{x_1} d\mu_1(x_1)$ be the disintegration of $\hat{\mu}$ over X. Integrating (2) with respect to $\hat{\mu}_{x_1}$ we get for μ_1-a.a. x_1

$$\omega_i^1(x_1) = \omega_i^1(x_1) c_i(x_1) \int_{X_2} \omega_i^2(x_2) d\hat{\mu}_{x_1}(x_2),$$

so

$$C^{-1} < c_i(x_1) = \left(\int_{X_2} \omega_i^2(x_2) d\hat{\mu}_{x_1}(x_2) \right)^{-1} < C,$$

since for μ_1-a.a. x_1 $C^{-1} < \omega^2(x_2) < C$ for μ_{x_1}-a.a. x_2. Substituting this inequality back into (1), the lemma follows. $\qquad\square$

If $\mathbf{Z} = (z, \mathcal{D}, \lambda, R)$ is any non-singular system and $z \in Z$ we denote by Ω_z the measure on Z defined by $\Omega_z(i) = \omega_{Z,i}(z)$. Given sequences $\xi, \eta \in A^{\mathbb{Z}}$, A finite, let $D(\xi, \eta) = \{i : \xi(i) \neq \eta(i)\}$. For $I \subset \mathbb{Z}$, I finite, define

$$\bar{d}_z^I(\xi, \eta) = \Omega_z(D(\xi, \eta) \cap I)/\Omega_1(I)$$

and then let

$$\bar{d}_z = \lim_{m,n \to \infty} \bar{d}_z^{[-m,n]}(\xi, \eta)$$

if the limit exists, equivalently

$$\bar{d}_z(\xi, \eta) = \lim_{m \to \infty} \bar{d}_z^{[-m,n]}(\xi, \eta) = \lim_{n \to \infty} \bar{d}_z^{[0,n]}(\xi, \eta)$$

if the two limits exist and agree. When we want to emphasize the measure λ we will write $\bar{d}_z = \bar{d}_{z,\lambda}$. It is clear that $\bar{d}_z$ satisfies the triangle inequality.

The proof of the following lemma is obvious.

Lemma 1.3: *Suppose $\bar{d}_z(\xi, \eta)$ exists. Call a sequence $I_1, I_2, \cdots$ of intervals in $\mathbb{Z}$ substantial if, denoting by $\sigma(I_n)$ the smallest symmetric interval containing I_n, there is a constant C such that*

$$\Omega_z(\sigma(I_n)) \leq C\Omega_z(I_n) \quad \forall n.$$

Suppose that $\lim_n \bar{d}_z^{I_n}(\xi, \eta) = 0$. Then $\bar{d}_z(\xi, \eta) = 0$.

By a *finite code* $\psi : A^{2k+1} \to B$, A and B finite, we mean any map $\psi = \psi_f$, determined by a choice of $f : A^{2k+1} \to B$, some k, via the formula

$$\psi_f(x)(i) = f(x[i - k, i + k]).$$

Such a ψ automatically commutes with the shifts. $2k + 1$ is called the *code length* of ψ. Now suppose $\mathbf{X}$ and $\mathbf{Y}$ are symbolic systems with alphabets A and B, $\mathbf{X}$ is ergodic and conservative and $\phi : \mathbf{X} \to \mathbf{Y}$ is a homomorphism. Then an application of the Hurewicz ergodic theorem shows that one can approximate ϕ within ε by a finite code ψ in the sense that

$$\bar{d}_x(\phi(x), \psi(y)) < \varepsilon \quad \text{for a.a.}x.$$

This is the natural analogue of a fact which is very well known when $\mathbf{X}$ is type II_1. We shall, however, need something a little more general in order to work in a relative product.

Proposition 1.4: *Suppose $\mathbf{X}_1$ and $\mathbf{Y}$ are symbolic systems with alphabets A and B, $\phi : A^{\mathbb{Z}} \to B^{\mathbb{Z}}$ is a homomorphism, $\hat{\mu}$ is a conservative non-singular joining of $\mathbf{X}_1$ with another system $\mathbf{X}_2$ and $\varepsilon > 0$. Then there is a finite code $\psi : A^{\mathbb{Z}} \to B^{\mathbb{Z}}$ such that*

$$\hat{\mu}\{(x_1, x_2) : \bar{d}_{(x_1, x_2)}(\phi(x_1), \psi(x_1)) < \varepsilon\} > 1 - \varepsilon.$$

(Here $\bar{d}_{(x_1, x_2)} = \bar{d}_{(x_1, x_2), \hat{\mu}}$.)

Proof: Approximating the sets $\{x : \phi(x)(0) = b\}, b \in B$, by cylinders of length $2k + 1$ in $A^{\mathbb{Z}}$ we obtain $f : A^{2k+1} \to B$ and hence $\psi = \psi_f$ such that, setting

$$D = D(\phi, \psi) = \{x \in X_1 : \phi(x)(0) \neq \psi(x)(0)\},$$

we have $\mu_1(D) < \varepsilon^2$. Since $\hat{\mu}$ is a joining we have

$$\hat{\mu}(D \times X_2) = \mu_1(D) < \varepsilon^2.$$

If $E(\cdot)$ denotes conditional expectation with respect to the σ-algebra of invariant sets for $(X_1 \times X_2, \hat{\mu}, T_1 \times T_2)$ it follows that

$$\hat{\mu}\{E(1_{D \times X_2}) < \varepsilon\} > 1 - \varepsilon. \tag{3}$$

Now, using the fact that ϕ and ψ commute with the shifts one sees that for $x_1 \in X_1$

$$\{i : T_1^i x_1 \in D\} = D(\phi(x_1), \psi(x_1)).$$

So, applying the Hurewicz ergodic theorem to $\hat{\mu}, T_1 \times T_2$ we get

$$
\begin{aligned}
E(1_{D \times X_2}) &= \lim_{n \to \infty} \frac{\sum_{i=0}^{n} \hat{\omega}_i(x_1, x_2) 1_{D \times X_2}(T_1^i x_1, T_2^i x_2)}{\sum_{i=0}^{n} \hat{\omega}_i(x_1, x_2)} \\
&= \lim_{n \to \infty} \frac{\Omega_{(x_1, x_2)}(\{i : T_1^i x_1 \in D\} \cap [0, n])}{\Omega_{(x_1, x_2)}[0, n]} \\
&= \lim_{n \to \infty} \bar{d}_{(x_1, x_2)}(\phi, \psi).
\end{aligned}
$$

Since the invariant σ-algebra of $T_1^{-1} \times T_2^{-1}$ is the same as that of $T_1 \times T_2$ we also get

$$E(1_{D \times X_2}) = \lim_{n \to \infty} \bar{d}_{(x_1, x_2)}^{[-n, 0]}(\phi, \psi).$$

In view of (3), this concludes the proof. $\qquad\square$

If $\mathbf{X} = (X, \mathcal{B}, \mu, T)$ is a finite measure-preserving system, $\mathcal{I}$ denotes its invariant σ-algebra, $\mathcal{F}$ is any factor algebra and h is an $\mathcal{F}$-measurable function then $E(h|\mathcal{I})$ is again $\mathcal{F}$-measurable, since

$$E(h|\mathcal{I}) = \lim \frac{1}{n} \sum_{i=0}^{n} h \circ T^i.$$

This observation, which is of course false for general non-singular $\mathbf{X}$, makes the proof of the following proposition almost trivial in the finite measure-preserving case.

Proposition 1.5: *Let $\mathbf{X} = (X, \mathcal{B}, \mu, T)$ be a (not necessarily ergodic) non-singular system and let $\mathcal{F}$ be a factor algebra which is ergodic, i.e. $TA = A, A \in \mathcal{F} \Rightarrow \mu(A) = 0$*

or 1. *Let $\mathcal{I}$ denote the σ-algebra of T-invariant sets. Suppose h is a positive $\mathcal{F}$-measurable function and that $\mu\{E(h|\mathcal{I}) = 0\} > 0$. Then $h = 0$.a.e.*

Proof: We will use the following facts about the ergodic decomposition of μ : there is a homomorphism $\phi : \mathbf{X} \to \mathbf{Y} = (Y, \mathcal{C}, \nu, id)$ such that $\phi^{-1}(\mathcal{C}) = \mathcal{I}$ and if $\mu = \int \mu_y d\nu(y)$ is the disintegration of μ over Y then for ν-a.a. y μ_y is non-singular and ergodic for T.

Let $D = \{y : \int_X h d\mu_y = 0\}$, so our hypothesis is that $\nu(D) > 0$, and we want to show $\nu(D) = 1$. Let

$$\hat{\mu} = \int_D \mu_y d\nu(y)$$

and let $\mu_{\mathcal{F}}$ and $\bar{\mu}_{\mathcal{F}}$ denote the restriction to $\mathcal{F}$. Clearly $\bar{\mu}_{\mathcal{F}}$ is absolutely continuous with respect to $\mu_{\mathcal{F}}$ so ergodicity of $\mu_{\mathcal{F}}$ implies that $\bar{\mu}_{\mathcal{F}}$ is equivalent to $\mu_{\mathcal{F}}$. By our definition of $\bar{\mu}$, $h = 0$ $\bar{\mu}_{\mathcal{F}}$-a.e., so $h = 0$ $\mu_{\mathcal{F}}$-a.e..

§2 A type III_λ version of Chacón's map

We will first describe this geometrically as a rank one cutting and stacking construction and then use a natural partition to code it to a symbolic system with alphabet $\{0, 1\}$. We fix a $\lambda \in (0, 1]$. When $\lambda = 1$ the construction we are about to give reduces to the classical Chacón map.

To start the construction let I_0 denote an interval of length 1, divide into intervals J_1, J_2 and J_4 of lengths a, b, a with $a/b = \lambda$ and let J_3 be an interval of length b abutting on I_0. Define $T : J_i \to J_{i+1}$ for $i = 1, 2, 3$ to be affine and leave T as yet undefined on J_4. This produces a tower ξ_1 for T of height 4 whose levels are intervals and T has constant Jacobian on each level, except the top one, where T is not yet defines. Note also that the union of the levels in an interval I_1.

After n steps of the construction we will have a tower ξ_n of height h_n whose levels are intervals with union I_n, also an interval. T will be an affine map on each level to the one above and as yet undefined on the top level. We extend the definition of T as follows. Divide the base B of the tower into intervals J_1, J_2, J_4 of lengths a, b, a with $a/b = \lambda$, let J_3 be an interval of length b abutting on I_n and extend T affinely so that

$$T : T^{h_n-1}J_1 \to J_2$$
$$T : T^{h_n-1}J_2 \to J_3$$
$$T : \qquad J_3 \to J_4,$$

to obtain a tower ξ_{n+1} of height $h_{n+1} = 3h_n + 1$. In the following picture the levels of ξ_n have been rescaled to appear of the same measure.

Note that $w_T = \lambda^{-1}$ on $T^{h_n-1}J_1, 1$ on $T^{h_n-1}J_2$ and λ on $J_3 = T^{h_n}J_2$.

In this way we obtain a sequence of towers ξ_n on intervals I_n, $I_n \uparrow X$, also an interval of finite Lebesgue measure, T is defined a.e. on I and is a non-singular automorphism. We let μ denote normalized Lebesgue measure on X, and $\mathcal{B}$ the Borel σ-algebra of X. Alternately $\mathbf{X} = (X, \mathcal{B}, \mu, T)$ can be viewed as a tower over the odometer on $\{0, 1, 2\}^{\mathbf{N}}$ with product measure $(a, b, a)^{\mathbf{N}} (2a + b = 1, \lambda = a/b)$ and height function $1_E, E = \{x : x(i) = 1 \text{ for some } i\}$.

We let P be the two-set partition which, viewed as function into $\{0, 1\}$, is defined by

$$P(x) = \begin{cases} 0 & \text{if } x \in I_0 \\ 1 & \text{if } x \in X - I_0. \end{cases}$$

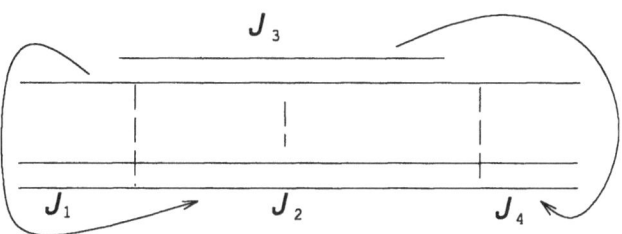

For $x \in X$ let $\theta(x) \in \{0,1\}^{\mathbb{Z}}$ denote the "P, T-name" of x:

$$\theta(x)(i) = P(T^i x).$$

Then θ is a homomorphism from $\mathbf{X}$ onto a symbolic system with alphabet $\{0,1\}$. We shall see in a moment that it is an isomorphism.

Define blocks of 0's and 1's inductively as follows:

$$B_0 = 0, \quad B_1 = 0010, \quad B_{n+1} = B_n B_n 1 B_n.$$

Clearly B_n is the partial P, T-name $\theta(x)[0, h_n - 1]$ of any point x in the base of ξ_n. The following lemma is easy to prove by induction (see [JK, Lemma 1]).

Lemma 2.1: *In each of the words $B_n B_n$ and $B_n 1 B_n$, B_n occurs as a subword only in the two obvious places.*

The following proposition is an easy consequence of Lemma 2.1.

Proposition 2.2: *P is a generating partition for T, i.e. θ is a.e. 1-1.*

This means that the cutting and stacking construction we have described is isomorphic via θ to a symbolic system with alphabet $\{0,1\}$ which we shall, by abuse of notation, also denote $\mathbf{X} = (X, \mathcal{B}, \mu, T)$. Henceforth we work with the symbolic version, referring back to the cutting and stacking construction when necessary to obtain various properties of a μ-typical $x \in X = \{0,1\}^{\mathbb{Z}}$, as in, for example, the following proposition.

Proposition 2.3: *There is a subset $X^* \subset X$ such that $\mu(X^*) = 1$ with the following properties:*

(A) *For each n, each $x \in X^*$ decomposes uniquely as a concatenation of n-blocks, B_n, some of which are separated by a single 1. Let us call such 1's in x (n-block) spacers. Any appearance of B_n in x must be one of the B_n's in the unique decomposition.*

(B) *For all $x \in X^*$ and for all sufficiently large $n = n(x)$ the 0th co-ordinate in x lies inside an n-block, which we call the time 0-n-block. The union over n of the intervals on which the time 0-n-blocks occur is $\mathbb{Z}$.*

(C) *If $x, y \in X^*$ and $x \notin \mathcal{O}(y) = \{T^n y : n \in \mathbb{Z}\}$ then for infinitely many n there are intervals I and J such that*

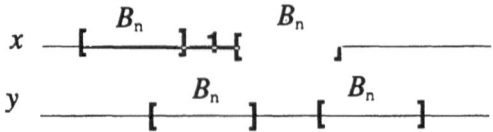

(a) $x(I) = B_n = y(J)$

(b) $|I \cap J| \geq \frac{1}{10}h_n = \frac{1}{10}|I| = \frac{1}{10}|J|$

(c) $I \cap J \subset [-10h_n, 10h_n]$

(d) $x(I)$ is followed by a spacer but $y(J)$ is not.

Roughly speaking, there are n-blocks in x and y, not too far from time 0 and overlapping substantially, one followed by spacer and the other not:
Let us call such a situation "broken n-blocks".

Proof:

(A) The existence of the decomposition is clear from the cutting and stacking description and the rest is just Lemma 2.1.

(B) This is just the fact that if $B_{k,n}$ denotes the union of k and bottom k levels of the nth tower ξ_n then, for fixed k, $B_{k,n}$ decreases with n and $\mu(B_{k,n}) \to 0$. Thus, for μ-a.a. x, $x \in B_{k,n}^c$ for n sufficiently large, in other words, in the symbolic description the 0th co-ordinate of x is at least $k + 1$ steps away from the ends of the time-0 n-block.

(C) Suppose x and y satisfy (a) and (b) and $x \notin \mathcal{O}(y)$. Note that any n-block in x (or y) nests as the first, second or third n-block in the $n+1$-block containing it. Let us say "the n-blocks in x and y are staggered" if each n-block in $x[-10h_n, 10h_n]$ overlaps two adjacent n-blocks in y both in intervals of length at least $\frac{1}{10}h_n$. Suppose that this is the case for some large n. Note that we must see the word $B_n 1 B_n 1 B_n$ somewhere in $x[-9h_n, 9h_n]$, since it appears in any copy of B_{n+2}. Thus, if there are no broken n-block pairs within $[-10h_n, 10h_n]$ we must see $B_n 1 B_n 1 B_n 1 B_n$ in $y[-10h_n, 10h_n]$. But this is impossible, since 0101010 never appears in a B_k.

Thus, we may assume that, for all n greater than some N, the n-blocks in x and y are not staggered. This means that at least one n-block in $x[-10h_n, 10h_n]$ lines up to within $\frac{1}{10}h(n)$ with an n-block in y, and if n is large it easily follows that all n-blocks in $x[-10h_n, 10h_n]$ line up to within $\frac{1}{9}h(n)$ with an n-block in y. In particular, this is true of the time 0 n-block in x, so wee see immediately that for $n > N$ the time 0 n-blocks in x and in y nest in the same way into the time 0 $n + 1$-blocks in x and y. Since the time 0 n-blocks asymptotically cover all of x and y this obviously implies that x is a shift of y. $\square$

Proposition 2.4: *T is ergodic.*

Proof: The standard proof of type II_1 rank one constructions (e.g. [F]) goes through. The key point is that the Jacobian is constant on levels. See also [R.S].

Proposition 2.5: *T is weakly mixing, i.e. $f \circ T = \lambda f$ with $f \in L^\infty$ implies f is a.e. constant. In particular T^k is ergodic $\forall k \neq 0$.*

Proof: Again the proof for the II_1-case ([F]) goes through.

§3 Consequences of rationality

In order to prove Theorem 0, in view of Lemma 1.1 we now fix a homomorphism $\phi : X \to Y, Y$ symbolic with alphabet $\{0, 1\}$. Let $\hat{\mu} = \mu \times_\phi \mu$ denote the relative product measure define in §1. In this section we obtain some properties of the system $(X \times X, \hat{\mu}, T \times T)$ which are all consequences of rationality via Lemma 1.2.

Proposition 3.1: *$\hat{\mu}$ is conservative for $T \times T$.*

Proof: Let $\omega = \omega_T$, $\hat{\omega} = \omega_{\hat{\mu}, T \times T}$. Note that $\omega(x) \in \{\lambda^{-1}, 1, \lambda\}$ a.e. Note also that the top and bottom levels of the nth tower ξ_n have the same measure, so $\omega_{h_n - 1} = 1$ on the base C_n on ξ_n. It follows that for $x \in C_n$ (i.e. $x[0, h_n - 1] = B_n$) $\omega_{h_n}(x) = \lambda^{-1}$ or $\omega_{h_n+1}(x) = \lambda$ according to whether $T^{h_n}x$ or $T^{h_n+1}x$ belongs to C_n. Since ω is constant on levels of ξ_n the cocycle relation then implies that for all $x \in X$ either $\omega_{h_n}(x) = \lambda^{-1}$, $\omega_{h_n+1}(x) = \lambda$, or $\omega_{h_n+1}(x) = 1$. (The last possibility occurs only when x lies in an n-block spacer, i.e. $x[-h_n, h_n] = B_n 1 B_n$, and $T^{h_n+1}x \in C_n$.) This implies that in any case $\lambda^2 \leq \omega_{h_n}(x) \leq \lambda^{-2}$. Now it follows from lemma 1.2 that

$$\lambda^6 \leq \hat{\omega}_{h(n)} \leq \lambda^{-6} \quad \hat{\mu}\text{-a.e.}$$

Thus $\sum_{i=1}^{\infty} \hat{\omega}_i = \infty$ $\hat{\mu}$-a.e., which is equivalent to conservativity.

Lemma 3.2.

(A) $\forall n \quad \lambda^6 \leq \hat{\mu}_{h_n} \leq \lambda^{-6} \quad \hat{\mu}$-a.e.

(B) $\forall k \quad \lambda^{3k} \leq \hat{\mu}_k < \lambda^{-3k} \quad \hat{\mu}$-a.e.

Proof: (a) is in the proof of Proposition 3.1. (b) follows from $\lambda \leq \omega \leq \lambda^{-1}$ and Lemma 1.2. □

For $(x_1, x_2) \in X \times X$ let $\Omega_{(x_1, x_2)} = \Omega_{\hat{\mu}, (x_1, x_2)}$ denote the measure on $\mathbb{Z}$ define in §1.

Proposition 3.3:

(A) $\forall C > 0 \ \exists C' > 0$ such that for $\hat{\mu}$-a.a. (x_1, x_2) whenever $I_1 \subset I_2$ are intervals in $\mathbb{Z}$ with $|I_1| < C|I_2|$ then $\Omega_{(x_1, x_2)}(I_2) < C'\Omega_{(x_1, x_2)}(I_1)$.

(B) $\forall k \in \mathbb{Z}^+$, $\varepsilon > 0$, $\exists L > 0$ such that for $\hat{\mu}$-a.a. (x_1, x_2) whenever $I_1 \subset I_2$ are intervals in $\mathbb{Z}$ with $|I_1| < k$ and $|I_2| > L$ then $\Omega_{(x_1, x_2)}(I_1) < \varepsilon\Omega_{(x_1, x_2)}(I_2)$.

(C) $\forall n$ and $i \in \mathbb{Z}$, $\lambda^6 < \Omega_{(x_1, x_2)}\{i\}/\Omega_{(x_1, x_2)}\{i + h_n\} < \lambda^{-6}$.

Proof: To see (a), since $h_{n+1}/h_n \leq 4$ for all n, there is an n such that $h_n \leq |I_1| \leq 4h_n$. Let $J \subset I_1$ be any interval of length h_n. Then I_2 can be covered by at most $4C + 1$ intervals $J + ih_n$ with $|i| < 4C + 1$. By Lemma 3.2.(a) we get

$$\Omega_{(x_1, x_2)}(I_2) \leq C'\Omega_{(x_1, x_2)}(J) \leq C'\Omega_{(x_1, x_2)}(I_1)$$

with $C' = (4C + 1)\lambda^{6(4c+1)}$.

To prove (b) we may assume $k = 1$. By Lemma 3.2.(a) since $h_n \leq 4^n$, $\sum_{i=0}^{4^n} \hat{\omega}_i > n\lambda^6$ $\hat{\mu}$-a.e. and $\sum_{i=-4^n}^{0} \hat{\omega}_i > n\lambda^6$ $\hat{\mu}$-a.e. This means that $\Omega_{(x_1,x_2)}\{j\}/\Omega_{(x_1,x_2)}[j, j+4^n] < \frac{1}{n}\lambda^{-6}$ and $\Omega_{(x_1,x_2)}\{j\}/\Omega_{(x_1,x_2)}[j - 4^n, j] < \frac{1}{n}\lambda^{-6}$, so (b) follows.
(c) is just a restatement of Lemma 3.2(a). $\qquad\square$

§4 Proof of Theorem 0

$\hat{\mu}$ continues to denote the relative product measure $\mu \times_\phi \mu$ as in §3. $\bar{d}_{(x_1,x_2)}$ denote $\bar{d}_{\hat{\mu},(x_1,x_2)}$.

Proposition 4.1: *For $\hat{\mu}$-a.a. (x_1, x_2) either $x_1 \in \mathcal{O}(x_2)$ or $\bar{d}_{(x_1,x_2)}(\phi(x_1), \phi(Tx_1)) = 0$.*

Proof: Use Proposition 1.4 to fix a sequence ϕ_m of finite codes with code lengths k_m such that

$$\hat{\mu}\left\{(x_1, x_2) : \bar{d}_{(x_1,x_2)}(\phi(x_1), \phi_m(x_1)) < 2^{-m}, \bar{d}_{(x_1,x_2)}(\phi(x_2), \phi_m(x_2)) < 2^{-m}\right\} > 1 - 2^{-m}.$$

Now fix a "$\hat{\mu}$-typical" (x_1, x_2) satisfying

- $\phi(x_1) = \phi(x_2)$ $\qquad\qquad\qquad\qquad\qquad\qquad\qquad\qquad$ (i)

- $\forall$ sufficiently large m $\bar{d}_{(x_1,x_2)}(\phi(x_1), \phi_m(x_1)) < 2^{-m}$,
  $$\bar{d}_{(x_1,x_2)}(\phi(x_2), \phi_m(x_2)) < 2^{-m},$$
  and $\bar{d}_{(x_1,x_2)}(\phi(Tx_1), \phi_m(Tx_1)) < 2^{-m}$ $\qquad\qquad$ (ii)

- $\bar{d}_{(x_1,x_2)}(\phi(x_1), \phi_m(Tx_1))$ exists. $\qquad\qquad\qquad\qquad\qquad$ (iii)

(Note that these conditions are satisfied for $\hat{\mu}$-a.a. (x_1, x_2).) Suppose moreover that $x_1 \notin \mathcal{O}(x_2)$; we shall show that $\bar{d}_{(x_1,x_2)}(\phi(x_1), \phi(Tx_1)) = 0$.

Fix a large $m = m(x_1, x_2)$ satisfying (ii) and then a much larger $n = n(m)$ and find "broken n-blocks" $x_1(I)$ and $x_2(J)$ in x_1 and x_2, as in Proposition 2.3(c). (We shall specify later how large n needs to be.) Let $I' = I \cap J$, $\alpha = x_1(I')$ and $\beta = x_2(I')$. Since $x_1(I)$ is followed by a spacer but $x_2(J)$ is not, we see that $x_2(I' + h_n) = \beta$, while $x_1(I' + h_n) = \alpha'$, where α' is essentially the shift of α, more precisely α' is either $0\alpha^*$ or $1\alpha^*$, where α^* denotes α with the last symbol deleted. Note that $\phi_m(\alpha)$, $\phi_m(\beta)$ and $\phi_m(\alpha')$ are meaningful as words of length $|I'| - 2k_m$. Let K denote the interval of length $|I'| - (2k_m + 1)$ obtained by deleting the leftmost $k_m + 1$ and rightmost k_m elements of I'. Once m and hence k_m are fixed, if n is large enough then I' is large so we can and do assume $|K| > \frac{1}{2}|I'|$. The following picture summarizes the situation we have described.

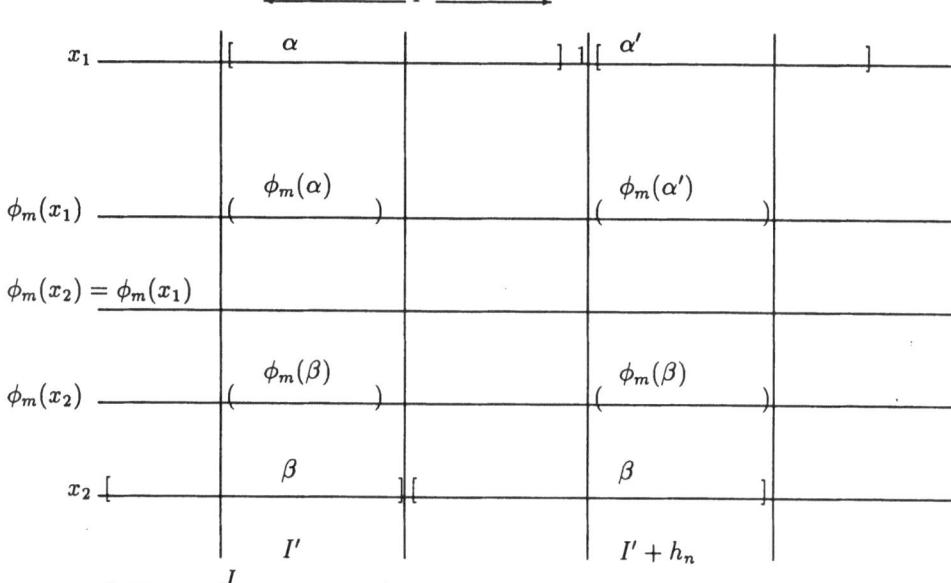

Now, suppressing the subscript in $\bar{d}_{(x_1,x_2)}$, we want to argue that $\bar{d}^K(\phi_m(\alpha),\phi_m(\beta))$ is small. If $[-N,N]$ is the smallest symmetric interval containing K then $2N+1 < 200|K|$, so by Lemma 3.3(a) $(C = 200)$ $\Omega_{(x_1,x_2)}[-N,N] < C'\Omega_{(x_1,x_2)}(K)$, whence $\bar{d}^K(\phi_m(\alpha),\phi_m(\beta)) < C'\bar{d}^{[-N,N]}(\phi_m(x_1),\phi_m(x_2))$. Moreover if n and hence N, is sufficiently large

$$\bar{d}^{[-N,N]}(\phi_m(x_1),\phi_m(x_2)) < \bar{d}^{[-N,N]}(\phi_m(x_1),\phi_m(x_2)) + \bar{d}^{[-N,N]}(\phi(x_2),\phi_m(x_2)) < 2 \cdot 2^{-m},$$

using (i) and (ii). Thus given $\varepsilon > 0$, if m was initially chosen large enough we have

$$\bar{d}^K(\phi_m(\alpha),\phi_m(\beta)) < C' \cdot 2 \cdot 2^{-m} < \varepsilon. \tag{iv}$$

By exactly the same reasoning applied on $K + h_n$

$$\bar{d}^{(K+h_n)}(\phi_m(\alpha'),\phi_m(\beta)) < C' \cdot 2 \cdot 2^{-m},$$

so by Lemma 3.3(c) we get

$$\bar{d}^K(\phi_m(\alpha'),\phi_m(\beta)) < \lambda^{-6} \cdot C' \cdot 2 \cdot 2^{-m} < \varepsilon,$$

again if m was initially chosen large enough. Combining this with (iv) we get

$$\bar{d}^K(\phi_m(\alpha),\phi_m(\alpha')) < 2\varepsilon,$$

in other words

$$\bar{d}^K(\phi_m(x_1),\phi_m(Tx_1)) < 2\varepsilon. \tag{v}$$

($\phi(Tx_1)$ and $\phi_m(\alpha')$ agree on K — this was the reason for deleting an extra point at the left end of I'.)

Now since $\bar{d}(\phi_m(x_1), \phi(x_1)) < 2^{-m}$, by the same reasoning as above

$$\bar{d}^K(\phi_m(x_1), \phi(x_1)) < C'2^{-m} < \varepsilon \tag{vi}$$

and similarly

$$\bar{d}^K(\phi_m(Tx_1), \phi(Tx_1)) < C'2^{-m} < \varepsilon. \tag{vii}$$

Combining (v),(vi) and (vii), if m is sufficiently large

$$\bar{d}^K(\phi(x_1), \phi(Tx_1)) < 4\varepsilon.$$

$K = K(m)$ depends on m and we have already observed that $K(m)$ is a substantial sequence in the sense of Lemma 1.3. Since we have just shown that

$$\lim_m \bar{d}^{K(m)}(\phi(x_1), \phi(Tx_1)) = 0$$

we conclude from Lemma 1.3 that $\bar{d}(\phi(x_1), \phi(Tx_1)) = 0$. $\square$

We are now in a position to conclude the proof of Theorem 0 quite easily. Suppose first that

$$\hat{\mu}\left\{(x_1, x_2) : \bar{d}_{(x_1,x_2)}(\phi(x_1), \phi(Tx_1)) = 0\right\} > 0. \tag{4}$$

Since ϕ and $\phi\circ T$ are both homomorphisms one sees exactly as in the proof of Proposition 1.4 that

$$\bar{d}_{(x_1,x_2)}(\phi(x_1), \phi(Tx_1)) = E_{\hat{\mu}}(1_{D\times X}|\mathcal{I}), \tag{5}$$

where $D = D_{\phi,\psi} = \{x : \phi(x)(0) \neq \phi(Tx)(0)\}$ and $\mathcal{I}$ denotes the σ-algebra of $T \times T$-invariant sets. In view of Proposition 1.5, (4) and (5) imply that $\hat{\mu}(D \times X) = \mu(D) = 0$, that is, $\phi(x)(0) = \phi(Tx)(0)$ for μ-a.a. x. By non-singularity we conclude that $\phi(x) = \phi(Tx)$ for μ-a.a. x and then ergodicity of T implies that ϕ is a.e. constant, so we are done.

If (4) does not hold, then according to Proposition 4.1, setting $\Delta_k = \{(x, T^k x) : x \in X\} \subset X \times X$, we must have

$$\hat{\mu}\left(\bigcup_{k\in\mathbb{Z}} \Delta_k\right) = 1.$$

If $\hat{\mu}(\Delta_k) > 0$ for some $k \neq 0$ it follows from the definition of the relative product $\hat{\mu} = \mu \times_\phi \mu$ that $\mu\{x \in X : \phi(T^k x) = \phi(x)\} > 0$. Since this set is T-invariant we conclude that it has full measure and then ergodicity fo T^k (Proposition 2.5) implies that ϕ is a.e. constant, so we are again done. The only remaining possibility is that $\hat{\mu}$ is supported on Δ_0, the diagonal in $X \times X$, which means that ϕ is a.e. 1-1 and concludes the proof.

References

[A]   .   J. Aaronson, *The intrinsic normalizing constants of transformations preserving infinite measures.* J. d'Analyse Math. **49**, (1987), 239-270.

[A,F]   T. Adams, N. Friedman, *Staircase mixing.* Preprint.

[A,N]   J. Aaronson, M. Nadkarni, L_∞ *eigenvalues ane L_2 spectra of non-singular transformations.* Proc. London Math. Soc. **55**, (3)(1987), 538-570.

[F]   N. Friedman, *Introduction to ergodic theory.* Van Nostrand, Princeton (1970).

[H]   T. Hamachi, *The normalizer group of an ergodic automorphism of type III and the commutant of an ergodic flow.* J. Functional Analysis **40**, (1981), 387-403.

[J1]   A. del Junco, *A simple measure-preserving transformation with trivial centralizer.* Pasific J. Math.**79**, (1978), 357-362.

[J2]   A. del Junco, *A family of counterexamples in ergodic theory.* Israel J. Math. **44**, (1983), 160-188.

[JK]   A. del Junco, M. Keane, *On generic points in the Cartesian square of Chacóns map.* Ergod Th. & Dynam. Sys. **5**, (1985), 59-69.

[O]   D. Ornstein, *On the root problem in ergodic theory.* Proc. Sixth Berkely Symp.Math. Stat. Prob. Vol. II. Univ. of California Press (1967), 347-356.

[R1]   D. Rudolph, *An example of a measure-preserving map with minimal self-joinings, and applications.* J. d'Analyse Math. **35**, (1979), 97-122.

[R2]   D. Rudolph, *Fundamentals of measurable dynamics.* Oxford Univ. Press (1990).

[R,S]   D. Rudolph, C. Silva, *Minimal self-joinings for nonsingular transformations.* Ergod Th. & Dynam. Sys. **9**, (1989), 759-800.

THE DYNAMICS OF SELF-SIMILAR SETS ON S^2 AND COMPLEX DYNAMICS

Atsushi Kameyama

Department of Mathematical Science
Faculty of Engineering Science
Osaka University
Toyonaka, Osaka, Japan

1. INTRODUCTION AND DEFINITIONS

We begin with the following theorem (see [3] and [4]).

Theorem. *Let X be a complete metric space and f_i ($1 \leq i \leq N, N > 1$) be contractions on X.*

Then there exists a unique compact subset K in X such that $K = \bigcup_{i=1}^{N} f_i(K)$.

Furthermore, there exists a coding map $\pi : \Sigma \to K$ such that π is surjective and the following diagram commutes for all i;

$$
\begin{array}{ccc}
\Sigma & \xrightarrow{\sigma_i} & \Sigma \\
\pi \downarrow & & \downarrow \pi \\
K & \xrightarrow{f_i} & K
\end{array}
$$

where Σ denotes the space of one-sided sequences of $\{1, 2, \ldots, N\}$ and σ_i maps $x_1 x_2 \ldots \in \Sigma$ to $i x_1 x_2 \ldots \in \Sigma$.

We call K the self-similar set. In this report, we will treat self-similar sets topologically. We will investigate the dynamics of self-similar sets on the 2-dimensional sphere. For the above reason, we regard only the coding map π and introduce a new definition of a self-similar set by forgetting the metric.

Definition 1. Let K be a compact set and f_i ($1 \leq i \leq N, N > 1$) be continuous maps of K into itself. A pair $(K, \{f_i\}_{i=1}^{N})$ is called a *self-similar system* if there exists a surjection π such that

$$
\begin{array}{ccc}
\Sigma & \xrightarrow{\sigma_i} & \Sigma \\
\pi \downarrow & & \downarrow \pi \\
K & \xrightarrow{f_i} & K
\end{array}
$$

Algorithms, Fractals, and Dynamics
Edited by Y. Takahashi, Plenum Press, New York, 1995

Fig. 1   The Sierpinski Gasket.

commutes for all i. We call K a *self-similar set*.

Now we recall the Sierpinski Gasket , which is the most famous self-similar set . Let X be a regular triangle and a_1, a_2, a_3 be the vertices of X. Let f_i be a similitude with ratio $1/2$ which fixes a_i $(i = 1, 2, 3)$. Then from the theorem we obtain a self-similar set K, which is called the Sierpinski Gasket (Fig. 1). Ushiki discovered a rational map g whose Julia set is homeomorphic to the Sierpinski gasket ([10]). See Fig. 2. We define Julia sets as follows.

Definition 2. For a rational map $g : \hat{C} \to \hat{C}$, the *Julia set* J_g is the set of points such that for any neibourhood U the family $\{g^n|U\}_{n>0}$ is not normal, where $\hat{C}$ denotes the Riemann sphere. The *Fotou set* is the complement of the Julia set.

The above example raises some questions.
Question 1. What kind of rational map has the Julia set homeomorphic to some self-similar set?
Question 2. What kind of self-similar set is homeomorphic to the Julia set for some rational map?
The answer to the first question is following ([6], [7]).

Definition 3. Suppose $f : S^2 \to S^2$ is a branched covering. We call f *postcritically finite* (pcf for short) if the *postcriticall set* $P_f = \{f^n(c)|c$ is a critical point, $n > 0\}$ is a finite set.

Theorem. *Suppose* $g : \hat{C} \to \hat{C}$ *is a pcf rational map of degree d at least two. If there exist d branches of g^{-1} on J_g, then J_g is self-similar.*

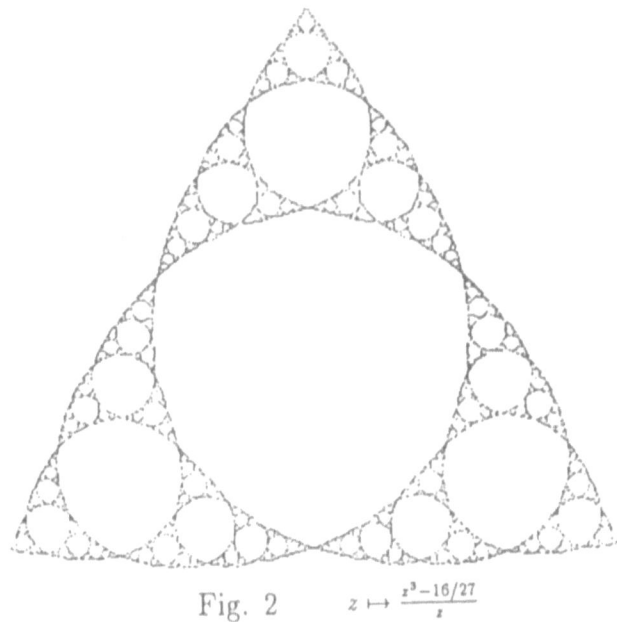

Fig. 2 $\qquad z \longmapsto \frac{z^3-16/27}{z}$

Corollary. *Suppose* $g : \mathbf{C} \to \mathbf{C}$ *is a polynomial map of degree* d *at least two. If any critical point is strictly preperiodic (i.e. eventually periodic and not periodic), then* J_g *is self-similar.*

Let $q_1, q_2, \ldots, q_d$ denotes the branches of g^{-1}. The self-similar system $(J_g, \{q_i\})$ satisfies the following.

(1) Each point in $J_i \cap J_j$ is a critical point for $i \neq j$, where $J_i = q_i(J_g)$. Therefore $\sharp(J_i \cap J_j) < \infty$. Actually $\sharp(J_i \cap J_j) \leq 2$.

(2) Since g is pcf, any point $c \in \bigcup_{i \neq j}(J_i \cap J_j)$ is strictly preperiodic.

We define a class of self-similr sets which includes self-similar Julia sets.

Definition 4. A self-similar system $(K, \{f_i\})$ is called *strictly preperiodic* if

(1) each f_i is injective,

(2) $K_i \cap K_j$ is at most finite for $i \neq j$, where $K_i = f_i(K)$,

(3) there exists a continuous map $g : K \to K$ such that

$$
\begin{array}{ccc}
\Sigma & \xrightarrow{\ \sigma\ } & \Sigma \\
\pi \downarrow & & \downarrow \pi \\
K & \xrightarrow[\ \ g\ \ ]{} & K
\end{array}
$$

commutes, where σ is the shift map and

(4) each point in $\bigcup_{i \neq j}(K_i \cap K_j)$ is strictly preperiodic by iteration of g.

We say (g, K) is the *dynamics* of $(K, \{f_i\})$.

Remark

◇ The third condition means that $f_i^{-1}(c) = f_j^{-1}(c)$ for any $c \in K_i \cap K_j$.

◇ A strictly preperiodic self-similar set K is metrizable ([5]).

◇ If K is connected, it is locally connected ([5]).

Definition 5. A strictly preperiodic self-similar system $(K, \{f_i\})$ is called spherical if K is a subset of S^2 and the dynamics of the system can be extended to a branched covering G on S^2.

Then it is proved that $G^{-1}(K) = K = G(K)$.

2. LOCALLY FINITENESS OF STRICTLY PREPERIODIC SELF-SIMILAR SETS

Theorem 1. *Let $(K, \{f_i\}_{i=1}^d)$ be a strictly preperiodic self-similar system. Suppose K is connected. Then there exists $N > 0$ such that for any point $x \in K$ and any connected neighborhood V, $V \setminus \{x\}$ has at most N connected components.*

Corollary. *Let $(K, \{f_i\}_{i=1}^d)$ be a spherical self-similar system. Suppose K is connected. Then there exists $N > 0$ such that for any point $x \in K$,*

$$\#\{U | U \text{ is a connected component of } S^2 \setminus K \text{ and } \partial U \ni x\} \leq N.$$

We prove the theorem. For $x \in K$ and $n > 0$, set

$$L_n(x) = \bigcup_{a:\pi(a)=x} K_{a_1 a_2 \ldots a_n},$$

where $K_{a_1 a_2 \ldots a_n} = f_{a_1} \circ f_{a_2} \circ \ldots \circ f_{a_n}(K)$. Then $\{L_n(x)\}_n$ forms a fundamental neighborhood system ([5]). If K is connected, so is $K_{a_1 a_2 \ldots a_n}$.

It is sufficient to show the following lemma.

Lemma. *Let $(K, \{f_i\}_{i=1}^d)$ be a strictly preperiodic self-similar system. Suppose K is connected. There exists $M > 0$ such that for any $n > 0$ and any $\alpha \in W_n$,*

$$\#\{\beta \in W_n | K_\alpha \cap K_\beta \neq \emptyset\} \leq M,$$

where $W_n = \{a_1 a_2 \ldots a_n | a_i \in \{1, 2, \ldots, d\}\}$.

Proof. For $a, b \in \Sigma$, if $\pi(a) = \pi(b)$ then there exist $x \in C = \bigcup_{i \neq j}(K_i \cap K_j)$ and $a^*, b^* \in \pi^{-1}(x)$ and $i_1 i_2 \ldots i_n \in W_n$ such that $a = i_1 i_2 \ldots i_n a^*, b = i_1 i_2 \ldots i_n b^*$.

It is sufficient to show that:

For any $a = a_1 a_2 \ldots \in \pi^{-1}(C)$ there exists N_a (independent of n) such that for any $\alpha = i_1 i_2 \ldots i_n \in W_n$,

$$\#\{k | 1 \leq k \leq n, \ i_k i_{k+1} \ldots i_n = a_1 a_2 \ldots a_{n-k+1}\} \leq N_a.$$

Indeed, it is clear that the statement of the lemma holds for $M = \sum_{\pi(a)=C} (l_a - 1) N_a$,

where $l_a = \#\{b \in \Sigma | a \sim b\}$.

We can write $a = \mu \bar{\lambda}$, where $\mu \in W_p$, $\lambda \in W_r$ (p and r are smallest possible) and $\bar{\lambda} = \lambda \lambda \ldots$.

Assume there exist $k_1 > k_2 \geq p + r$ such that

$$a_1 a_2 \ldots a_{k_s} = i_{n-k_s+1} i_{n-k_s+2} \ldots i_n \quad (i = 1, 2).$$

120

If $k_1 - k_2$ is a multiple of r, it contradicts minimality of p. Otherwise, it contradicts minimality of r. Therefore,

$$\#\{k | 1 \leq k \leq n, i_k i_{k+1} \ldots i_n = a_1 a_2 \ldots a_{n-k+1}\} \leq p + r.$$

Set $N_a = p + r$. $\qquad\qquad\qquad\qquad\qquad\qquad\qquad\qquad\qquad\qquad\qquad\qquad\qquad\qquad$ $\square$

Remark

$\diamond$ The theorem is true for self-similar sets that satisfy only (1), (2) and (4) in Definition 5, if not strictly preperiodic.

$\diamond$ Doubtless the choice of M in the proof is not best possible. It is an interesting problem to find the best estimate.

$\diamond$ See Fig. 3 and 4, which are examples with the branching number four and three.

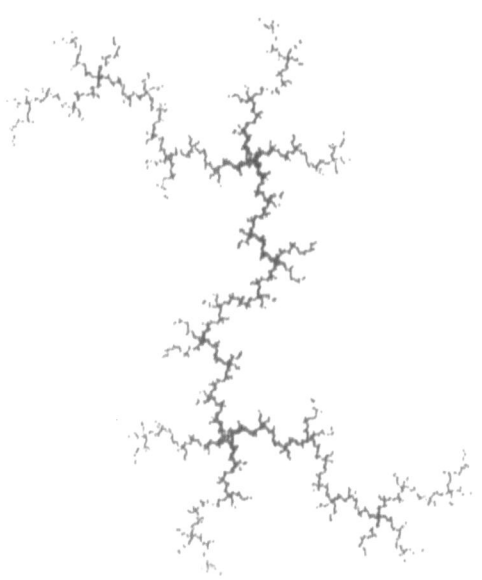

Fig. 3 $\quad \mapsto z^2 + c$.

$c = 0.3950141 + 0.5556246i$

3. THE DYNAMICS OF SPHERICAL SELF-SIMILAR SETS

Suppose $(K, \{f_i\})$ is a spherical self-similar system with K connected. Let $\mathcal{U}$ be the set of connected components of $S^2 \setminus K$. For $U \in \mathcal{U}$, the image $G(U)$ also belongs to $\mathcal{U}$.

Theorem 2. *Every component $U \in \mathcal{U}$ is eventually periodic.*

Proof. Take $U \in \mathcal{U}$. Since K is connected, U is homeomorphic to the unit disc and ∂U is connected. It is clear that $G(\partial U) = \partial G(U)$.

We can take two points x and $y \in \partial U$. For some $n > 0$, there exist distinct i and j such that $g^n(x) \in K_i$ and $g^n(y) \in K_j$. Since $\partial G^n(U) \cap K_i \neq \emptyset$ and $\partial G^n(U) \cap K_j \neq \emptyset$, we have $\partial G^n(U) \cap C \neq \emptyset$, where $C = \bigcup_{i \neq j}(K_i \cap K_j)$. Let c be one element of $\partial G^n(U) \cap C$.

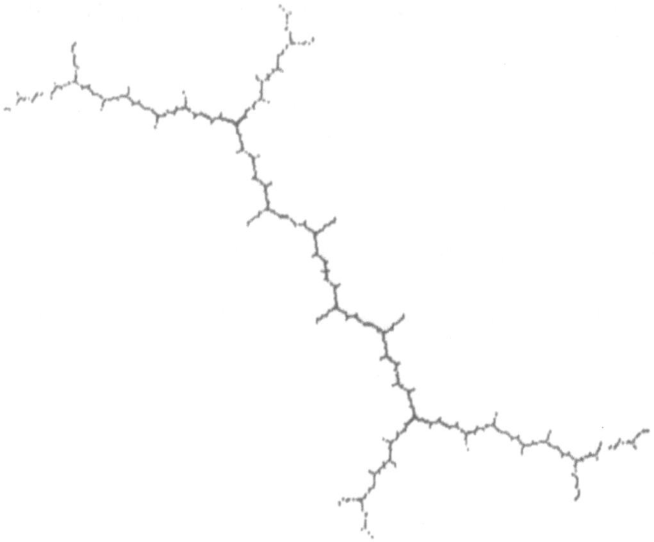

Fig. 4 $z \mapsto z^2 + c$

$c = -0.155788 + 1.1122171i$.

Recall that points in C are eventually periodic. Therefore, for some $m > 0$, $p = g^m(c)$ is periodic. Let k be the period of p. We obtain:

$$\partial G^{n+m+tk}(U) \ni p \quad \text{for } t = 0, 1, 2 \dots .$$

From the corollary of Theorem 1, we can see that $\{\partial G^{n+m+tk}(U)\}_{t=0,1,2\dots}$ is finite. So U is eventually periodic. □

Corollary. *For a spherical self-similar system $(K, \{f_i\})$ with K connected, the branched covering $G : S^2 \to S^2$ is pcf.*

4. A FURTHER STUDY

In this section we give results without proofs. Compare the result in the previous section with Sullivan's theorem ([9], [1]).

Theorem(Sullivan). *Every connected component of the Fotou set of a rational map is eventually periodic.*

It suggests that the dynamics of a spherical self-similar system resembles that of a rational map.

Actually, we can prove the following:

Theorem. *Suppose $(K, \{f_i\}_{i=1}^d)$ be a spherical self-similar system with K connected. Then there exists a rational map $R : \hat{C} \to \hat{C}$ of degree d such that the Julia set J_R is homeomorphic to K, and $R|J_R$ and g are topologically conjugate.*

Theorem. *Suppose $(K, \{f_i\}_{i=1}^d)$ be a strictly preperiodic self-similar system with K connected and simply connected. Then there exists a polynomial map $P : \mathbb{C} \to \mathbb{C}$ of degree $D = d^n$ such that the Julia set J_P is homeomorphic to K, and $P|J_P$ and g^n are topologically conjugate.*

The proofs are in [6] and [7]. They use Thurston's Theorem ([2]).

References

1. A. F. Beardon, *Iteration of Rational Functions*, Graduated Text in Mathematics, Vol.132, Springer-Verlag, 1991.
2. A. Douady and J. H. Hubbard, *A Proof of Thurston's Topological Characterization of Rational Maps*, preprint.
3. M. Hata, *On the Structure of Self-similar Sets*, Japan J. Appl. Math. **2** (1985), 381–414.
4. J. E. Hutchinson, *Fractals and Self-similarity*, Indiana Univ. Math. J. **30** (1981), 713–747.
5. A. Kameyama, *Self-similar Sets from the Topological Point of View*, to appear in Japan J. Ind. Appl. Math.
6. ———— , *Julia Sets and Self-similar Sets*, to appear.
7. ———— in preparation.
8. J. Milnor, *Dynamics in One Complex Variable: Introductory Lectures*, preprint, SUNY, Stony Brook, New York, (1990).
9. D. Sullivan, *Quasiconformal Homeomorphisms and Dynamics I: Solution of the Fotou-Julia Problem on Wandering domains*, Ann. of Math. **122** (1985), 401–418.
10. S. Ushiki, *Julia sets with Polynomial Symmetries in "Proceedings of the International*, Conference on Dynamical Systems and Related Topics," Ed. K. Shiraiwa, Advanced Series in Dynamical Systems, Vol. 9, World Scientific, (1991).

THE ESSENCE OF THE LAW OF LARGE NUMBERS

Michael Keane

CWI P.O. Box 94079
1090 GB Amsterdam
Electronic mail: keane@cwi.nl

The law of large numbers, not really a law but a mathematical theorem, is at the same time a justification for application of statistics and an essential tool for the mathematical theory of probability. As such, it must be taught to many students. The traditional method for this, using independent and identically distributed random variables, was developed by Kolmogorov in the 1930's, and explains well what happens, and much more, at this level of generality. However, it has recently come to light that the reason for the validity of this theorem in its general setting, that of stationarity, is much simpler than was first thought. In this short article, I shall try to explain to the general audience towards whom this collection is directed, the essence of the law of large numbers. A complete treatment should certainly include many references and interesting historical comments, and I apologize for their absence here.

Let me start with the *basic law of large numbers* by considering, very simply, an infinite sequence

$$x_0, \ x_1, \ x_2, \ldots$$

each of whose elements is either 0 or 1. Perhaps it will help (or hinder!) to think of x_n as the result of the nth trial of an uncertain experiment, with $x_n = 1$ designating success and $x_n = 0$ failure. Let

$$a_n = \frac{x_0 + x_1 + \ldots + x_{n-1}}{n} \qquad (n \geq 1)$$

denote then the average numbers of successes up to time n. It is very easy to see mathematically that for some sequences x,

$$\lim_{n \to \infty} a_n$$

exists, while for other sequences x, this is not the case. One can only affirm with certainty that

$$\lim_{n \to \infty} (a_{n+1} - a_n) = 0,$$

but nothing impedes the averages a_n from oscillating more and more slowly as n grows. Thus it seems that further discussion is useless, and that uncertainty here must be accepted.

Phenomenologically, however, we are faced with the fact that in certain situations, such limits seem to exist, and the society makes seemingly understandable statements concerning the percentage of smokers dying of cancer, the probability of rain tomorrow, or an industrial average yield. We are confronted with the question as to whether nature produces sequences whose averages do converge, and why. Of course, this is not a mathematical question, and in order to say something mathematically sensible, one must adopt a model.

The currently accepted model, and it is difficult to see how it could be replaced by something else, is that for a given situation in which such sequences x appear, in principle all sequences are possible, but there is also a mass distribution with total mass 1 over the set of sequences, which assigns to each "event" which might occur a *probability*, this being the total mass of those sequences for which the event occurs. If an event, for instance the existence of $\lim_{n\to\infty} a_n$, has probability 1, then one says that the event will occur *almost surely*.

The determination of such a mass distribution in different practical situations is one of the most important tasks for probabilists, and requires a good mixture of mathematics, other sciences, and good old common sense. First principles are of utmost importance, as determining such an object by experimentation resembles very much a cat chasing its own tail! One of the basic properties of such a mass distribution, already alluded to briefly above, is that of *stationarity*. We say that the probability measure (= mass distribution) is *stationary* if the events have time-homogeneous probabilities. That is, shifting any event forwards or backwards in time does not change its probability.

Perhaps a brief remark on mass distributions is in order. There is a branch of mathematics, measure theory, which deals extensively with the specification and manipulation of such objects. However, one can understand well most arguments and principles by using the intuitive notion, which is my intention here.

Now we can state the

BASIC LAW OF LARGE NUMBERS:

If $x = (x_0, x_1, \ldots)$ is a stationary sequence of zeroes and ones, then $lim_{n\to\infty} a_n$ exists almost surely.

Just to be sure that you are (mathematically) still with me: A unit mass distribution on sequences of zeroes and ones is given; it is stationary. Then the set of all sequences x for which $\lim_{n\to\infty} a_n$ exists has total mass 1. The set of sequences for which this limit does not exist has mass 0. Remember, this is a theorem, and I want to explain the proof.

To understand the proof will require the level of first-year university analysis, given the intuitive acceptance of the mass distribution notion. We begin by defining

$$\bar{a} := \limsup_{n\to\infty} a_n;$$

this always exists, and $0 \le \bar{a} \le 1$. It is also clear that if we had started observing x at

a later time point, the value $\bar{a}$ would be the same:

$$\bar{a} = \limsup_{n \to \infty} \frac{x_k + x_{k+1} + \ldots + x_{k+n-1}}{n}$$

for any $k \geq 0$ and any sequence x.

Next, we need a way to measure how close we are to the lim sup, $\bar{a}$. Thus, let $\epsilon > 0$ be a fixed positive number, and for each $k \geq 0$, define

$$N_k := \min\left\{ n \geq 1 : \frac{x_k + x_{k+1} + \ldots + x_{k+n-1}}{n} \geq \bar{a} - \epsilon \right\}.$$

By the definition of lim sup, the set on the right is non-empty and hence N_k is finite for each k. The crucial point we need to address concerns the size of the numbers N_k; to make our idea clear, let us examine the simplest case first.

CASE 1. Suppose that for each $\epsilon > 0$ there exists a (large) positive integer M such that for each k, $N_k \leq M$ almost surely. (That is, the set of sequences x for which $N_k \leq M$ has total mass 1.)

REMARK: Note that by our assumption of stationarity the events $N_k \leq M$ for different k all have the same probability.

If now x is such a sequence that for each k, $N_k \leq M$, we claim that $\lim_{n \to \infty} a_n$ exists. The idea is that, as n gets larger, a_n can only change more and more slowly, and that then wandering is impossible because the lim sup is reached again and again within M steps. Formally, one proceeds as follows. Fix $\epsilon > 0$ and choose any $n > M/\epsilon$. Then starting at the beginning of x, break x up into pieces of lengths at most M such that the average of x over each piece is at least $\bar{a} - \epsilon$. Stop at the piece containing the coordinate n. Then it is clear that

$$x_0 + x_1 + \ldots + x_{n-1} \geq (n - M)(\bar{a} - \epsilon),$$

so that

$$a_n = \frac{x_0 + x_1 + \ldots + x_{n-1}}{n} \geq (1 - \epsilon)(\bar{a} - \epsilon) \geq \bar{a} - 2\epsilon$$

for each $n > M/\epsilon$; it follows that $\lim_{n \to \infty} a_n = \bar{a}$ exists.

REMARK: Note that only the last piece is of importance; it must not become too long.

Actually, the same type of argument works in the general case, when combined with an idea coming originally from non-standard analysis.

CASE 2: General case. By the remark after Case 1, it remains true that the events $N_k \leq M$ all have the same probability, for any k and fixed M. Since N_k is finite for each x, we may not be able to find an M, for $\epsilon > 0$ given, such that these events have probability 1, but we certainly can choose M so large that for any k, the probability of $N_k \leq M$ is less than ϵ.

Fix now such an integer M, given $\epsilon > 0$. Next, we want to make the same inequality work for us, but we are impeded whenever $N_k > M$. So let us change x at those places to insure quick arrival at the lim sup.

Namely, define

$$x_k^* := \begin{cases} x_k & \text{if} \quad N_k \leq M \\ \\ 1 & \text{if} \quad N_k > M. \end{cases} \qquad (k \geq 0)$$

Then clearly $x_k^* \geq x_k$ for each k, so that if we set

$$N_k^* := \min\left\{ n \geq 1 : \frac{x_k^* + \ldots x_{k+n-1}^*}{n} \geq \bar{a} - \epsilon \right\}$$

(same $\bar{a}$), then $N_k^* \leq N_k$, and moreover if k is such that

$$N_k > M,$$

then we have

$$N_k^* = 1,$$

since setting $x_k^* = 1$ insures immediate arrival above $\bar{a} - \epsilon < 1$.

Now we are almost ready. As above, breaking x^* up into pieces yields for $n > M/\epsilon$.

$$x_0^* + x_1^* + \ldots + x_{n-1}^* \geq (n - M)(\bar{a} - \epsilon),$$

but now we cannot conclude anything about the sequence x because we have replaced it by x^*.

Instead, we now need to use our mass distribution to calculate the average value of each side of the inequality over all sequences x, called by probability theory the *expectation* and denoted by $\mathbb{E}(\cdot)$. Let

$$\mathbb{E}(x_0) =: p$$

and

$$\mathbb{E}(x_0^*) =: p^*;$$

by stationarity, $\mathbb{E}(x_k^*) = p^*$ for all k, and by the choice of M, we have

$$p^* \leq p + \epsilon.$$

Of course, p is just the probability that $x_k = 1$, and p^* the probability that $x_k^* = 1$, for any k. Now, taking expectations of each side of the inequality results in

$$n(p + \epsilon) \geq np^* \geq (n - M)(\mathbb{E}(\bar{a}) - \epsilon) \qquad (n \geq M/\epsilon).$$

Now divide by n, send n to infinity and then ϵ to zero, giving

$$\mathbb{E}(\bar{a}) = \mathbb{E}\left(\limsup_{n \to \infty} \frac{x_0 + \ldots + x_{n-1}}{n}\right) \leq p.$$

Finally, apply the entire argument above to the "mirrored" $0 - 1$−sequence $y_k = 1 - x_k$; an easy calculation (exercise!) shows that

$$\mathbb{E}\left(\liminf_{n \to \infty} \frac{x_0 + \ldots + x_{n-1}}{n}\right) \geq p.$$

But for any sequence x, certainly

$$\liminf_{n \to \infty} \frac{x_0 + \ldots + x_{n-1}}{n} \leq \limsup_{n \to \infty} \frac{x_0 + \ldots + x_{n-1}}{n};$$

it is an elementary fact of expectations or averaging that the three inequalities then must be equalities, the last one almost surely. Hence lim sup = lim inf for a set of sequences of total mass one, i.e. the limit exists almost everywhere. This concludes the proof of the basic law of large numbers.

In concluding, we state without proof that this method can be widely extended with minor, straight-forward modifications to the most general laws of large numbers based on stationarity. The above proof should, however, in my opinion be included in basic probability courses, since it so clearly shows the nature of the interplay of stationarity assumptions and the existence of statistical limits.

ROTATION INVARIANCE AND CHARACTERIZATION OF A CLASS OF SELF-SIMILAR DIFFUSION PROCESSES ON THE SIERPINSKI GASKET

Takashi Kumagai

Department of Mathematics
Osaka University
Toyonaka, Osaka 560, Japan

Introduction

In [B.P], Barlow-Perkins succeeded in the characterization of the Brownian motion on the Sierpinski gasket. They proved that the diffusion on the gasket which has local translation and reflection invariance is a constant time change of the Brownian motion. On the other hand, Kumagai [Kum] introduced a class of Feller diffusions which is invariant under the operation of local rotation. These diffusions are called p-stream diffusions on the Sierpinski gasket, which contains Brownian motion as a typical case. They were constructed as a limit of a sequence of random walks which has some consistency (called decimation property). In this paper, we will characterize these Feller diffusions. In fact, the non-degenerate self-similar Feller diffusion which has local rotation invariance is a constant time change of some p-stream diffusion. In general, the problem of this type is essentially reduced to show the uniqueness of the fixed point for some non-linear map. In Section 1, we briefly introduce the p-stream diffusions and give some properties of them. In Section 2, we characterize these diffusions. In Section 3, we give some remarks for the existence of non-symmetric Feller diffusions on some fractals.

§1 Construction and some properties of p-stream diffusions on the Sierpinski Gasket

In this section, we briefly introduce the construction and properties of a class of Feller diffusions on the Sierpinski gasket which is locally rotation invariant and non-symmetric. As the detailed results are written in [Kum], we omit proofs. First, we give some notation.

Notation. We follow most of the notation of Barlow-Perkins [B.P].

Algorithms, Fractals, and Dynamics
Edited by Y. Takahashi, Plenum Press, New York, 1995

1) We denote by G_0 all the vertices of the pre-Sierpinski gasket consisting of regular triangles with side 1 (c.f. Figure 1.1).
Let $G_n = 2^{-n}G_0$, $n \in \mathbb{Z}$, be the vertices of the pre-Sierpinski gasket consisting of regular triangles with side 2^{-n}. Set

$$G_\infty = \cup_{n \in \mathbb{Z}} G_n.$$

Then, its closure in $\mathbb{R}^2$

$$G = Cl(G_\infty)$$

is the Sierpinski gasket.

2) Let $X = (X(t))$ be a G-valued process.
For $m \in \mathbb{Z}$, set

$$T^m(X) = T_0^m(X) = \inf\{t \geq 0 : X(t) \in G_m\},$$
$$T_{i+1}^m(X) = \inf\{t > T_i^m(X) : X(t) \in G_m - \{X(T_i^m(X))\}\}, \quad i \geq 0,$$
$$W_i^m(X) = T_i^m(X) - T_{i-1}^m(X), \quad i \geq 1,$$
$$\tau_A(X) = \inf\{t \geq 0 : X(t) \in A\} \quad \text{for} \quad A \subset G.$$

3) For each $x \in G_m$, we can locate x and its four G_m-neighborhoods $N_m^i(x), 1 \leq i \leq 4$ as in Figure 1.2 by rotation. Here, G_m-neighborhoods of x are points with distance 2^{-m} from x and are connected with x by G_m-bonds.
Write $N_m(x) = \{N_m^1(x), N_m^2(x), N_m^3(x), N_m^4(x)\}$.

4) Let J_0 be the regular triangle (includes its inner part) with edges $(0,0)$, $(1,0)$, $(\frac{1}{2}, \frac{\sqrt{3}}{2})$. A G_n-triangle is a closed set of points in G that lie inside a regular triangle which is a translation of $2^{-n}J_0$ and whose vertices are three neighboring points in G_n. Let $\mathcal{F}_n$ denote the set of G_n-triangles.

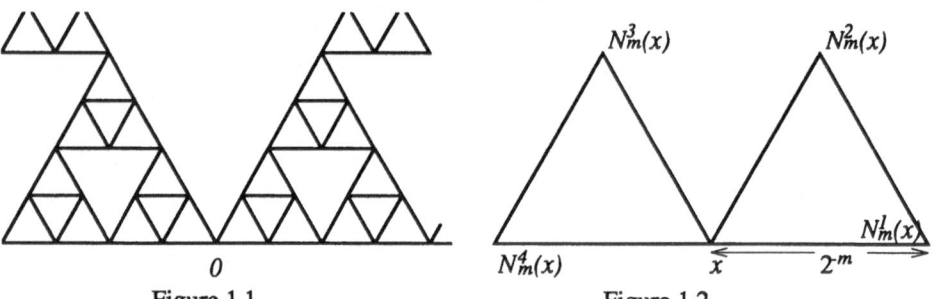

Figure 1.1   Figure 1.2

For each $0 < p < 1$, let $\{Y_r\}_{r=0}^\infty$ be a r.w. (r.w. in abbreviation) on G_0 with the following transition probabilities:

$$P(Y_{r+1} = y | Y_r = x) = \begin{cases} p^2 & \text{if} \quad y = N_0^1(x) \\ p(1-p) & \text{if} \quad y = N_0^2(x) \\ p(1-p) & \text{if} \quad y = N_0^3(x) \\ (1-p)^2 & \text{if} \quad y = N_0^4(x) \\ 0 & \text{otherwise} \end{cases} \quad \text{for} \quad x, y \in G_0.$$

We call this the p-stream r.w.

Proposition 1.1. *(Decimation property)* *For each* $m \in \mathbb{N}$, $Y(T_i^{-m}(Y))$, $i \geq 0$ *is a p-stream r.w. on G_{-m}.*

Remark. Assume that we have a r.w. $\{\tilde{Y}_r\}_{r=0}^{\infty}$ on G_0 with the transition probabilities given by $\mathbf{p} = (p_1, p_2, p_3, p_4)$: i.e.

$$P(\tilde{Y}_{r+1} = y | \tilde{Y}_r = x) = \begin{cases} p_i & \text{if } y = N_0^i(x) \quad (1 \leq i \leq 4) \\ 0 & \text{otherwise} \end{cases} \quad \text{for } x, y \in G_0.$$

Set $f_i(\mathbf{p}) = P^0(\tilde{Y}(T_1^{-1}(\tilde{Y})) \in N_{-1}^i(0))(1 \leq i \leq 4)$, and $f(\mathbf{p}) = (f_1(\mathbf{p}), f_2(\mathbf{p}), f_3(\mathbf{p}), f_4(\mathbf{p}))$. Then, it is easy to verify that $\tilde{Y}(T_1^{-1}(\tilde{Y}))$ has the transition probability given by $f(\mathbf{p})$. We call the map f non-linear decimation map. Proposition 1.1 assert that $(p^2, p(1-p), p(1-p), (1-p)^2)$ is an fixed point of the map.

Let $\{X(n, x) : x \in G_n\}$ be a family of p-stream r.w. which satisfy the following properties:

(1) $\{X(n, x) : n \in \mathbb{Z}\}$ is a p-stream r.w. on G_n starting at x,

(2) If $m \leq n$ and $x \in G_n$, then $X(m, x)(i) = X(n, x)(T_i^m(X(n, x)))$, $i \geq 0$.

Notation

1) For each $x \in G_m$, we call a path from x to $N_m^i(x)$ a path of type $< i >$, $1 \leq i \leq 4$. Let η_i be the number of times which Y (starting at 0) has passed paths of type $< i >$ before $T_1^{-1}(Y)$, $1 \leq i \leq 4$.

2) For the G_n-valued r.w. X_n, let $T_i^{m<l>}(X_n)$ $(1 \leq l \leq 4)$ be the number of times which X_n has passed $< l >$ type paths before the time $T_i^m(X_n)$, and set $W_i^{m<l>}(X_n) = T_i^{m<l>}(X_n) - T_{i-1}^{m<l>}(X_n)$.

Proposition 1.2. *If $x \in G_m$ and $i \in \mathbb{N}$ are fixed, then the four dimensional process*

$$r \to Z_r = (Z_r^{<l>}, 1 \leq i \leq 4),$$

where $Z_r^{<l>} := W_i^{m<l>}(X(m+r, x))$, $1 \leq l \leq 4$, $r \geq 0$ is a multi four-type branching process with the generating functions

$$f_0^l(s_1, s_2, s_3, s_4) = E^0(s_1^{\eta_1} s_2^{\eta_2} s_3^{\eta_3} s_4^{\eta_4} | Y(T_1^{-1}(Y)) = N_{-1}^l(0)).$$

Moreover, the largest eigenvalue of 4×4 matrix $M = (\frac{\partial f^i}{\partial s_j}(1,1,1,1))$ is $-\frac{2(p^2-p-1)}{2p^2-2p+1} = -1 + \frac{3}{2(p-\frac{1}{2})^2 + \frac{1}{2}}$, (denote it λ_p).

Using the theory of supercritical multi-type branching process (c.f. Athreya-Ney [A.N]), we can prove the following.

Proposition 1.3. *For $x \in G_n$, let $X_n(x)(j \cdot \lambda_p^{-n}) = X(n, x)(j)$ and extend $X_n(x)(t)$ to $t \in [0, \infty)$ by the linear interpolation. Then, for $x \in G_\infty$, $X_n(x)$ converges a.s. in $C([0, \infty), G)$ as $n \to \infty$ to a process, $X(x)$.*

By taking the sequence $\{X(n,x) : x \in G_n\}$ adequately, we can extend $X(x)$ to $x \in G$, which we call the p-stream diffusion.

Theorem 1.4. *Let $\Omega = C([0,\infty), G)$, $\mathcal{F} = $ Borel σ-field on Ω augmented in a usual manner, $X_t(\omega) = \omega(t)$ for $\omega \in \Omega$ and $P^x = $ law of $X(x)$ on Ω. Then, $(\Omega, \mathcal{F}, X_t, P^x)$ is a Feller diffusion process. I.e. it is a diffusion process such that $P_t : C_\infty(G) \to C_\infty(G)$. Here $C_\infty(G)$ is a set of continuous functions on G vanishing at ∞.*

Proposition 1.5. *The p-stream diffusion is recurrent. I.e. it hits every non-empty open set with probability one.*

A diffusion process on a locally compact state space S with the transition semigroup P_t is called symmetric if there exists a positive Borel measure m such that

$$\int_S P_t f(x) \cdot g(x) m(dx) = \int_S f(x) \cdot P_t g(x) m(dx), \quad \text{for } f, g \in C_b(S).$$

Here $C_b(S)$ is a set of continuous bounded functions on S.

Proposition 1.6. *The p-stream diffusion is not symmetric for $p \neq \frac{1}{2}$. (In the case of $p = \frac{1}{2}$, it is the Brownian motion.)*

Theorem 1.7. *There are constants $c_{1.1}, c_{1.2}, c_{1.3}$ and $c_{1.4}$ such that*

$$c_{1.1} \exp\{-c_{1.2}(\delta t^{-\gamma})^{\frac{1}{1-\gamma}}\} \leq P^x(|X_t - X_0| \geq \delta)$$
$$\leq P^x(\sup_{s \leq t} |X_s - X_0| \geq \delta) \leq c_{1.3} \exp\{-c_{1.4}(\delta t^{-\gamma})^{\frac{1}{1-\gamma}}\}$$

for all $x \in G$ and $t, \delta \in (0, \infty)$. Here $\gamma = \frac{\log 2}{\log \lambda_p}$.

Proposition 1.8. *There are constants $\{c_{1.5}(x) > 0 : x \in G\}$ such that for all $x \in G$,*

$$\limsup_{t \downarrow 0} \frac{|X_t - X_0|}{t^\gamma (\log \log t^{-1})^{1-\gamma}} = c_{1.5}(x) \quad P^x - a.s.$$

Proposition 1.9. *For $p \neq p'$, $P^x_{(p)}$ and $P^x_{(p')}$ are mutually singular for $x \in G$, where $P^x_{(p)}$ is a law of the p-stream diffusion starting at x on the path space.*

§2 Characterization of the p-stream diffusions

In [B.P], Barlow-Perkins gave a wonderful characterization of the Brownian motion on the Sierpinski gasket. Roughly saying, they showed that the Brownian motion is uniquely determined (up to a constant time scaling $t \to ct$) by the properties of local translation and reflection invariance. The p-stream diffusions are invariant under local translation and rotation but not reflection invariant if p is not $\frac{1}{2}$. In fact, this property plus self-similarity characterize p-stream diffusions.

Consider the following properties on G-valued processes.

[P_1] (**Feller diffusion**) $(\Omega, \tilde{\mathcal{F}}, Y_t, \tilde{P}^x)$ is a G-valued Feller diffusion defined on the continuous paths space.

[P_2] (**Self-similarity**) For any $n \in \mathbb{Z}$ and any triangles Δ in $\mathcal{F}_n$, let F_1, F_2, F_3 be usual contractions operated on Δ. If we denote Y_Δ the diffusion killed at $\partial \Delta$, then there exists positive constant λ such that

$$F_i(Y_\Delta(x)(t)) \overset{d}{\sim} Y_{F_i(\Delta)}(F_i(x))(\lambda t) \qquad \text{for } x \in \Delta, \ 1 \le i \le 3.$$

($X \overset{d}{\sim} Y$ means that processes X and Y have the same distribution.)

[P_3] (**Local translation and rotation invariance**) For any $n \in \mathbb{Z}$ and any pair of adjacent triangles in $\mathcal{F}_n$, Δ_1 and Δ_2, which intersect at $x \in G_n$, let $\Psi : \Delta_1 \cup \Delta_2 \to \Psi(\Delta_1) \cup \Psi(\Delta_2)$ be the composition of a translation and rotation of $\pm\frac{2\pi}{3}$ or 0 which maps $\Delta_1 \cup \Delta_2$ onto another pair of adjacent triangles in $\mathcal{F}_n$. Then

$$\Psi(Y_{\Delta_1 \cup \Delta_2}(x)(t)) \overset{d}{\sim} Y_{\Psi(\Delta_1 \cup \Delta_2)}(\Psi(x))(t).$$

[P_4] (**Non-degeneracy**) Let $a_1 = (1,0), a_2 = (\frac{1}{2}, \frac{\sqrt{3}}{2}), a_3 = (-1,0), a_4 = (-\frac{1}{2}, \frac{\sqrt{3}}{2})$. If $P^0(\min_i \tau_{a_i}(Y) < \infty) > 0$, then

$$P^0(\tau_{a_i}(Y) < \min_{j \ne i} \tau_{a_j}(Y)) > 0 \text{ for } \ 1 \le i \le 4.$$

Theorem 2.1. $(\Omega, \tilde{\mathcal{F}}, Y_t, \tilde{P}^x)$ *satisfies* [P_1], [P_2],[P_3],[P_4] *if and only if there is a* $c \in [0, \infty)$ *and* $0 < p < 1$ *such that for every* $x \in G, \tilde{P}^x$ *is the* $P^x_{(p)}$-*law of* $X(c \cdot)$. *Here* $X(t)(\omega) = \omega(t)$.

For the proof, we prepare some lemmas.

Lemma 2.2. *Let* $x \in G_n$ *and suppose* $\tilde{P}^x(W_1^n(Y) = \infty) = 0$.

(a) $\{W_i^n(Y) : i \in \mathbb{N}\}$ *are i.i.d.* $[0, \infty)$-*valued random variables, whose common law does not depend on* x.

(b) *Let* $\hat{Y}_n(i) = Y(T_i^n(Y))$. *Then, for some* $0 < p < 1$, $\hat{Y}_n$ *is a* p-*stream r.w. on* G_n.

Proof. (a) is an easy consequence of the strong Markov property and [P_3].
For (b), consider the induced r.w. $\hat{Y}_n(x_n)(i)$ on G_n starting at $x_n \in G_n$. From the assumption $\tilde{P}^x(W_1^n(Y) = \infty) = 0$ and [P_3], $\tilde{P}^{x_n}(W_i^n(Y) = \infty) = 0$ for all $i \in \mathbb{N}$. Let

$$P(\hat{Y}_n(x_n)(1) = y) = \begin{cases} p_n(x_n) & \text{if} \quad y = N_n^1(x_n) \\ q_n(x_n) & \text{if} \quad y = N_n^2(x_n) \\ r_n(x_n) & \text{if} \quad y = N_n^3(x_n) \\ s_n(x_n) & \text{if} \quad y = N_n^4(x_n) \\ 0 & \text{otherwise.} \end{cases}$$

Then, from [P_3], $p_n(x_n)$, $q_n(x_n)$, $r_n(x_n)$, $s_n(x_n)$ are independent of x_n. From [P_2], they are also independent of n (therefore we denote them p, q, r, s). From [P_4], $p, q, r, s > 0$. Clearly this r.w. has a decimation property $\hat{Y}_n(T_i^{n-1}(\hat{Y}_n)) = \hat{Y}_{n-1}(i)$. Thus we are looking for all the sequence of random walks which has decimation property and of which the transition probabilities are the same. (Equivalently, we are looking for all the fixed points of non-linear decimation map.) By easy calculations, we can see that only the sequence of p-stream random walks ($1 < p < 1$) satisfies the conditions. This concludes the proof. $\square$

Remark

1) If one does not assume $[P_4]$, then some non-degenerate random walks appear. For example when $p = s = \frac{1}{2}, q = r = 0$, which is mentioned in [G]. The author was informed of this by O.Jones. Using numerical plotting, O.Jones has checked that the p-stream random walks correspond to the attractive fixed points of the non-linear decimation map.

2) If one does not assume $[P_3]$, then there appear other diffusion processes. The examples are mentioned in [O].

Lemma 2.3(Ito [I]: §44 (xi)). *Let* $(\Omega, \mathcal{F}, Z(t), P^x)$ *be a Feller diffusion on* G. *If* $a \in G$ *is not a trap, then for any* $\epsilon > 0$, *there exists an open set* U $(a \in U)$ *such that* $E^y(\tau_{U^c}(Z)) < \epsilon$ *for all* $y \in U$.

Proof. As a is not a trap, there exists $f \in \mathcal{D}(\mathcal{A})$ such that $\mathcal{A}f(a) \neq 0$. Taking $-f$ if necessary, $\mathcal{A}f(a) > 0$. As $\mathcal{A}f$ is continuous, there exist $\alpha > 0$ and U, neighborhood of a, such that $\mathcal{A}f(b) > \alpha$ if $b \in U$. Now, by the Dynkin's formula,

$$f(b) = E^b[\int_0^{\tau_{U^c} \wedge n} (-\mathcal{A}f)(Z(t))dt] + E^b[f(Z(\tau_{U^c} \wedge n))].$$

On the other hand, as $Z(t)$ is a diffusion, $Z(\tau_{U^c} \wedge n)(b) \in Cl\ U$ ($Z(\cdot)(b)$ starts at b). By the continuity of f, choosing U small enough, we have $f(Z(\tau_{U^c} \wedge n)) < f(b) + \alpha\epsilon$. Using the above facts, we have $f(b) < -\alpha E^b(\tau_{U^c} \wedge n) + f(b) + \alpha\epsilon$, therefore $E^b(\tau_{U^c} \wedge n) < \epsilon$. Letting $n \to \infty$ and using Fatou's lemma, we obtain the result. $\square$

The next lemma is proved in the same way as [Lemma 8.3:B.P].

Lemma 2.4. *If* n, Δ_1, Δ_2 *and* Ψ *are as in [P_3] and* $y \in (\Delta_1 \cup \Delta_2) \cap G_\infty$, *then* $[P_3]$ *holds if* x *is replaced by* y.

Lemma 2.5. *Under the assumptions* $[P_1],[P_3]$, *if the is no trap, then there is an* $n_1 \in \mathbb{N}$ *such that*

$$\sup_{y \in G} \tilde{P}^y(T_1^{n_1}(Y) > n_1) < 1.$$

Proof. Take the sequence which attains the supremum and denote it $\{x_n^{n_1}\}$. Also, define $a_n^{n_1} = \tilde{P}^{x_n^{n_1}}(T_1^{n_1}(Y) > n_1)$. First, we claim that for each n, there exists $\{y_n^{n_1}\} \subset G_\infty$ such that

(2.1) $$\tilde{P}^{y_n^{n_1}}(T_1^{n_1}(Y) > n_1) \geq a_n^{n_1} - \frac{1}{n}.$$

To prove this, fix n, n_1. From Lemma 2.3, $\forall \epsilon > 0, \exists l \in \mathbb{Z}$ such that $\tilde{E}^{x_n^{n_1}}(\tau_{G_l}(Y)) < \frac{\epsilon}{2}$. Therefore

$$(2.2) \qquad \tilde{P}^{x_n^{n_1}}(\tau_{G_l}(Y) > 1) \leq \frac{\epsilon}{2}.$$

If $\exists \epsilon > 0$ such that $\tilde{P}^y(T_1^{n_1}(Y) > n_1 - 1) \leq a_n^{n_1} - \epsilon$ for $\forall y \in G_l$, then by (2.2),

$$\tilde{P}^{x_n^{n_1}}(T_1^{n_1}(Y) > n_1) \leq \tilde{P}^{x_n^{n_1}}(\tau_{G_l}(Y) > 1) + \max_{y \in G_l} \tilde{P}^y(T_1^{n_1}(Y) > n_1 - 1)$$

$$\leq \frac{\epsilon}{2} + a_n^{n_1} - \epsilon = a_n^{n_1} - \frac{\epsilon}{2},$$

which contradicts the definition of $a_n^{n_1}$. Thus, $\forall \epsilon > 0$, $\exists y \in G_l$ such that $\tilde{P}^y(T_1^{n_1}(Y) > n_1 - 1) \geq a_n^{n_1} - \epsilon$. As $T_1^{n_1-1}(Y) \geq T_1^{n_1}(Y)$, letting $\epsilon = \frac{1}{n}$ we have proved (2.1). Now if the statement of this lemma does not hold, then by (2.1) and the diagonal method, we see that there exists $\{x_n^n\} \subset G_\infty$ such that $\tilde{P}^{x_n^n}(T^n(Y) > n) \to 1$. Using Lemma 2.4, we can assume that $\{x_n^n\}$ is in a compact set. Take a converging subsequence $x_{n_k} \to x$ (,say). On the other hands, by Lemma 2.3, there exists U, neighborhood of x, such that $\tilde{E}^y(\tau_{U^c}(Y)) < \epsilon$ for all $y \in U$. Using Chebyshev's inequality, we have contradiction. $\square$

The next lemma is obtained by a standard argument of the diffusion processes.

Lemma 2.6. Let $(\Omega, \mathcal{F}, Z(t), P^x)$ be a Feller diffusion on G.
If $P^x(\exists t \geq 0, \epsilon > 0$ such that $Z(s) = Z(t)\forall s \in [t, t + \epsilon)) > 0$, for some $x \in G$, then $P^y(Z(t) = Z(0)\forall t \geq 0) = 1$ for some $y \in G$.

Proof of the theorem 2.1. The sufficiency part is clear. So we will prove the necessity.
If there is a trap in G_∞, then, by $[P_3]$, all the $x \in G_\infty$ should be traps and by the topological structure of the Sierpinski gasket, all the $x \in G$ are traps. Thus, the theorem holds with $c = 0$. If there is a trap in $G - G_\infty$, then by $[P_2]$, there is a sequence of traps which converges to some $y \in G_\infty$. By Lemma 2.3, y is a trap and we have the same conclusion. Thus we can assume that there is no trap in the Feller diffusion $(\Omega, \tilde{\mathcal{F}}, Y_t, \tilde{P}^x)$.
By Lemma 2.5, there is an $n_1 \in \mathbb{N}$ such that

$$\sup_{y \in G} \tilde{P}^y(T_1^{n_1}(Y) > n_1) < 1.$$

Using the Markov property inductively, we see that for some positive constants θ, T,

$$(2.3) \qquad \tilde{P}^y(T_1^{n_1}(Y) > t) \leq \exp(-\theta t) \quad \text{for } t \geq T \quad \text{and all } y \in G.$$

(2.3) ensures the integrability of $T_1^{n_1}$.
Fix $x \in G_{n_0}$ where $n_0 \geq n_1$. If $n \geq n_0$, then $T_1^n(Y) < \infty$ $\tilde{P}^x$-a.s. and therefore $\hat{Y}_n(i) = Y(T_i^n(Y))$ is a p-stream r.w. on G_n for some $0 < p < 1$ by Lemma 2.2. Define

$$(2.4) \qquad \tau_n(t) = T_{[\lambda_p^n t]}^n(Y), \quad Y_n(t) = Y(\tau_n(t)) = \hat{Y}_n([\lambda_p^n t]).$$

As $\{\hat{Y}_n : n \geq n_0\}$ satisfies $\hat{Y}_m(i) = \hat{Y}_n(T_i^m(Y))$ for $n \geq m$, by the same argument as [Sect. 2:Kum], we have

$$\lim_{n \to \infty} Y_n = X \quad \text{in } C([0, \infty), G) \ \tilde{P}^x\text{-a.s.},$$

where $\tilde{P}^x(X \in \cdot) = P^x_{(p)}(\cdot)$ on $C([0,\infty), G)$.
If we define
$$N_n = W_1^{n_0}(\hat{Y}_{n_0+n}) = \sum_{i=1}^{4} W_1^{n_0 <i>}(\hat{Y}_{n_0+n}),$$

then by the same reason as [Lemma 2.5:Kum], $\{W_1^{n_0<i>}(\hat{Y}_{n_0+n})\}$ is a multi four-type branching process. By the definition of N_n,

$$W_1^{n_0}(Y) = \sum_{i=0}^{N_n} W_i^{n_0+n}(Y), \qquad n \in \mathbb{N}.$$

The summands in this equation are *i.i.d.* but not independent of N_n which makes a contrast to the case of Barlow-Perkins. The rest of the proof is essentially the same as that of Theorem 8.1 in Barlow-Perkins [B.P] except this point. So we just sketch it emphasizing the difference. Let

$$\begin{aligned}
\mu_{n_0,k} &= E^x(W_1^{n_0}(Y)|Y(T_1^{n_0}) \in N_{n_0}^k(Y(0))) \\
\tilde{\mu}_{n_0,k} &= E^x(\{W_1^{n_0}(Y)\}^2|Y(T_1^{n_0}) \in N_{n_0}^k(Y(0))) \\
\sigma_{n_0,k} &= \tilde{\mu}_{n_0,k} - \mu_{n_0,k}^2 \qquad \text{for } 1 \le k \le 4.
\end{aligned}$$

Also, let

$$\begin{aligned}
\mu_{n_0} &= E^x(W_1^{n_0}(Y)) = p^2\mu_{n_0,1} + p(1-p)\mu_{n_0,2} + p(1-p)\mu_{n_0,3} + (1-p)^2\mu_{n_0,4} \\
\tilde{\mu}_{n_0} &= E^x(\{W_1^{n_0}(Y)\}^2), \quad \sigma_{n_0} = \tilde{\mu}_{n_0} - \mu_{n_0}^2.
\end{aligned}$$

All of these values are finite from (2.3). If we define $\vec{\mu}_{n_0}$ be the four-dimensional vector which has $\mu_{n_0,k}$ in the k-th column and define $\vec{\tilde{\mu}}_{n_0}$, $\vec{\sigma}_{n_0}$ in the same way, we easily obtain

$$\begin{aligned}
\vec{\mu}_{n_0} &= \vec{\mu}_{n_0+n} M^n \\
\vec{\tilde{\mu}}_{n_0} &\ge \vec{\sigma}_{n_0+n} M^n,
\end{aligned}$$

where M is a 4×4-matrix mentioned in Proposition 1.2. By the general theory of supercritical multi-type branching process, we know that $\frac{1}{\lambda_p^n} M^n$ converges to some positive matrix. Thus we have

(2.5) $$\mu_{n_0+n} \cdot \lambda_p^n \overset{n \to \infty}{\longrightarrow} c_{2.1} \qquad \tilde{P}^x\text{-a.s.}$$

(2.6) $$\sigma_{n_0+n} \le c_{2.2}\lambda_p^{-n}\tilde{\mu}_{n_0}.$$

Here the constants are positive and independent of n. Our goal is to show that $\tau_n(t)$ converges in probability to $c^{-1}t$ where c is a positive constant.

Using the *i.i.d.* property of $W_i^{n_0+n}$,

(2.7) $$\tilde{E}^x(\tau_{n_0+n}(t)) = [\lambda_p^{n_0+n} t]\mu_{n_0+n},$$

which converges to $c^{-1}t$ ($c > 0$) by (2.5). Thus, it is enough to check that $\tau_{n_0+n}(t) - \tilde{E}^x(\tau_{n_0+n}(t)) \overset{\tilde{P}^x}{\longrightarrow} 0$. (Here $X \overset{\tilde{P}^x}{\longrightarrow} Y$ means X converges to Y in probability.) By (2.5) and (2.7), we see that there exist $c_{2.3}, c_{2.4} > 0$ such that

(2.8) $$c_{2.3}\lambda_p^{n_0} t \le \tilde{E}^x(\tau_{n_0+n}(t)) \le c_{2.4}\lambda_p^{n_0} t.$$

By (2.6) and (2.8),

$$(2.9) \qquad \tilde{E}^x(\{\tau_{n_0+n}(t)\}^2) \le c_{2.2}\lambda_p^{-n}\tilde{\mu}_{n_0} + (c_{2.4}\lambda_p^{n_0}t)^2.$$

We claim that, for fixed t,

$$(2.10) \qquad \max_{i \le [\lambda_p^{n_0+n}t]} W_i^{n_0+n}(Y) \xrightarrow{\tilde{P}^x} 0 \quad \text{as} \quad n \to \infty.$$

If $\epsilon, t > 0$, then

$$\tilde{P}^x(\max_{i \le [\lambda_p^{n_0+n}t]} W_i^{n_0+n}(Y) > \epsilon)$$

$$\le \tilde{P}^x(T_{[\lambda_p^{n_0+n}t]}^{n_0+n}(Y) > M) + \tilde{P}^x(\inf_{t \le M} \sup_{u \in [t, t+\epsilon]} |Y(u) - Y(t)| \le 2^{-(n_0+n)}).$$

The last term converges to zero as $n \to \infty$ by Lemma 2.6 and the assumption that this Feller diffusion does not have any traps. The first term is bounded by $M^{-1}c_{2.4}\lambda_p^{n_0}t$ by (2.8). Hence it can be arbitrarily small, uniformly in n, by taking M large. This proves (2.10).

For $n \in \mathbb{N}$ and $\epsilon, t > 0$, we have

$$
\begin{aligned}
(2.11) \qquad & \tau_{n_0+n}(t) - \tilde{E}^x(\tau_{n_0+n}(t)) \\
& = \sum_{i=1}^{[\lambda_p^{n_0+n}t]} (W_i^{n_0+n}(Y) - W_i^{n_0+n}(Y) \wedge \epsilon) \\
& + \sum_{i=1}^{[\lambda_p^{n_0+n}t]} (W_i^{n_0+n}(Y) \wedge \epsilon - \tilde{E}^x(W_i^{n_0+n}(Y) \wedge \epsilon)) \\
& - \sum_{i=1}^{[\lambda_p^{n_0+n}t]} \tilde{E}^x(W_i^{n_0+n}(Y) - W_i^{n_0+n}(Y) \wedge \epsilon).
\end{aligned}
$$

From (2.9) and (2.10), the first term is $\mathbb{L}^2$-bounded uniformly in n and converges in probability to zero as $n \to \infty$. Thus the last term approaches zero as $n \to \infty$. The middle term goes to zero because the square of the $\mathbb{L}^2$-norm of it is bounded by $\epsilon c_{2.5}\lambda_p^{-n}\lambda_p^{n_0+n}t$. Therefore we have proved

$$\tau_n(t) \xrightarrow{\tilde{P}^x} c^{-1}t \quad \text{as} \quad n \to \infty, \qquad \text{for all} \quad t \ge 0.$$

Letting $n \to \infty$ in (2.4), we obtain

$$(2.12) \qquad X(t) = Y(c^{-1}t) \qquad \text{for all} \quad t \ge 0 \qquad \tilde{P}^x\text{-a.s.}$$

The choice of c is independent of $x \in G_{n_0}$ for each n_0 and hence cannot depend on n_0. Thus, (2.12) is valid for all $x \in G_\infty$. In general, we claim that

$$(2.13) \qquad \tilde{E}^x(f(Y(t))) = E^x(f(X(ct))) \qquad \text{for } f \in C_b(G),$$

which is a direct consequence of (2.12) and the Feller property. $\qquad \square$

§3 Remarks for other fractals

So far we have considered the two-dimensional gasket. Generally, the question of existence and uniqueness of the diffusions which is invariant under the operation of some group on G_0 is essentially reduced to that of a sequence of decimation random walks. Here we mention some remarks about this question on some fractals.

1) Higher dimensional Sierpinski gasket

Consider the d-dimensional ($d \geq 3$) gasket E with length 1. Let $\partial E = G_0$. Then a r.w. which satisfies $[P_2],[P_3]$, $[P_4]$ is in fact a simple r.w. (this was calculated by Y.Towa). So, by the same argument discussed in Section 2, the process on E satisfying $[P_1],[P_2],[P_3],[P_4]$ is a constant time change of the Brownian motion. This tells us that the invariance of the transition probability on G_0 under the operation of $\mathcal{A}_d$ (alternating group) is so strong that only the simple r.w. satisfies it. But there could exist a (not simple) r.w. which is invariant under the operation of some smaller subgroup. If d is prime, then we can prove the existence of one parameter r.w. invariant under the operation of cyclic subgroup in the same way as Proposition 3.2.

2) Nested fractals

Nested fractals, introduced by Lindstrøm [L], is a class of finitely ramified fractals which has strong symmetry. Pentakun (Figure 1) is a typical example. Let E be Pentakun with length 1. If we define $F_i(x) = \alpha^{-1}(x - a_i) + a_i$ ($1 \leq i \leq 5$), where a_i are the vertices of the pentagon with length 1 and $\alpha = \frac{3+\sqrt{5}}{2}$, then $E = \cup_{i=1}^5 F_i(E)$. Let $\Omega = \{1, \cdots, 5\}^{\mathbb{N}}$. It is easy to verify that a map $\pi : \Omega \to E$ defined by $\pi(\omega) = \lim_{n\to\infty} F_{\omega_1}(\cdots(F_{\omega_n}(a_1))\cdots)$ exists and is continuous. Further, E has the following property.

Property 3.1 Let $\tau = (1, 2, \cdots, 5) \in S_5$ and let $\vec{\tau} : \Omega \to \Omega$ be the mapping which operate τ on each coordinate. Then, there exists a continuous surjection $\tilde{\tau} : E \to E$ with order 5 such that

$$\pi \circ \vec{\tau} = \tilde{\tau} \circ \pi.$$

Using this property, we have the following proposition.

Proposition 3.2. *There exists one parameter family of non-symmetric Feller diffusion processes on E.*

Proof. We should show the existence of one-parameter family of decimation r.w. on E. Then the rest of the argument is the same as [Kum]. We define $G_0 = \{a_1, \cdots, a_5\}$ and $G_n = \cup_{i_1,\cdots,i_n=1}^5 F_{i_1} \circ \cdots \circ F_{i_n}(G_0)$. Let $B = \{(x_1, \cdots, x_4)|x_i > 0, \sum_{i=1}^5 x_i = 1\}$, $\hat{B} = \{(x_1, \cdots, x_4)|x_i \geq 0, \sum_{i=1}^4 x_i = 1\}$ and $(p_1, p_2, p_3, p_4) \in B$. Define

$$P_{\mathbf{p}}(i, j) = \begin{cases} p_{i-j} & \text{if} \quad i > j \\ p_{5+i-j} & \text{if} \quad j > i \\ 0 & \text{if} \quad i = j \end{cases} \qquad \text{for } 1 \leq i, j \leq 5.$$

For $x \in G_n$, define $\rho_n(x) = \{C : C$ is a n-cell (i.e. a set of the form $F_{i_1} \circ \cdots \circ F_{i_n}(G_0))$ containing x.$\}$. Then the range of ρ_n is $\{1, 2\}$. For an arbitrary choice of s ($0 < s < 1$),

define a r.w. on G_n with the transition probability $P_n^{s,\mathbf{P}}(x,y)$ as follows:
If x,y are G_n-neighborhoods and they could be expressed as $x = s_1 \cdots s_{m-1} k_0 \dot{\alpha}$, $y = s_1 \cdots s_{m-1} k_0 \, \alpha, \alpha, \cdots, \alpha \dot{\beta}$ (here $i = iiii\cdots$), then $P_n^{s,\mathbf{P}}(x,y) = P_{\mathbf{p}}(\alpha,\beta)$ when

$$\underbrace{\quad}_{n-m \text{ times}}$$

$\rho_n(x) = 1$. When $\rho_n(x) = 2$ and the other expression of x is $x = s_1 \cdots s_{m-1} k_1 \dot{\alpha}'$, then

$$P_n^{s,\mathbf{P}}(x,y) = \begin{cases} sP_{\mathbf{p}}(\alpha,\beta) & \text{if} \quad k_0 - k_1 = 1 \ (\bmod \ 5) \\ (1-s)P_{\mathbf{p}}(\alpha,\beta) & \text{if} \quad k_0 - k_1 = 4 \ (\bmod \ 5). \end{cases}$$

Otherwise $P_n^{s,\mathbf{P}}(x,y) = 0$.

Thus, the transition probability of G_n is defined by s and $\mathbf{p}$. By the same idea as Proposition 1.1, we can induce r.w. on G_{n-1}. Because of the Property 3.1 and the self-similarity of Pentakun, the induced r.w. has a transition probability $P_{n-1}^{\tilde{s},\tilde{\mathbf{P}}}(x,y)$ for some $0 < \tilde{s} < 1, \tilde{\mathbf{p}} \in B$ (it is easily checked that the induced variables are non zero). First, we show that $s = \tilde{s}$ for any choice of $\mathbf{p}$. Let x be an element of G_m with $\rho_m(x) = 2$. Locate the $(m+1)$-cells connected to x (Figure 2) and assume without loss of generality that the probability for the r.w. on G_{m+1} starting at x to hit an element of the $(m+1)$-cell (except x) on the right hand side before hitting that on the left hand side is s. Let I_1^R be the probability for this r.w. to arrive at an element of the right m-cell (except x) before coming back to x. Define I_1^L in the same way. Also, let I_2 be the the probability with which the r.w. comes back to x before arriving at one of the elements of the m-cells (except x).
Then, by the strong Markov property,

$$\tilde{s} = I_1^R + I_1^R I_2 + I_1^R I_2^2 + \cdots = \frac{I_1^R}{1 - I_2},$$

$$\tilde{s} = \frac{I_1^L}{1 - I_2}.$$

By the Property 3.1 and the self-similarity of Pentakun, $I_1^R : I_1^L = s : (1-s)$. Thus we see that $s = \tilde{s}$.

So, we should find a fixed point of a non-linear map $f_s : \mathbf{p} \to \tilde{\mathbf{p}}$.
By Shauder's fixed point theorem, $f_s : \hat{B} \to \hat{B}$ has a fixed point. Because $\sharp G_0$ is prime, the fixed point should be in the interior and we have finished the proof. $\qquad\square$

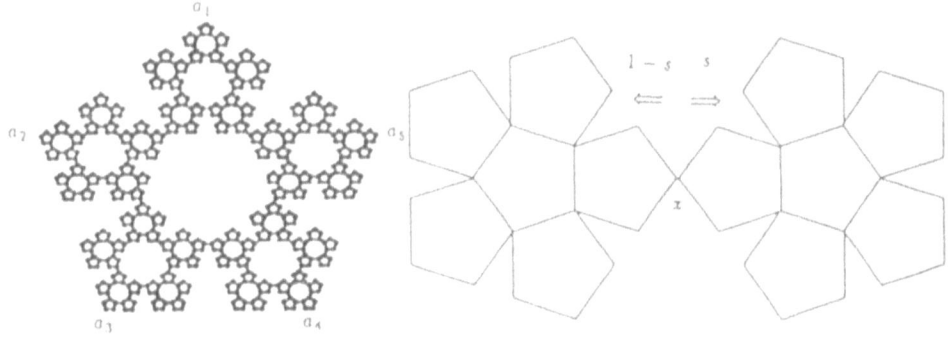

Figure 1   Figure 2

By the same proof, we see that D-dimensional Sierpinski gasket has at least one parameter family of Feller diffusion processes if D is prime. The statement of Proposition 3.2 might be extended to the following:

Let E be a nested fractal which is not contractible (i.e. E has a subset which is homeomorphic to a circle) and has Property 3.1. If $\sharp G_0$ is prime, then there exists one parameter family of non-symmetric diffusion processes on E.

Acknowledgments

The author thanks to Dr. M.T.Barlow and Dr. B.M.Hambly for fruitful discussions during his visit to Cambridge. Also he is grateful to Hayashibara company for giving him an opportunity to meet people working in the area of applied mathematics.

References

[A.N]   Athreya,K.B. & Ney,P.E., *Branching processes*, Springer, New York, (1972).

[B.P]   Barlow,M.T. & Perkins,E.A., *Brownian motion on the Sierpinski gasket*, Prob. Th. Rel. Fields, **79** (1988), 543-623.

[G]   Goldstein,S., *Random walks and diffusions on fractals*, In: Kensten,H. (ed.) Percolation theory and ergodic theory of infinite particle systems, IMA Math. Appl. 8, Springer, New York, (1987), 121-128.

[Ki]   Kigami,J., *A harmonic calculus on the Sierpinski spaces*, Japan J. Appl. Math. **6** (1989), 259-290.

[I]   Ito,K., *Stochastic Processes II*, Iwanami-Kōza Gendai Ōyō-Sūgaku (in Japanese).

[Kum]   Kumagai,T., *Construction and some properties of a class of non-symmetric diffusion processes on the Sierpinski gasket*, In: K.D.Elworthy & N.Ikeda (eds.) Asymptotic Problems in Probability Theory: stochastic models and diffusions on fractals, Pitman , 219-247 (1993).

[Kus]   Kusuoka,S., *A diffusion process on a fractal*, In: Itô,K. & Ikeda,N. (eds.) Probabilistic methods in Mathematical Physics, Kinokuniya, Tokyo (1987), 251-274.

[L]   Lindstrøm,T., *Brownian motion on nested fractals*, Mem. Amer. Math. Soc. **420** (1990).

[O]   Osada,H., *Self-similar diffusions on a class of infinitely ramified fractals*, Preprint, (1992)

CHAOTIC PHENOMENA IN DENSITY-WAVE OSCILLATIONS

R.T.Lahey, Jr., C.J.Chang, F.Bonetto, and D.A.Drew

Center for Multiphase Research
Rensselaer Polytechnic Institute
Troy, NY 12180-3590

INTRODUCTION

The physics of density-wave instabilities is now rather well understood. Density-wave oscillations are caused by the lag introduced into a thermal-hydraulic system by the finite speed of propagation of density perturbations. In principle, density-wave oscillations can occur in both diabatic and adiabatic two-phase systems, and in diabatic single-phase systems.

In order to gain some insight into density-wave instability mechanisms, let us consider the special case of air/water flow in an adiabtic channel (Svanholm and Friedly, 1983). While this is a simple case, it does serve to illuminate the basic physics involved in density-wave oscillations.

Figure-1 is a schematic of a constant head tank of water (i.e. $H = H_1(t) + H_2(t) = $ constant), which has a drain line connected near the bottom. The drain line discharges to atmosphere, thus the system is subjected to a constant pressure drop boundary condition, $\Delta_p = \rho_l g H$. It is convenient, but not essential, to assume that the drain line has no hydraulic losses except at the inlet and the exit. Moreover, since the air flow is presumed to be introduced into the drain line through a choked flow orifice, the air flow rate is constant.

The hydraulic head loss across the inlet orifice is given by,

$$H_1 = \frac{\Delta p_1}{\rho_l g} = K_1 \frac{v_1^2}{2g} \tag{i}$$

Similarly, the loss in hydraulic head due to the exit loss is given by,

$$H_2 = \frac{\Delta p_2}{\rho_l g} = K_2 \frac{\bar{\rho}_2}{\rho_l} \frac{v_2^2}{2g} \tag{ii}$$

where $\bar{\rho}_2$ is the two-phase density and v_2 the two-phase velocity at the exit plane (ie, station-2). For the conditions of interest here, it is a good assumption that the mass

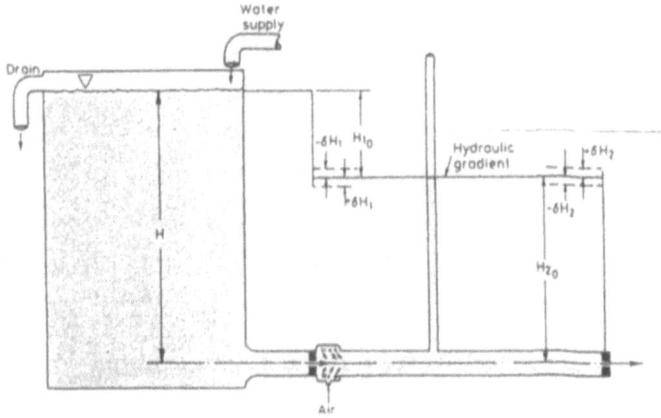

FIGURE 1. Air/Water Flow Channel (Svanholm and Friedly, 1983).

flow rate contributed by the air is negligible. In contrast, the air flow controls the volumetric flow rate at the exit, and thus the exit velocity, v_2.

Let us assume we have a positive perturbation in the inlet velocity, δv_1. In accordance with Eq.(i), this will lead to an increase in head loss across the inlet orifice, δH_1. Also, since the air flow is held constant, and the inlet flow of liquid has increased, the density of the flowing two-phase mixture will be increased by $\delta \bar{\rho}$. This density perturbation will propagate through the channel at the kinetic wave speed, as shown by the lower (z, t) trajectory in Figure 2. When the density perturbation reaches the local loss at the exit, the increased velocity will cause an increase in head loss across the exit orifice, δH_2. Since the total head is constant (i.e., $H_1 + H_2$ =constant), this perturbation in head loss across the exit orifice will, neglecting some effects, lead to a negative perturbation in the inlet head, $-\delta H_1$, and hence velocity, $-\delta v_1$. This causes a negative perturbation in mixture density, $-\delta \bar{\rho}$, to propagate through the channel, leading to a $-\delta H_2$ at the exit and thus inducing a (positive) δH_1 at the inlet. Clearly this feedback mechanism can lead to system instability if the magnitude of the induced δH_1 equals or exceeds that of the previous perturbation in H_1. Indeed, the condition for neutral stability (i.e. when the amplitude of the oscillations neither diverge nor converge) is when, $|\delta H_1| = |\delta H_2|$.

Admittedly, this discussion has been for a fairly simple system. Moreover, we have neglected consideration of some nontrivial effects, such as distributed frictional losses and the inertia of the single-phase liquid in the drain line, in our consideration of system response. Nevertheless, as we shall see subsequently, the essence of density-wave instability phenomena has been captured by this simple example.

Let us now consider the case of interest in this paper. In particular, let us consider a boiling channel which has subcooled inlet and is subjected to a constant, parallel-channel-type, pressure drop boundary condition.

A perturbation in the inlet velocity, δj_{in}, due, for instance, to changing the setting of a butterfly valve at the inlet, will create a propagating enthalpy perturbation, δh, in the single phase region. The point at which bulk boiling begins will be perturbed by the arrival of this enthalpy wave. In the two-phase region there will be propagating void fraction perturbation (and thus, a density-wave), $\delta \alpha$, due to the perturbation in

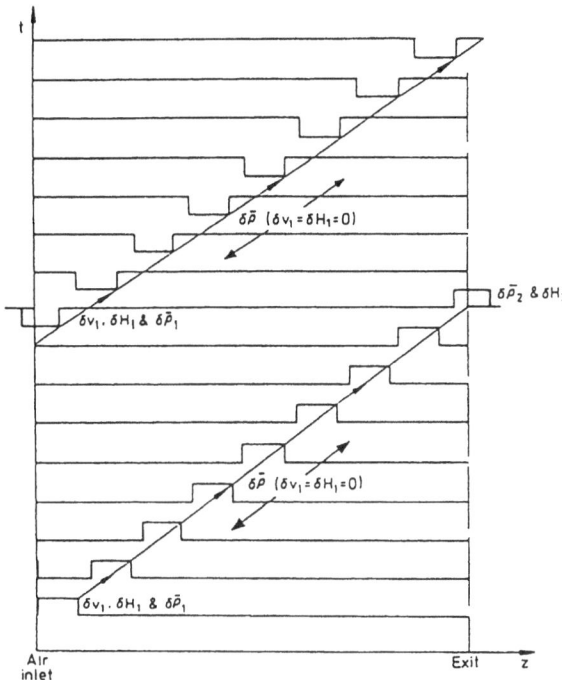

FIGURE 2. Density-wave trajectories in an air/water flow channel (Svanholm & Friedly, 1983).

the boiling inception point.

Due to the change in flow rate and nonboiling length, $(\delta\lambda)$, there will be a perturbation in the two-phase pressure drop, $\delta(\Delta p_{2\phi})$. Since the total pressure drop across the channel is externally imposed, there will necessarily be an equal and opposite perturbation in the single-phase pressure drop, $\delta(\Delta p_{1\phi})$. Because of the lags associated with the finite speed of propagation of the enthalpy and void fraction perturbations, the resultant pressure drop perturbation in the two-phase region, and the corresponding feedback perturbation in the single-phase pressure drop, will normally be out-of-phase with the inlet velocity perturbation. Depending on the various lags which take place, the resultant pressure drop perturbation in the single-phase region may either reinforce or attenuate subsequent imposed inlet velocity perturbations (due to changes in position of the butterfly valve).

In order to better appreciate these concepts, let us consider Fig. 3. As can be seen, at time zero we open the butterfly valve such that there is a positive step perturbation of inlet velocity in the (parallel) channel on which a constant pressure drop in impressed. This is a somewhat contrived case but it serves to give insight into the physics involved in density-wave instabilities. If all other parameters are held constant, this will be produce a negative void fraction perturbation in the two-phase region $(\delta\alpha < 0)$. Moreover, as can be easily understood from steady-state considerations, the length of the single-phase region will increase and that of the two-phase region decrease. After one transport time, T_{tr}, the system may come to equilibrium at the new operating conditions, $\Delta p_{1\phi}$, and

$\Delta p_{2\phi}$.

The transport time, T_{tr}, that it takes a perturbation in inlet velocity to pass through the channel can be approximated by,

$$T_{tr} = T_{tr_{2\phi}} + T_{tr_{1\phi}} \simeq \int_{\lambda_o}^{L_H} \frac{dz}{C_K(z)} + \lambda_0/j_{in} \qquad \text{(iii)}$$

where, C_K is the speed at which void perturbations propagate (i.e., kinematic wave speed). It should be noted that in Eq. (iii) we have assumed that the transport delay associated with the single-phase enthalpy wave is given by, $T_{tr_{1\phi}} = \lambda_0/j_{in}$. This is only valid for the case of a heater having negligible heat capacity. In practice, the transport time in the single-phase region, $T_{tr_{1\phi}}$, will be somewhat longer because of the attenuation of the thermal wave by the heat capacity of the conduit. Nevertheless, for our purpose here, Eq.(iii) is a reasonable approximation.

Let us now return to Figure 3. We see that for a boiling parallel channel on which

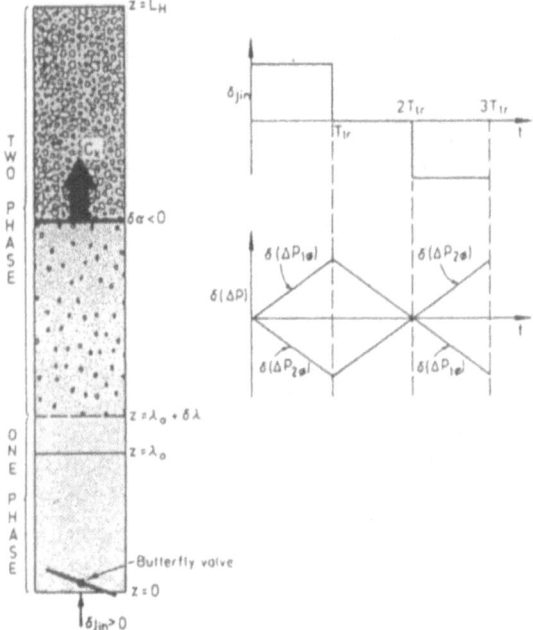

FIGURE 3. Changes in boling channel presure drops due to step changes in inlet velocity.

a constant pressure drop is impressed, if we have a step inclease in inlet velocity, and the velocity remains constant for a time T_{tr}, the single and two-phase pressure drops may change monotonically (although not necessarily linearly, as shown) until a new operating state is reached at time $t = T_{tr}$. Finally, if we slightly close the butterfly valve, and thus change the inlet velocity by a step reduction just equal to the initial step increase, the system will achieve a new operating state at time $t = 3T_{tr}$.

It should be noted that for each perturbation in inlet velocity, it takes one transport time to achieve a new equilibrium operating state. Clearly there is considerable lag between the inlet velocity perturbation and the perturbation in the single and two-phase pressure drops.

Let us now consider the implications of a sinusoidal variation in inlet velocity. It should be noted that the boiling system under considerations can behave as a resonant system. That is, when the resonant frequency, f_r, is reached, the oscillation can become self-excited. This occurs when the period of the oscillation equals twice the transport time. That is,

$$f_r = \frac{1}{2T_{tr}} \tag{iv}$$

To better understand this phenomena, let us consider Figure 4. It can be seen that we have imposed a harmonic variation in inlet velocity. Case #1 is a hypothetical case in which there is no transport delay through the system (i.e., $T_{tr} = 0$). For this special case the pressure drop perturbations are always in-phase with the inlet velocity perturbation. This case cannot actually occur, since in practice there is always some lag.

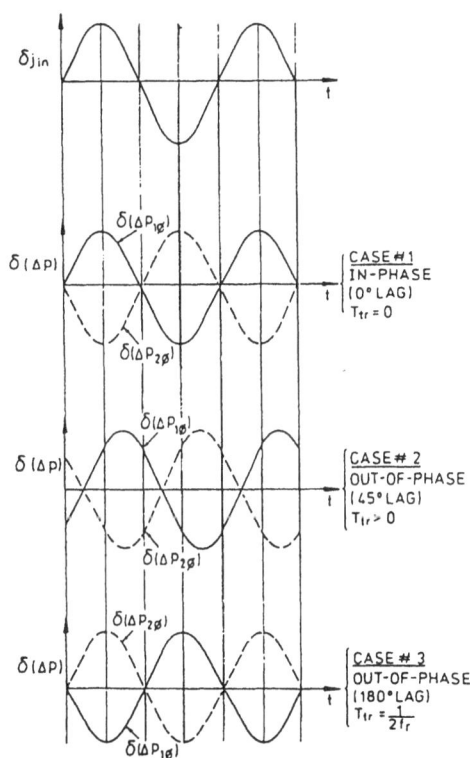

FIGURE 4. Density-wave instability phenomena (harmonic inlet velocity).

In Case #2, we have assumed a finite amount of lag. In particular, we have assumed operating conditions such that the pressure drop perturbations lag the inlet velocity perturbation by 45°. For such conditions, the system is quite stable. In contrast, in Case #3, we have assumed operating conditions such that the pressure drop perturbations and the inlet velocity perturbation are 180° out-of-phase. For these conditions the system will be self-excited, and oscillate at its resonant frequency, f_r. That is, the resultant perturbation in the single-phase pressure drop will induce a harmonic perturbation in the inlet velocity, and thus, unlike cases 1 and 2, no external means are necessary to cause the system to oscillate since it has become self-excited.

DISCUSSION - THE MODEL

The partial differential equations which describe the conservation of mass, energy and momentum are given by,

MASS

$$\frac{\partial \rho}{\partial t} + \frac{\partial}{\partial z}[\rho u] = 0 \tag{1}$$

ENERGY

$$\frac{\partial(\rho h)}{\partial t} + \frac{\partial(\rho h u)}{\partial z} = \frac{q'' P_H}{A_{X-S}} + \frac{\partial p}{\partial t} \tag{2}$$

MOMENTUM

$$\frac{\partial(\rho u)}{\partial t} + \frac{\partial}{\partial z}(\rho u^2) = -\frac{\partial p}{\partial z} - \rho g - \left[\frac{f}{D_H} + \sum_{i=1}^{N} K_i \delta(z - z_i)\right] \frac{\rho u^2}{2} \tag{3}$$

The corresponding equations of state are:

Single-Phase

$$\rho = \rho_f , \qquad \text{for } h < h_f \tag{4a}$$

Tow-Phase

$$\rho = \rho = \left[v_f + \frac{v_{fg}}{h_{fg}}(h - h_f)\right]^{-1} \qquad , h > h_f \tag{4b}$$

In addition, we may also have conversation equations for the heated wall dynamics and, for the case of nuclear reactor feedback, the neutron kinetics.

Heater Wall

$$\frac{dT_H}{dt} = \frac{q'''}{c_{pH}\rho_H} - \frac{P_H H(T_H - T_b)}{c_{pH}\rho_H A_H} \tag{5}$$

Neutron (Point) Kinetics

$$\frac{dn}{dt} = \frac{(\Delta k - \beta)}{l}n + \lambda_d C \tag{6a}$$

$$\frac{dC}{dt} = \frac{\beta}{l}n + \lambda_d C \tag{6b}$$

The emerging science of chaos is based on the theory of systems of ordinary differential equations. Thus, in order to use the relatively rich literature associated with chaos theory, we must first transform the system of coupled nonlinear partial differential equations (PDE), Eqs.(1)-(3), into an equivalent nonlinear system of coupled ordinary differential equations (ODE).

One way to accomplish this is to divide up the heated channel into a number of nodes (having moving boundaries), and then to spatially integrate the PEDs across each node under the assumption of linear nodal enthalpy profiles.

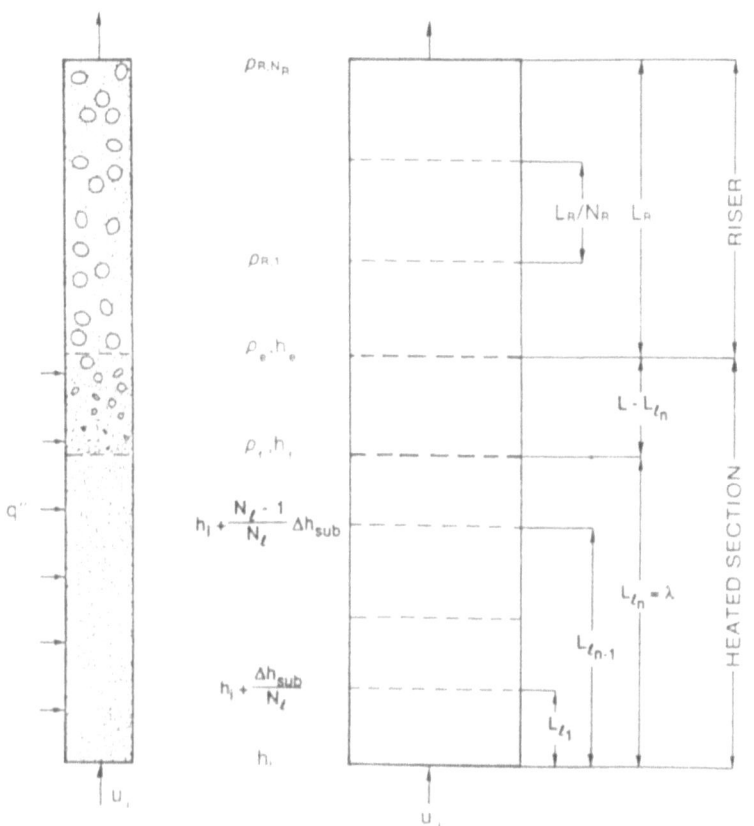

FIGURE 5. Schematic diagram of the boiling channel.

The nodal structure used is shown in Fig.-5, and the resultant non-dimensional nodal model is given by:

Nodal Energy Balance in the Single-Phase Region $(N_l$ nodes)

$$\frac{dL^*_{ln}}{dt^*} = 2u^*_i - 2N_l(L^*_{ln} - L^*_{ln-1})Q'''^*_{ln} - \frac{dL^*_{ln-1}}{dt^*} \tag{7}$$

Mass Balance in Diabatic Region of Channel

$$\frac{dM^*_{CH}}{dt^*} = u^*_i - \rho^*_e u^*_e \tag{8}$$

Mass Balance in Riser Nodes $(N_r$ nodes)

$$\frac{dM^*_{Rn}}{dt^*} = \frac{u^*_e}{A_R}(\rho^*_{Rn-1} - \rho_{Rn}) \tag{9}$$

Channel Momentum Equation

$$\frac{dW^*}{dt^*} = Eu - \Delta p^*_a - \Delta p^*_f - \Delta p^*_g \tag{10}$$

The heated wall dynamics equations are given by a non-dimensional lumped parameter energy balance as:

Heater Dynamics, The Single-Phase Region

$$\frac{dT^*_{H_{ln}}}{dt^*} = \frac{(T^*_{H_{ln+1}} - T^*_{H_{ln}})}{2(L^*_{ln} - l^*_{ln-1})}\frac{dL^*_{ln}}{dt^*} + \frac{(T^*_{H_{ln}} - T^*_{H_{ln-1}})}{2(L^*_{ln} - l^*_{ln-1})}\frac{dL^*_{ln-1}}{dt^*} \\ + \frac{\left[Q'''^* - K_{DB}u^{*0.8}_i\left(T^*_{H_{ln}} - \frac{(2N_l - 2n + 1)}{2N_l}h^*_i\right)\right]}{N_{MH}} \tag{11}$$

where, using the Dittus-Boelter correlation:

$$K_{DB} = 0.023\frac{k_f}{D^{0.2}_H}\frac{Pr^{0.4}}{\nu^{0.8}_f}\frac{P_H L_H u^{0.8}_{ref}}{c_{pf}\rho_f u_{io} A_{xs}} \tag{12a}$$

$$N_{MH} = \frac{N_{pch}}{N_{sub}}\frac{c_{pH}\rho_H}{c_{pf}\rho_f}A^*_H \tag{12b}$$

Heater Dynamics, The Two-Phase Region

$$\frac{dT^*_{H_{2\phi}}}{dt^*} = \frac{(T^*_{H_{2\phi}} - T^*_{H_{lN_l}})}{2(1 - \lambda^*)}\frac{d\lambda^*}{dt^*} + \frac{Q'''^* - K_{JL}T^*_{H_{2\phi}}}{N_{MH}} \tag{13}$$

where, using the Jens-Lottes correlation:

$$K_{JL} = \frac{\exp(4p/900)}{1296 \times 10^4}\frac{10^6}{3600}\frac{P_H L_H T^3_{ref}}{c_{pf}\rho_f u_{io} A_{xs}u_{io}} \tag{14}$$

The non-dimensional point kinetics neutronics equations [Weaver, 1963] are given by

Neutron Density

$$\frac{dn^*}{dt^*} = N_l \left[\left(\frac{\Delta k}{\beta} - 1 \right) n^* + C^* \right] \tag{15}$$

Precursor Density

$$\frac{dC^*}{dt^*} = N_\lambda [n^* - C^*] \tag{16}$$

where,

$$N_l \triangleq \frac{\beta}{l} t_{ref} \tag{17a}$$

$$N_\lambda \triangleq \lambda t_{ref} \tag{17b}$$

DISCUSSION - MODEL EVALUATION

The nodal model has been numerically evaluated for two nodes in the single-phase region ($N_l = 2$), one in the two-phase region of the heater and three in the unheated riser ($N_R = 3$). The model thus has a total of seven state variables when constant heat flux conditions are assumed (ie, when heater wall dynamics and neutron feedback effects were neglected). The parameters used in these calculations are given in Table-I.

TABLE 1.

$N_{sub} = 100$	$K_e = 10$
$Fr = 0.0016$	$K_{re} = 15$
$b = 0.0002$	$L_R/L_H = 3$
$K_i = 42$	$A_R/A_H = 4$

Figure-6 shows that the boiling channel experienced a limit cycle for a case in which $N_{pch} = 109$. As the phase change number (N_{pch}) was reduced the steady-state position of the boiling boundary moved toward the exit of the heater. As can be seen in Figure-7, for $N_{pch} = 107$ period doubling was observed. At $N_{pch} = 106.7$ a T^2 torus appeared, as can be seen in Figures-8 and 9. The corresponding Poincaré sections were calculated using the method presented in Appendix-I, and are shown in Figure-10.

The phase change number was further reduced until at $N_{pch} = 106.545$ a strange attractor was found. Figures 11 and 12 show phase space representations of the strange attractor and Figure-13 gives the corresponding Poincaré sections.

The most definitive measure of a strange attractor is the evaluation of Lyapunov exponents. The largest Lyapunov exponent (Λ_{max}) is given by [Wolf et al, 1985],

$$\Lambda_{max} = \frac{1}{t_m - t_0} \sum_{i=1}^{m} \log_2 \left[\frac{L(t_i)}{L(t_{i-1})} \right]$$

Note that, if in each interval,

$$L(t_i) = L(t_{i-1}) 2^{\Lambda_i (t_i - t_{i-1})}$$

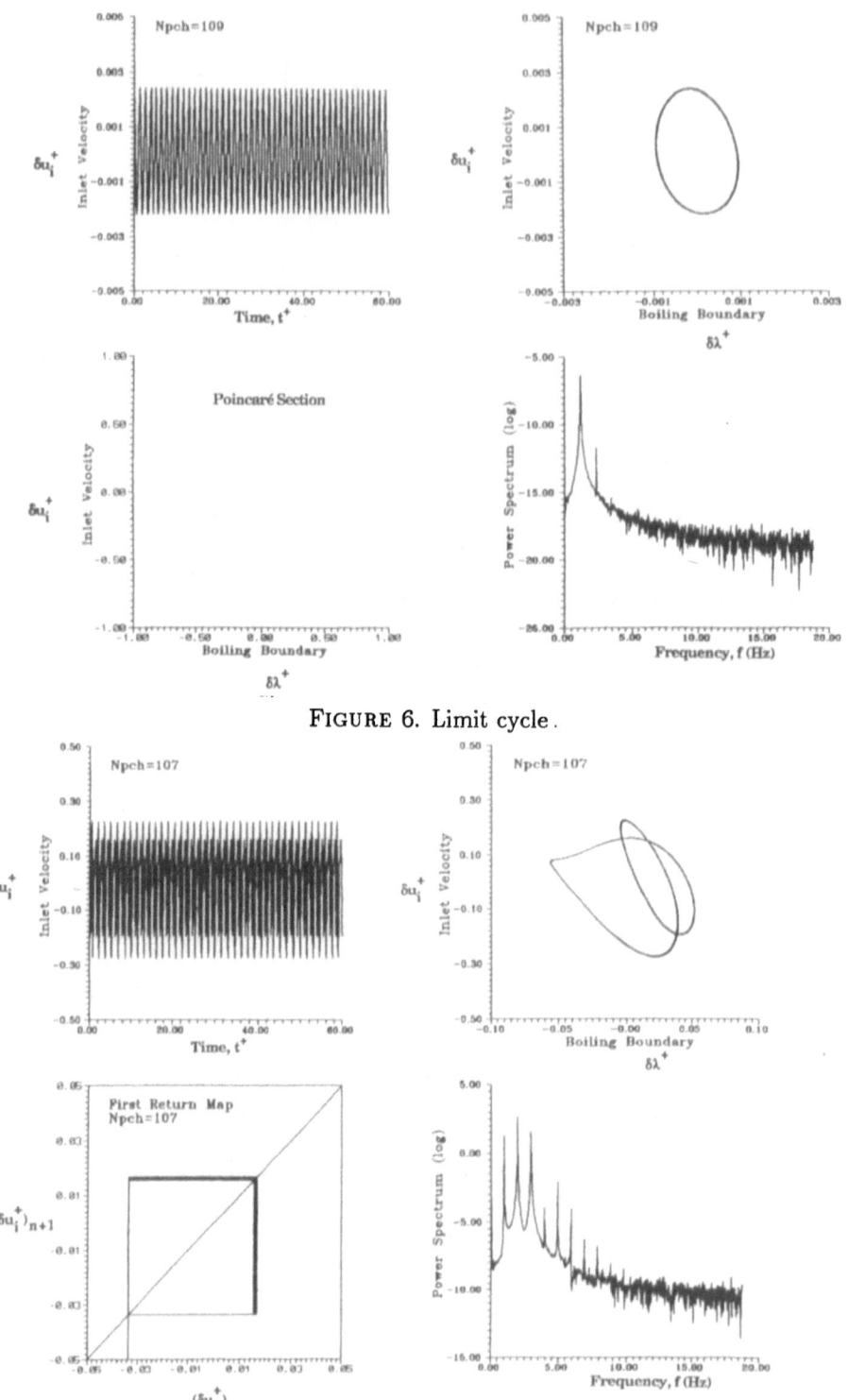

FIGURE 6. Limit cycle.

FIGURE 7. Period doubling.

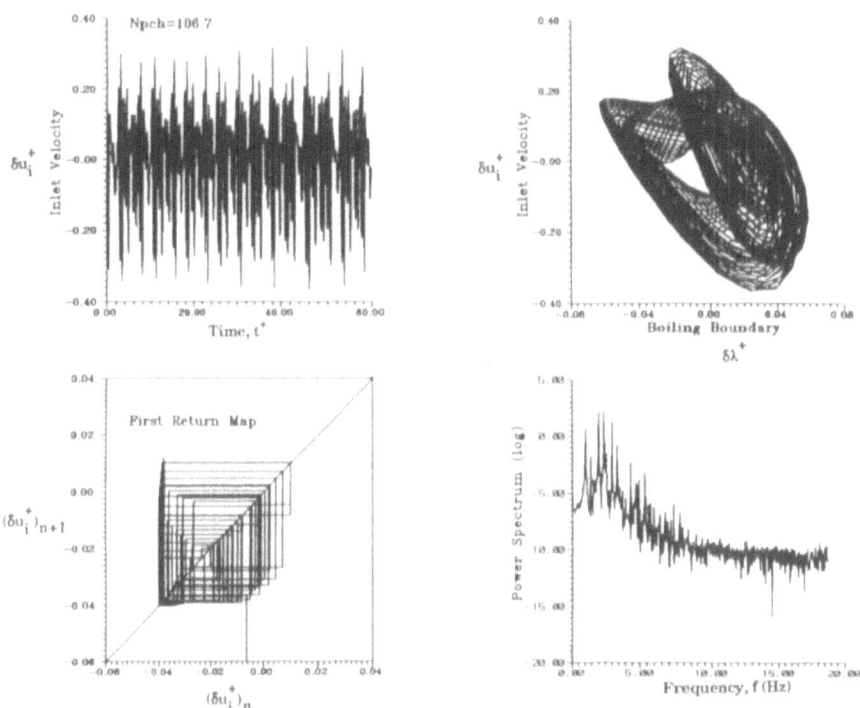

FIGURE 8. $\mathbb{T}^2$ torus.

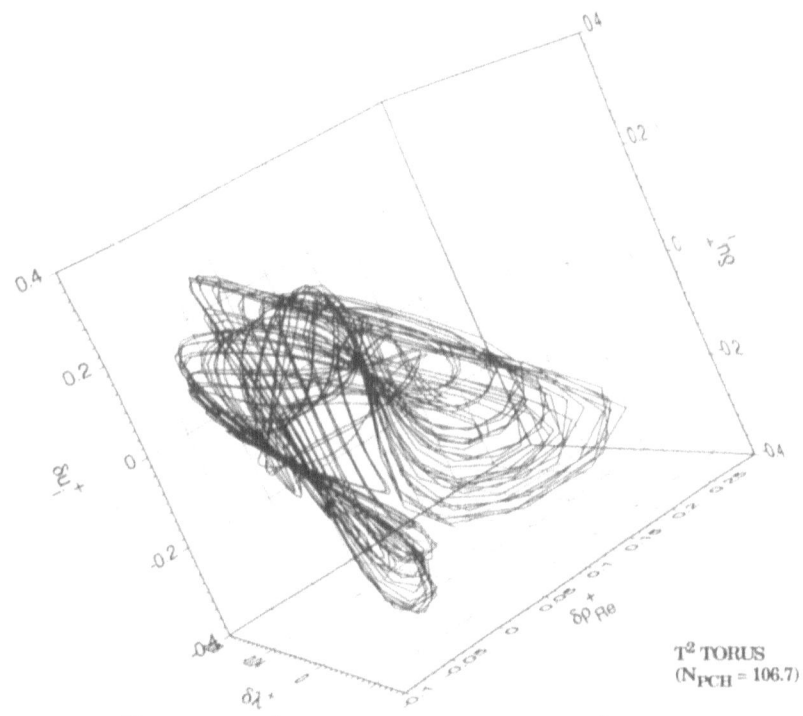

FIGURE 9. Three-dimensional display of $\mathbb{T}^2$ torus.

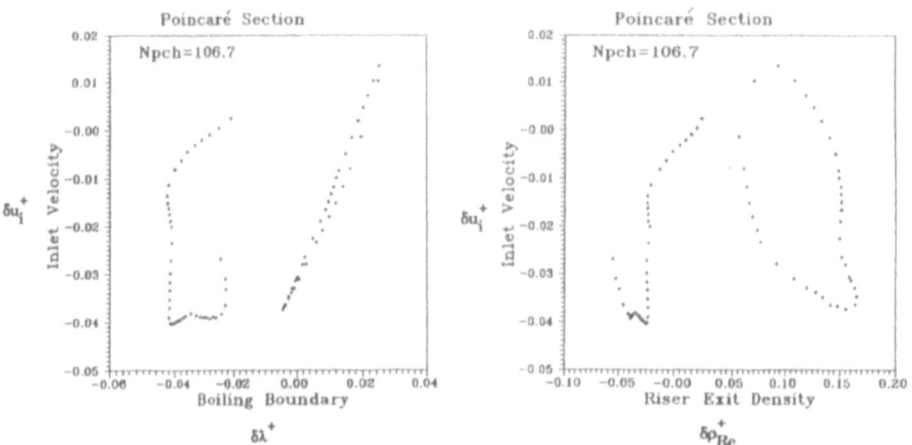

FIGURE 10. Poincarè sections of $\mathbb{T}^2$ torus.

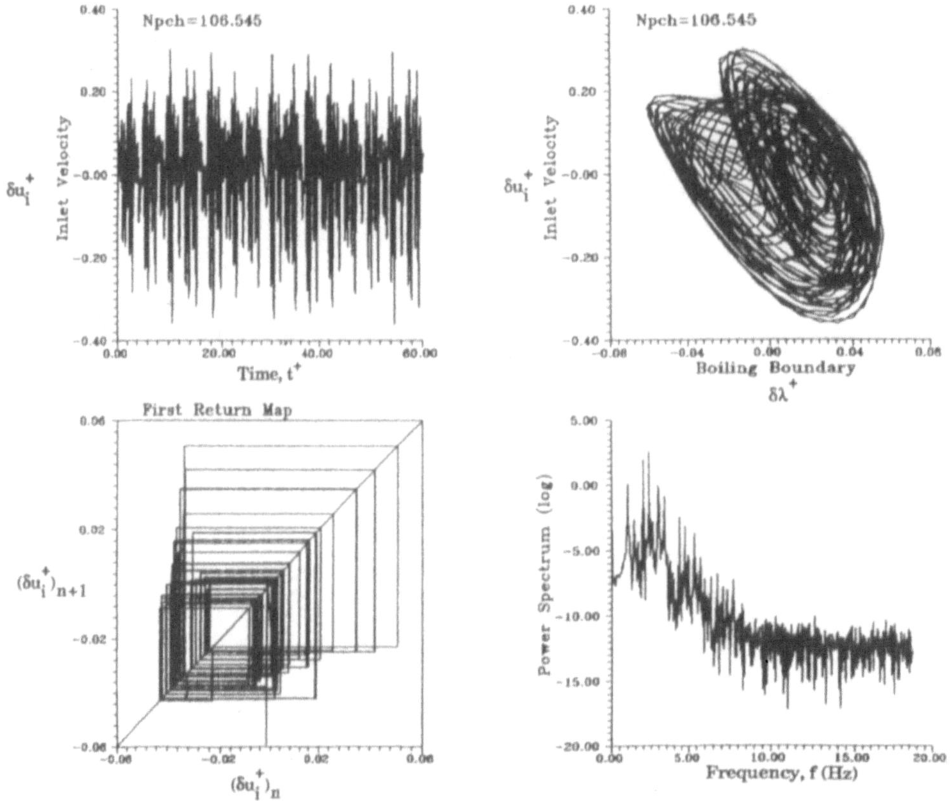

First Return Map

FIGURE 11. Strange attractor.

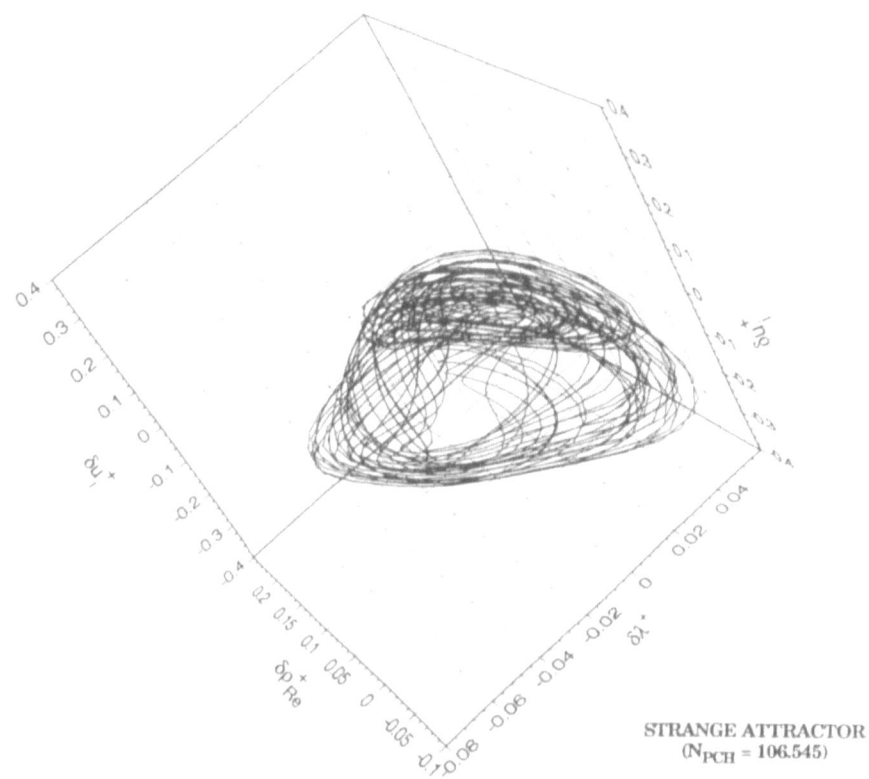

FIGURE 12. Three dimensional display of strange attractor.

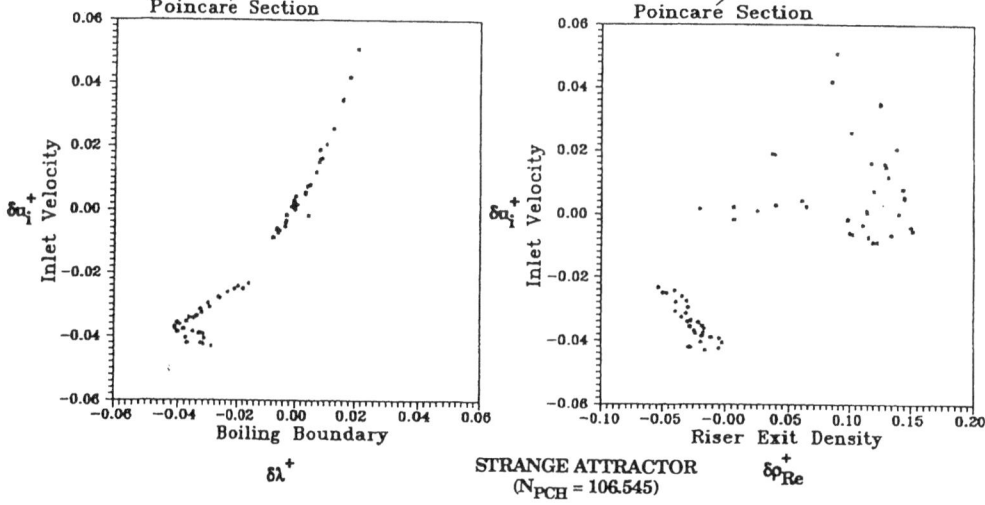

FIGURE 13. Strange Attractor.

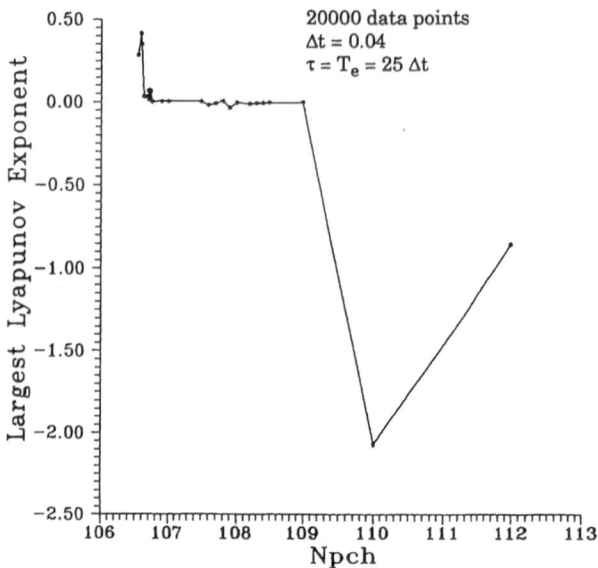

FIGURE 14. Largest Lyapunov exponent.

then,

$$\log_2 \left(\frac{L(t_i)}{L(t_{i-1})} \right) = \Lambda_i(t_i - t_{i-1}) \triangleq \Lambda_i \Delta t_i$$

Hence, renormalizing the process after each time step, we have:

$$\Lambda_{\max} = \bar{\Lambda} = \frac{1}{(t_m - t_0)} \sum_{i=1}^{m} \Lambda_i \Delta t_i$$

Figure-14 gives the maximum Lyapunov exponent ($\Lambda_{\max}$) for various phase change numbers. We note that $\Lambda_{\max} = 0$ implies a periodic limit cycle while $\Lambda_{\max} > 0$ denotes an aperiodic strange attractor.

ACKNOWLEDGMENTS

The authors wish to acknowledge the financial support given this study by the Institute of Nuclear Energy Research (INER) of the Republic of China, and the work of, and helpful discussions with, Professor N.Takenaka (Kobe University).

References

1. Svanholm, K. & Friedly, J.C.,1983, *An elementary introduction t the problem of density-wave oscillations,* Proceedings of the HTFS Research Symposium,Bath, England.
2. Weaver, L.E., 1963 *System Analysis of Nuclear Reactor Dynamics,* ANS Monograph, Rowen and Littlefield, Inc., New York.
3. Wolf et al, 1985, *Determining Lyapunov exponents from a time series,* Physicaa 16D, 285-317

NOMENCLATURE

$A =$ Flow area

$D_H =$ Hydraulic diameter

$h_k =$ Enthalpy of phase-k

$L(t_{i-1}) =$ The distance between two point at time t_{i-1}

$\Delta p =$ Pressure drop

$t =$ time

$v_k =$ Specific volume of phase-k

$\rho_k =$ Density of phase-k

$E_u = \dfrac{\Delta p_{total}}{\rho_f u_{ref}^2}$

$h^* = \dfrac{h - h_f^2}{h_{ref}}$

$N_{sub} = \dfrac{(h_f - h_{io})}{h_{fg}} \dfrac{v_{fg}}{v_f}$

$Q'''^* = \dfrac{q'''}{q_0'''}$

$u^* = \dfrac{u}{u_{ref}}$

$\rho^* = \dfrac{\rho}{\rho_{ref}}$

where,

$h_{ref} = \dfrac{q_0''' L_H}{\rho_f u_{io}}$

$u_{ref} = \dfrac{q_0''' L_H}{\rho_f (h_f - h_{io})}$

$b =$ Thermal expansion coefficient

$g =$ Gravity

$L_H =$ Heated length

$L'(t_i) =$ The evolved distance between two points after time step Δt_i

$q''' =$ Internal heat generation rate

$u =$ Velocity

$z =$ axial position

$A^* = \dfrac{A}{A_{X-S}}$

$F_r = \dfrac{u_{ref}^2}{g L_H}$

$N_{pch} = \dfrac{q_0''' L_H}{\rho_f u_{io} h_{fg}} \dfrac{v_{fg}}{v_f}$

$p^* = \dfrac{p - p_i}{\Delta p_{total}}$

$t^* = \dfrac{t}{t_{ref}}$

$z^* = \dfrac{z}{L_H}$

$\Lambda = \dfrac{f L_H}{2 D_H}$

$t_{ref} = \dfrac{\rho_f (h_f - h_{io})}{q_0'''}$

APPENDIX-I

POINCARÉ SECTIONS IN HYPERSPACE

For a higher order dimensional (i.e., $N > 3$) phase hyperspace, it is difficult to locate the positions of cutting plane (ie, the Poincaré section) which the solution trajectories pass through. The easiest way to locate a cutting plane perpendicular to the trajectories in phase hyperspace.

Given two points, $\underline{y}_1, \underline{y}_2$, which are typical values obtained from the time series data, the particular plane which is perpendicular to $\underline{y}_2 - \underline{y}_1$, and contains $\underline{y}_2$, can be calculated. In particular, the normal to the plane perpendicular to $\underline{y}_2 - \underline{y}_1$ is given by:

$$\underline{a} = C(\underline{y}_2 - \underline{y}_1) \tag{A.I.1}$$

where C is proportionality constant, and $\underline{a}$ is a vector prependicular to the cutting plane.

The equation of the cutting plane is given by:

$$\underline{a} \cdot \underline{x} = 1 \qquad \text{(A.I.2)}$$

Since the point $\underline{y_2}$ is asuumed to be located in the plane,

$$\underline{a} \cdot \underline{y_2} = 1 = C(\underline{y_2} - \underline{y_1}) \cdot \underline{y_2} \qquad \text{(A.I.3)}$$

Therefore,

$$C = \frac{1}{\underline{y_2} \cdot (\underline{y_2} - \underline{y_1})} \qquad \text{(A.I.4)}$$

and from Eq.(A.I.1),

$$\underline{a} = C(\underline{y_2} - \underline{y_1}) = \frac{(\underline{y_2} - \underline{y_1})}{\underline{y_2} \cdot (\underline{y_2} - \underline{y_1})} \qquad \text{(A.I.5)}$$

Similar to 3-D example shown in Figure I.1, the positions where the cutting plane

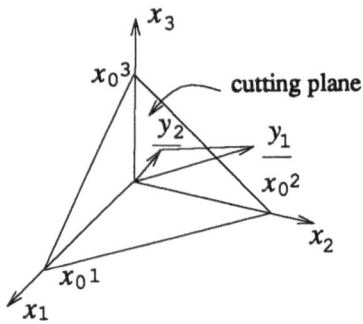

FIGURE I.1. A Typical Cutting Plane in Three Dimension.

intersects an N-dimensional coordinate system are determined from,

$$a_1 x_1 + a_2 x_2 + a_3 x_3 + a_4 x_4 + \cdots = 1 \qquad \text{(A.I.6)}$$

where only one x_i is not equal to zero. Thus,

$$x_{0_i} = \frac{1}{a_i} \qquad \text{(A.I.7)}$$

In order to find the points of intersection of the solution flow (ie, trajectories) with the cutting plane, let us consider two points, $\underline{\hat{y}_1}$ and $\underline{\hat{y}_2}$, which belongs to different hemi-hyperspace, as shown in Figure I.2.

As noted before, the equation of the cutting plane is,

$$\underline{a} \cdot \underline{x} = 1 \qquad \text{(A.I.8)}$$

The straight line containing $\underline{\hat{y}_1}$, and $\underline{\hat{y}_2}$ is given by,

$$\underline{x} = (\underline{\hat{y}_2} - \underline{\hat{y}_1})t + \underline{\hat{y}_1} \qquad \text{(A.I.9)}$$

158

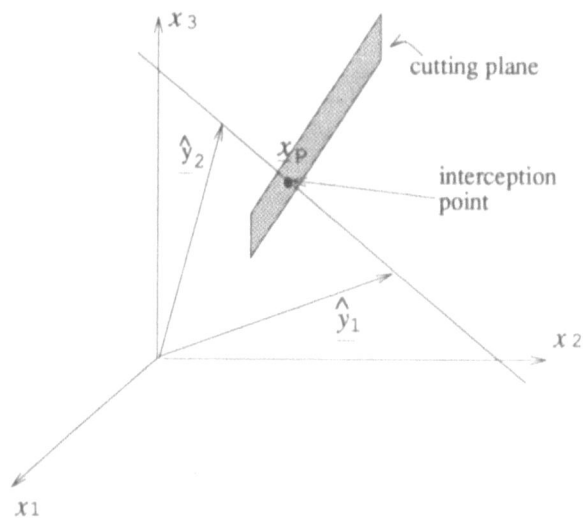

<figure>
FIGURE I.2. Interception Point $(\underline{x}_p)$.
</figure>

where, $0 \leq t \leq 1$.

Substituting Eqs. (A.I.9) into Eq. (A.I.8) we obtain,

$$\underline{a} \cdot (\underline{\hat{y}}_2 - \underline{\hat{y}}_1)t_p + \underline{a} \cdot \underline{\hat{y}}_1 = 1 \qquad (\text{A.I.10})$$

where, t_p denote that $\underline{x} = \underline{x}_p$.

Solving Eq. (A.I.10) for t_p,

$$t_p = \frac{(1 - \underline{a} \cdot \underline{\hat{y}}_1)}{\underline{a} \cdot (\underline{\hat{y}}_2 - \underline{\hat{y}}_2)} \qquad (\text{A.I.11})$$

and the interception point can be obtained as,

$$\underline{x}_p = (\underline{\hat{y}}_2 - \underline{\hat{y}}_1)t_p + \underline{\hat{y}}_1 \qquad (\text{A.I.12})$$

In summary, using Eqs. (A.I.5) and (A.I.7) we may determine a_i and x_{0_i}. Substituting $\underline{a}$ into Eq. (A.I.11), t_p can be computed for given $\underline{\hat{y}}_1$ and $\underline{\hat{y}}_2$. Finally the intersection point, $\underline{x}_p$, can be calculated using Eq.(A.I.12).

FREDHOLM MATRIX AND ZETA FUNCTIONS FOR 1-DIMENSIONAL MAPPINGS

Makoto Mori

National Defense Academy
Yokosuka-shi, Kanagawa 239, Japan

We consider a mapping $F : I \to I$ (I is a bounded interval) which satisfies:

(F1) F is a piecewise C^2, that is, there exists a partition of I into finite subintervals $\{\langle a \rangle\}_{a \in \mathcal{A}}$ and

(1) F can extend to $\overline{\langle a \rangle}$ in C^2,

(2) F is strictly monotone on each $\langle a \rangle$.

(F2) The lower Lyapunov number $\xi > 0$:

$$\xi = \liminf_{n \to \infty} \frac{1}{n} \operatorname*{ess\,inf}_{x \in I} \log |F^{n\prime}(x)|.$$

Remark. (i) From (F2), it follows

$$\operatorname*{ess\,inf}_{x \in I} |F'(x)| > 0.$$

(ii) We can consider transformation with a countable alphabet $\mathcal{A}$, for which

$$\overline{F(\langle a \rangle)} = I$$

except finite number of $a \in \mathcal{A}$.

Definition. The Perron–Frobenius operator $P : L^1 \to L^1$ is defined by

$$\int Pf(x)g(x)\,dx = \int f(x)g(F(x))\,dx$$

for $g \in L^\infty$. In other words

$$Pf(x) = \sum_{y : F(y) = x} f(y)|F'(y)|^{-1}.$$

Hereafter, we denote the restriction of P to BV also by P.

Aim: Our aim are the followings:

(A1) To study the spectrum of P.

(A2) To study the relation between the spectrum of P and the zeta function $\zeta(z)$:

$$\zeta(z) = \exp\left[\sum_{n=1}^{\infty} \frac{z^n}{n} \sum_{p:F^n(p)=p} |F^{n\prime}(p)|\right].$$

We want to express these (A1) and (A2) in terms of *Fredholm Matrix* $\Phi(z)$, which we will define by constructing renewal equation. Roughly speaking,

(1) For $|z| < e^{\xi}$, z^{-1} belongs to the spectrum of P if and only if $\Phi(z)$ has an eigenvalue 1.

(2) $\det(I - \Phi(z))$ is analytic in $|z| < e^{\xi}$ and

$$\zeta(z) = \frac{1}{\det(I - \Phi(z))}.$$

Therefore the zeta function has a meromorphic extension to the domain $|z| < e^{\xi}$, and in this domain z^{-1} belongs to the spectrum of P if and only if $\zeta(z)$ has a singularity at z.

Tools: We use the following tools to solve our problem.

(T1) Renewal Equation.

(T2) Signed Symbolic Dynamics.

(T3) Formal Piecewise Linear Transformations.

Renewal Equation. First, we will explain renewal equation.

Example 1. piecewise linear Markov cases: Let

$$\begin{aligned}
&\mathcal{A} = \{a, b\}, \\
&I = [0,1] = \langle a\rangle \cup \langle b\rangle, \quad (\langle a\rangle \cap \langle b\rangle = \emptyset) \\
&F(\langle a\rangle) = I, \\
&F(\langle b\rangle) = \langle a\rangle, \\
&F'(x) = \begin{cases} \lambda_a & x \in \langle a\rangle, \\ \lambda_b & x \in \langle b\rangle. \end{cases}
\end{aligned}$$

We consider generating functions:

$$s_g^c(z) = \sum_{n=0}^{\infty} z^n \int 1_{\langle c\rangle}(x) g(F^n(x)) \, dx,$$

where $g \in L^{\infty}$, $c = a$ or b and $1_{\langle c\rangle}$ is the indicator function of $\langle c\rangle$. Then

$$\begin{aligned}
s_g^c(z) &= \sum_{n=0}^{\infty} z^n \int P^n 1_{\langle c\rangle}(x) g(x) \, dx \\
&= \int (I - zP)^{-1} 1_{\langle c\rangle}(x) g(x) \, dx.
\end{aligned}$$

This roughly shows that the spectrum problem of P now turns into the problem of singulariteis of $s_g^c(z)$ $(c = a, b)$.

On the other hand, we get the relations:

$$s_g^a(z) = \chi_g^a + \lambda_a^{-1} z(s_g^a(z) + s_g^b(z)),$$
$$s_g^b(z) = \chi_g^b + \lambda_b^{-1} z s_g^a(z),$$

where

$$\chi_g^c = \int 1_{(c)}(x) g(x) \, dx.$$

Taking

$$s_g(z) = \begin{pmatrix} s_g^a(z) \\ s_g^b(z) \end{pmatrix} \quad \text{and} \quad \chi_g = \begin{pmatrix} \chi_g^a \\ \chi_g^b \end{pmatrix},$$

we get a *Renewal Equation* of the form:

$$s_g(z) = (I - \Phi(z))^{-1} \chi_g,$$

where

$$\Phi(z) = \begin{pmatrix} \lambda_a^{-1} z & \lambda_a^{-1} z \\ \lambda_b^{-1} z & 0 \end{pmatrix}.$$

This shows

PROPOSITION 1 z^{-1} *belongs to the spectrum of P if and only if* $\det(I - \Phi(z)) = 0$.

Therefore we may call $\Phi(z)$ the *Fredholm Matrix*.

Since $\Phi(z)$ is essentially a structure matrix, $\text{tr}(\Phi(z))^n$ corresponds to the sum over the periodic orbits with period n, Therefore, we get

$$\begin{aligned}
\zeta(z) &= \exp\left[\sum_{n=1}^{\infty} \frac{z^n}{n} \text{tr}(\Phi(z))^n\right] \\
&= \exp[-\text{tr}\log(I - \Phi(z))] \\
&= \frac{1}{\det(I - \Phi(z))}.
\end{aligned}$$

Example 2. Now we consider the β–expansion. Let $\lambda = \lambda_a = \lambda_b$ satisfy $1 < \lambda \leq 2$. Set

$$\begin{aligned}
\mathcal{A} &= \{a, b\}, \\
\langle a \rangle &= [0, 1/\lambda) \text{ and } \langle b \rangle = [1/\lambda, 1], \\
F(x) &= \lambda x \mod 1.
\end{aligned}$$

In this case, $s_g^a(z)$ is the same as before, but

$$\begin{aligned}
s_g^b(z) &= \int 1_{(b)}(x) g(x) \, dx + \lambda_b^{-1} z s_g^{J_1}(z) \\
&= \int 1_{(b)}(x) g(x) \, dx + (\lambda_b^{-1} z(\int_{J_1} (x) g(x) \, dx + \lambda_a^{-1}(s_g^a(z) + s_g^{J_2}(z)) \\
&= \cdots \\
&= \chi_g^b(z) + \Phi(z)_{b,a} s_g^a(z),
\end{aligned}$$

where
$$J_i = [0, F^i(1)] \cap \langle b^+[i+1]\rangle,$$
and $b^+[1]b^+[2]\cdots$ is the expansion of the point 1, that is,
$$F^n(1) \in \langle b^+[n+1]\rangle.$$

Note that $\chi_g^b(z)$ and $\Phi(z)_{b,a}$ are analytic function in $|z| < e^{\mathfrak{e}}$. Thus taking $\Phi(z)_{b,b} = 0$, this also gives the renewal equation of the form
$$s_g(z) = (I - \Phi(z))^{-1}\chi_g(z),$$
and
$$\zeta(z) = \frac{1}{\det(I - \Phi(z))}.$$

We can also prove the same results in a same way for transformations for which only one endpoint does not satisfy Markov condition such as unimodal linear transformation (for the proof refer [3] and for a certain extension, refer [4]).

Signed Symbolic Dynamics. We now consider transformations for which more than two endpoints do not satisfy the Markov condition. It seems impossible to trace the both endpoints of the subintervals at a time. Thus we introduce the signed symbolic dynamics. Let
$$\delta[L] = \begin{cases} 1 & \text{if } L \text{ is true,} \\ 0 & \text{otherwise,} \end{cases}$$
$$\sigma[L] = \delta[L] - \frac{1}{2}.$$

Then for points c, d $(c < d)$, we get
$$\sigma[x > c] + \sigma[x < d] = 1_{(c,d)}(x). \tag{$*$}$$

Now we call endpoints of the subinterval $\langle a \rangle$ by a^- and a^+. We identify a^- (a^+) with the infinite sequence of symbols $a^-[1]a^-[2]\cdots$ $(a^+[1]a^+[2]\cdots)$, respectively, where

$$\lim_{x\downarrow a^-} F^n(x) \in \langle a^-[n]\rangle,$$
$$\lim_{x\uparrow a^+} F^n(x) \in \langle a^+[n]\rangle.$$

We also denote a finite sequence $\tilde{a}[m]\cdots\tilde{a}[n]$ by $\tilde{a}[m,n]$ and $F^{n-1}(\tilde{a})$ by $\tilde{a}[n,\infty)$, where $\tilde{a} = a^-$ or a^+ and $m < n$. We divide $s_g^a(z)$, using the equation $(*)$
$$s_g^a(z) = s_g^{a^+}(z) + s_g^{a^-}(z),$$
where $s_g^{a^\sigma}(z)$ depends only on the orbit of a^σ $(\sigma \in \{+, -\})$. Then we can construct a renewal equation of the form:
$$s_g(z) = (I - \Phi(z))^{-1}\chi_g(z)$$

for

$$s_g(z) = \left(s_g^{a^\sigma}(z)\right)_{a\in\mathcal{A},\sigma\in\{+,-\}},$$
$$\chi_g(z) = \left(\chi_g^{a^\sigma}(z)\right)_{a\in\mathcal{A},\sigma\in\{+,-\}}.$$

Concerning the definitions and the proof, see [5].

Examples

(E1) β–expansion.

$$F(x) = \lambda x \mod.1,$$
$$= \begin{cases} \lambda x & \text{if } x \in \langle a \rangle = [0, 1/\lambda), \\ \lambda x - 1 & \text{if } x \in \langle b \rangle = [1/\lambda, 1]. \end{cases}$$

For $1 < \lambda \le 2$,

$$\Phi(\lambda z) = \begin{pmatrix} z/2 & z/2 & z/2 & z/2 \\ z/2 & z/2 & z/2 & z/2 \\ z/2 & z/2 & z/2 & z/2 \\ z/\{2(1-z)\} & a(z) & a(z) & -z/\{2(1-z)\} \end{pmatrix},$$

where

$$a(z) = \sum_{n=1}^{\infty} \{b^+[n+1] - 1/2\} z^n.$$

Then

$$\det(I - \Phi(\lambda z)) = 1 - \sum_{n=1}^{\infty} b^+[n] z^n,$$

and the density function of the invariant measure is

$$\rho(x) = C^{-1} \sum_{n=0}^{\infty} \lambda^{-n} \delta[x < b^+[n+1, \infty)]$$

where C is the normalizing constant.

(E2) unimodal linear transformation.

$$F(x) = \begin{cases} -\lambda x + 1 & x \in \langle a \rangle = [0, 1/\lambda), \\ \lambda x - 1 & x \in \langle b \rangle = [1/\lambda, 1], \end{cases}$$

where $1 < \lambda \le 2$.

$$\Phi(\lambda z) = \begin{pmatrix} z/2 & z/2 & z/2 & z/2 \\ z/2 & z/2 & z/2 & z/2 \\ z/2 & z/2 & z/2 & z/2 \\ b(z) & c(z) & c(z) & -b(z) \end{pmatrix},$$

where

$$b(z) = \sum_{n=1}^{\infty} \sigma[\text{sgn } b^+[1,n] = +]z^n,$$

$$c(z) = \sum_{n=1}^{\infty} \left\{ \delta[\text{sgn } b^+[1,n] = + \text{ and } b^+[n] = b] \right.$$

$$\left. + \delta[\text{sgn } b^+[1,n] = - \text{ and } b^+[n] = a] - \frac{1}{2} \right\} z^n,$$

$$\text{sgn } a_1 \cdots a_n = \begin{cases} + & \text{if } F^{n\prime}(x) > 0 \text{ for } x \in \langle a_1 \cdots a_n \rangle, \\ - & \text{otherwise,} \end{cases}$$

$$\langle a_1 \cdots a_n \rangle = \bigcap_{i=1}^{n} F^{-i+1} \langle a_i \rangle.$$

Then

$$\det(I - \Phi(\lambda z)) = 1 - \sum_{n=1}^{\infty} \text{sgn } b^+[1,n]z^n,$$

and the density function of the invariant measure is

$$\rho(x) = C^{-1} \sum_{n=0}^{\infty} \lambda^{-n} \delta[x < b^+[n+1, \infty)] \text{ sgn } b^+[1,n].$$

(E3) linear mod.1 transformation.

$$F(x) = \lambda x - r \quad (\text{mod. 1}),$$
$$= \begin{cases} \lambda x - r & \text{if } x \in \langle a \rangle = [0, (1+r)/\lambda), \\ \lambda x - r - 1 & \text{if } x \in \langle b \rangle = [(1+r)/\lambda, 1], \end{cases}$$

where $0 < r < 1$. Then for $1 < \lambda \le 2$

$$\Phi(\lambda z) = \begin{pmatrix} z/\{2(1-z)\} & d(z) & d(z) & -z/\{2(1-z)\} \\ z/2 & z/2 & z/2 & z/2 \\ z/2 & z/2 & z/2 & z/2 \\ z/\{2(1-z)\} & e(z) & e(z) & -z/\{2(1-z)\} \end{pmatrix},$$

where

$$d(z) = \sum_{n=1}^{\infty} \sigma[a^-[n+1] = a]z^n,$$

$$e(z) = \sum_{n=1}^{\infty} \sigma[b^+[n+1] = b]z^n.$$

Then

$$\det(I - \Phi(\lambda z)) = (1-z)^{-1}\{1 - \sum_{n=1}^{\infty}(b^+[n] - a^-[n])z^n\},$$

and the density function of the invariant measure is

$$\rho(x) = C^{-1} \sum_{n=0}^{\infty} \lambda^{-n} \{\delta[x < b^+[n+1, \infty)] - \delta[x < a^-[n+1, \infty)]\}.$$

Formal Piecewise Linear Transformation. We want to approximate a piecewise C^2 transformation F by piecewise linear transformations. But it may change the structure of the symbolic dynamics.

Thus we will introduce "formal piecewise linear transformations" F_N on the symbolic dynamics where F is realized. Let for an infinite sequence of alphabets α

$$F'_N(\alpha) = \frac{F(\alpha[1, N]^+) - F(\alpha[1, N]^-)}{\text{the length of } \langle \alpha[1, N] \rangle}.$$

When we consider the derivative F'_N on the symbolic dynamics, we express the shift operator by F_N. Then approximating F by F_N:

$$F'(\alpha)^{-1} = F'_1(\alpha)^{-1} + \sum_{N=1}^{\infty} (F'_{N+1}(\alpha)^{-1} - F'_N(\alpha)^{-1}),$$

we can construct a formal renewal equation

$$s_g(z) = \chi_g(z) + \Phi(z)s_g(z)$$

on the set of signed words $\tilde{\mathcal{W}}$, where

$$
\begin{aligned}
s_g(z) &= \left(s_g^{\tilde{w}}(z) \right)_{\tilde{w} \in \tilde{\mathcal{W}}}, \\
\chi_g(z) &= \left(\chi_g^{\tilde{w}}(z) \right)_{\tilde{w} \in \tilde{\mathcal{W}}}
\end{aligned}
$$

and $\Phi(z)$ is a countable dimensional matrix. By $\Phi_N(z)$, we denote the restriction of $\Phi(z)$ to the set of signed words with length less and equal to N. Then this matrix corresponds to the Fredholm matrix of the formal piecewise linear transformation F_N. Hence we get:

THEOREM 1 *Assume that $|z| < e^{\xi}$. Then z^{-1} belongs to the spectrum of P if and only if $(I - \Phi_N(z))^{-1}$ are bounded.*

Using this fact, we also get:

THEOREM 2 *Assume that $|z| < e^{\xi}$. Then z^{-1} belongs to the spectrum of P if and only if there exists $\{z_N\}$ which converges to z and $\det(I - \Phi_N(z_N)) = 0$.*

Moreover, we can show

$$\zeta(z) = \lim_{N \to \infty} \frac{1}{\det(I - \Phi_N(z))},$$

by showing the uniform boundedness of $\det(I - \Phi_N(z))$ in wider sence in $|z| < e^{\xi}$.

Hence combining the results:

THEOREM 3 *The zeta function has a meromorphic extension to the domain $|z| < e^{\xi}$, and in this domain, z^{-1} belongs to the spectrum of P if and only if $\zeta(z)$ has a singularity at z.*

These theorems are proved in [6], and concerning Theorem 3 refer also [1].

References

1. V.Baladi and G.Keller, *Zeta functions and transfer operators for piecewise monotone transformations*, Commun. Math. Phys. **127**, (1990), 459-478.
2. F.Hofbauer and G.Keller, *Zeta functions and transfer–operators for piecewise linear transformations*, J. Reine Angew. Math. **352**, (1984), 100-113.
3. M.Mori, *On the decay of correlation for piecewise monotonic mappings I*, Tokyo J. Math. **8**, (1985), 389-414.
4. _____, *On the decay of correlation for piecewise monotonic mappings II*, Tokyo J. Math. **9**, (1986), 135-161.
5. _____, *Fredholm determinant for piecewise linear transformations*, Osaka J. Math. **27**, (1990), 81-116.
6. _____, *Fredholm determinant for piecewise monotonic transformations*, Osaka J. Math. **29**, (1992), 497-529.
7. Y.Takahashi, *Fredholm determinant of unimodal linear maps*, Sci. Papers of Coll. Gen. Ed. Univ. Tokyo **31**, (1981), 61-87.
8. _____, *One dimansional maps and power spectrum*, Recent Studies on Turbulent Phenomena (1985), 99-116, Association for Science Documents Information, Tokyo.

ON THE LENGTH SPECTRUM OF THE BOUNDED SCATTERING BILLIARDS TABLE

Takehiko Morita

Department of Mathematics
Osaka University
Toyonaka, Osaka 560 Japan

INTRODUCTION

Let Q be a bounded connected plane domain with piecewise smooth boundary ∂Q. We assume the following conditions throughout the article.

(A.1) ∂Q consists of a finitely many smooth components $\Gamma_j, j = 1.2, ..., L$.

(A.2)   Each Γ_j is an arc of a smooth, simply closed, convex curve (possibly the whole of it).

(A.3)   For $i \neq j$, Γ_i and Γ_j have at most one point in common and the interior angle (with respect to Q) made by these two curves is strictly positive. Moreover, if Γ_i does not intersect the other smooth components, Γ_i itself is assumed to be a simply closed curve. The Q will be called a *bounded scattering billiards table*. Regarding $\overline{Q} = Q \cup \partial Q$ as a manifold with boundary, we consider the 'geodesic flow' on the unit tangent bundle $\overline{Q} \times S^1 = \{(q,v); q \in \overline{Q}, v \in \mathbb{R}^2 \text{ with } |v| = 1\}$. The flow is assumed to obey the law of reflections at the boundary. It is denoted by S_t and is called the *billiards flow* after Sinai [S]. Since the reflection vector and the incidental vector at the boundary are identified, it is natural to employ $M = \pi^{-1}(Q) \cup M_-$ as the phase space of the billiards flow, where $\pi : \overline{Q} \times S^1 \to \overline{Q}$ is the natural projection and $M_- = \{x = (q,v) \in \pi^{-1}(\partial Q); (n(q), v) \leq 0\}$. Here $n(q)$ denotes the unit inner normal of Q at q and $(\cdot, \cdot)$ is the Euclidean inner product. It is well known that M admits a canonical Liouville measure m induced from the Lebesgue measure on $\overline{Q} \times S^1$ (see [C.F.S]).

The distribution of the length spectrum (the distribution of the periods of the closed orbits of the billiards flow) of the billiards table has been investigated by those mathematicians and physicist who are working in Ergodic Theory, Scattering Theory, Quantum Chaos and so on, in this decade (e.g.[B.S.C], [E], [G], [I], [M], [St]). One of the typical asymptotic formulae for the length spectrum is the prime number type theorem

Algorithms, Fractals, and Dynamics
Edited by Y. Takahashi, Plenum Press, New York, 1995

which asserts that

$$\#\{\tau; \exp(h\ell(\tau)) \leq t\} \frac{\log t}{t} \to 1 \qquad (t \to \infty) \qquad (1)$$

holds for some $h > 0$, where τ and $\ell(\tau)$ denote a closed orbit of the flow and its period, respectively. In the case when Q is the so-called open billiards table without eclipse, the author proves the prime number type theorem in [M]. But in the present billiards table no such theorems exist because of many technical difficulties. For example, the Markov partition for the present billiards flow is necessarily countably infinite. The construction of the Markov partition itself is of course a very difficult problem although we find it in [B.S.C] under the finite multiplicity condition (see (A.4) in the next section). At this stage we are only able to show the following insufficient fact.

Theorem 1. *We impose the finite multiplicity condition on the bounded scattering billiards table. Let η be a Markov patrition constructed in [B.S.C]. Then there exist a sequence of positive numbers $\{h_n\}_{n=1}^{\infty}$ and a sequence of subsets $\{M_n\}_{n=1}^{\infty}$ of M satisfying the following: (1) $h_n \leq h_{n+1} \leq \sup h_n < \infty$ for any $n \geq 1$. (2) $M_n \subset M_{n+1}$ for any $n \geq 1$ and $m(M_n) \uparrow m(M)$ $(n \to \infty)$. (3) For each n, the prime number type theorem*

$$\#\{\tau; \tau \subset M_n, \exp(h_n\ell(\tau)) \leq t\} \frac{\log t}{t} \to 1 \quad (t \to \infty) \qquad (2)$$

holds. (4) $\bigcup_{n=1}^{\infty} M_n$ contains all but a finite number of the non-degenerate closed orbits of the billiards flow (see the next section for the definition of the non-degeneracy).

The main purpose of this article is to explain why the above theorem holds by showing an estimate for the first collision time of the flow to M_-.

FUNDAMENTAL FACTS AND RESULTS

One of the most traditional ways to analyze the qualitative behavior of the flow dynamics is the method of Poincaré maps. For $x \in M_-$, we define the first collision time by

$$t_+(x) = \inf\{t > 0; \pi(S_t x) \in \partial Q\}, \qquad (3)$$

where $t_+(x)$ is regarded as $+\infty$ if the set in (3) is empty. The last collision time is defined similarly by

$$t_-(x) = \sup\{t < 0; \pi(S_t x) \in \partial Q\}. \qquad (4)$$

The first collision map (*billiards map*) and the last collision map are defined by

$$Tx = S_{t_+(x)}x \text{ and } T^{-1}x = S_{t_-(x)}x, \qquad (5)$$

respectively. They are well-defined on the set of ν-full measure containing all periodic points and the notation T^{-1} is compatible with the inverse mapping of T. In the above ν denotes the canonical T-invariant area measure on M_- induced from the measure m (see [C.F.S]). The measure-theoretical dynamical system (M_-, T, ν) plays a role of the Poincaré map and the billiards flow S_t can be represented as a special flow over M_- with ceiling function t_+. Now we reach the following questions.

1. (Q.1) Does the discrete dynamical system (M_-, T) have a 'nice'coding?

2. (Q.2) Does the collision time t_+ have 'nice'representation on the coding space in (Q.1)?

We have to note that we can prove the prime number type theorem (1) in the case of the so-called open billiards without eclipse in [M] by answering these questions. In the present billiards table [B.S.C] gives an answer to (Q.1) by constructing a Markov partition. We have to explain what the Markov partition in [B.S.C] is for our convenience. First we introduce convenient local coordinates systems for M_-. Choose a point $q(j) \in \Gamma_j$ for each j. For any $x = (q, v) \in M_-$ so that q belongs to the interior of the arc Γ_j, we put $j(x) = j$, $r(x) =$ the arclength between q and $q(j)$ measured clockwise along the arc Γ_j, and $\varphi(x) =$ the angle between the incidental vector v and the unit inner normal $n(q)$ measured anticlockwise. Clearly (j, r, φ) gives a local coordinates in the neighborhood of x. In what follows we often write $x = (j, r, \varphi)$ or simply $x = (r, \varphi)$. Put $S_0 = \{x = (q, v) \in M_- ; (n(q).v) = 0\}$ and $V_0 = \{x = (q, v); q \in \Gamma_i \cap \Gamma_j \text{ for some } i \neq j\}$. Moreover, we put $R_0 = S_0 \cup V_0$, $R_k = T^k R_0$, and $R_{k,\ell} = \bigcup_{j=k}^{\ell}$ for $k, \ell \in \mathbb{Z}$ with $k < \ell$. It is not hard to see that the singularity set of T^n and that of T^{-n} are $R_{-n,0}$ and $R_{0,n}$, respectively for $n > 0$. In particular, the set $R_{-\infty,\infty}$ consists of countably many smooth curves which will be called *discontinuity curves*. The number of the discontinuity curves passing through $x \in M_-$ is called the *multiplicity* of x. We impose the following generic condition on our billiards table.

(A.4) The multiplicity is uniformly bounded in x.

To define the Markov partition we need the notions of the *local stable manifold (LSM)* and the *local unstable manifold (LUM)*. A LSM is defined as a curve in M_- without endpoints such that T^n are continuous on it for all $n \geq 1$. A LUM is a curve in M_- without endpoints such that T^{-n} are continuous on it for all $n \geq 1$. It is well known that ν-almost every point x has a LSM and a LUM passing through it. We denote by $\gamma^s(x)$ and $\gamma^u(x)$ the maximal smooth components of them. For $x, y \in M_-$, we put $[x, y] = \gamma^u(x) \cap \gamma^s(y)$. Choose $A \subset \gamma^s(x)$ and $B \subset \gamma^u(y)$ so that $[x', y']$ is defined for any $x' \in A$ and $y' \in B$. We pot $[A, B] = \{[x', y']; x' \in A, y' \in B\}$. A subset R of M_- is called a *rectangle* if $[x, y] \in R$ holds for any $x, y \in R$. A rectangle R is said to be *non-degenerate* if $\nu(R) > 0$. For any subset A of M_-, we set $\gamma_A^u = \gamma^u \cap A$ and $\gamma_A^s = \gamma^s \cap A$. We say that two rectangles R_1 and R_2 *intersect regularly* if $R_1 \cap R_2 = [\gamma_{R_2}^s(x), \gamma_{R_1}^u(y)]$ for any $x, y \in R_1 \cap R_2$.

A finite or countable family $\eta = \{R_j\}$ of M_- consisting of closed rectangles is called a *Markov partition* for T if it satisfies the following: (1) $\nu(M_- - \bigcup_j R_j) = 0$ and $\nu(R_i \cap R_j) = 0$ for $i \neq j$. (2) For each j there exists a connected domain $U(R_j)$ containing R_j on which the map T and T^{-1} are continuous. (3) For ν-a.e. $x \in M_-$ $R(x)$ and $TR(T^{-1}x)$ intersect regularly, where we denote by $R(x)$ the rectangle in η containing x.

Our argument depends heavily on the following result in [B.S.C].

Theorem 2. *Assume that the conditions (A.1), (A.2), (A.3) and (A.4) are fulfilled. Then for any $\varepsilon > 0$, there exists a Markov partition for T whose elements have diameter less than ε.*

We note that the Markov partition for our billiards map T is necessarily infinite and possibly contains the degenerate elements.

Let η be a Markov partition for T. A periodic point of T is said to be *non-degenerate* if it is contained in a non-degenerate element in η. A closed orbit τ of the billiards flow S_t is said to be non-degenerate if the set $\tau \cap M_-$ consists of non-degenerate periodic points of T.

In appearance, the non-degeneracy depends on the choice of the Markov partition. We expect that the set of non-degenerate periodic points of T for distinct Markov partitions must coincide up to a finite number of exceptional members. But we do not know whether it is true or not.

In the sequel we consider a fixed Markov partition η constructed in [B.S.C]. We are interested in only non-degenerate rectangles in η. Therefore we need not construct the shift space (the coding space in (Q.1)) so carefully as in Section 7 of [B.S.C]. We define an infinite matrix with $0 - 1$ entries A so that $A(i, j) = 1$ if and only if R_i and $T^{-1}R_j$ intersect regularly. We can not guarantee $\nu(R_i \cap T^{-1}R_j) > 0$ even if both R_i and R_j are non-degenerate. This is the reason why we define the matrix A as above. Put

$$\Sigma = \{\xi = \{\xi_n\}_{n=-\infty}^{\infty} ; A(\xi_n, \xi_{n+1}) = 1 \text{ for any } n \in \mathbb{Z}\}. \tag{6}$$

Then we obtain a discrete dynamical system (Σ, σ), where $\sigma : \Sigma \to \Sigma$ is the shift transformation defined as $(\sigma\xi)_n = \xi_{n+1}$ for $n \in \mathbb{Z}$. We prepare the following lemma.

Lemma 1. *The map* $\Phi : \Sigma \to \Sigma$ *defined by* $\Phi(\xi) = \bigcap_{m=-\infty}^{\infty} T^{-n}R_{\xi_n}$ *gives a one to one correspondence up to a finite number of points between the sets* $Per(\Sigma)$ *and* $NPer(T)$, *where* $Per(\Sigma)$ *and* $NPer(T)$ *denote the set of periodic points of* σ *and that of non-degenerate periodic points of* T, *respectively.*

Proof. To begin with we claim that in Section 5 of [B.S.C], it is proved that T has at most a finite number of periodic orbits contained in the distinct rectangles in the same time. Let ξ be in $Per(\Sigma)$. Clearly, $\Phi(\xi)$ is a non-degenerate periodic point of T. Combining this with the above claim we see that $\Phi|_{Per(\Sigma)} : Per(\Sigma) \to NPer(T)$ is injective except for finitely many elements.

It remains to show that $\Phi|_{Per(\Sigma)}$ is surjective up to finitely many points. Let $x \in NPer(T)$ be such that any point in its orbit is contained exactly one element in η. From the above claim it suffices to show that $TR(x)$ and $R(Tx)$ intersect regularly. But this is proved in Section 5 of [B.S.C] too. Now the proof of the lemma is complete. $\quad\square$

The following proposition is a discrete version of Theorem 1 in Introduction.

Proposition 1. *There exists a sequence of topologically transitive subshifts of finite type* $\{\Sigma_n\}_{n=1}^{\infty}$ *such that (1)* $\Sigma_n \subset \Sigma_{n+1}$ *and* $\Sigma_n \subset \Sigma$ *for any* $n \geq 1$, *and (2)* $\bigcup_{n=1}^{\infty} Per(\Sigma_n) = Per(\Sigma)$:

Proof. It suffices to show that for any N, there is a subset of positive integers K_N containing $\{1, 2, ..., N\}$ such that the $\sharp K_N \times \sharp K_N$-matrix A_N is irreducible, where A_N is the restriction of A to the indices in K_N. Since the measure-theoretical dynamical system (M_-, T, ν) is mixing, there is a positive integer k_N such that $\nu(T^{-k}R_i \cap R_j) > 0$ for all $1 \leq i, j \leq N$ whenever $k \geq k_N$. For each $(i, j) \in \{1, 2, ..., N\} \times \{1, 2, ..., N\}$, we select a chain of rectangles $R_0^{(i,j)}, R_1^{(i,j)}, ..., R_{k_N}^{(i,j)}$ in η satisfying $R_0^{(i,j)} = R_i$, $R_{k_N}^{(i,j)} = R_j$, and $\nu(R_0^{(i,j)} \cap T^{-1}R_1^{(i,j)} \cap \cdots \cap R_{k_N}^{(i,j)}) > 0$. If we put

$$K_N = \{ \text{ the indices appearing in the above procedure } \},$$

it is easy to see that it satisfies the desired condition. Thus we obtain the proposition. □

The next task is to obtain an answer to the question (Q.2). For this purpose we define d_θ-metric on the sift space Σ. For $0 < \theta < 1$ we set $d_\theta(\xi, \xi') = \theta^n$ if $\xi_j = \xi'_j$ for all j with $|j| < n$ and $\xi_{-n} \neq \xi'_{-n}$ or $\xi_n \neq \xi'_n$.

For a periodic point $\xi \in \Sigma$, we set

$$f(\xi) = t_+(\Phi(\xi)). \tag{7}$$

One of the main results in this article is the following,

Proposition 2. f extends uniquely to Σ as a Lipschitz function with respect to d_θ for some $0 < \theta < 1$.

The proof will be given in the next section. In the rest of this section we give:

Proof of Theorem 1. The assertion (1) follows from the rough estimate in [St]. The assertions (2) and (4) are easy consequences of Lemma 1 and Proposition 1. Therefore it remains to prove the prime number type theorem.

In virtue of the results in [P.P], we have only to show that the spacial flow over the topologically mixing component of Σ with ceiling function $f_n = f + f \circ \sigma + f \circ \sigma^{d_n - 1}$ is topologically weakly mixing, where d_n is the number of the topologically mixing components of Σ_n. To this end we show that $f_n|_{\Sigma_n}$ can not have the form

$$f_n = aK + g \circ \sigma^{d_n} - g \tag{8}$$

for any $a > 0$, any integer valued function K and any real valued function g. If $f_n|_{\Sigma_n}$ could have the form (8) for some $a > 0$, K, and g as above, the length spectrum of the billiards flow corresponding to Σ_n would be a subset of $a\mathbb{Z}$. We can choose distinct words $\alpha - (\alpha_0\alpha_1 \ldots \alpha_{s-1})$ and $\beta - (\alpha_0\beta_1 \ldots \beta_{t-1})$ such that $A(\alpha_0, \alpha_1) = \cdots = A(\alpha_{s-2}, \alpha_{s-1}) = A(\alpha_{s-1}, \alpha_0) = A(\alpha_0, \beta_1) = \cdots = A(\beta_{t-2}, \beta_{t-1}) = A(\beta_{s-1}, \alpha_0)$. We use the following conventions. For words $w^{(i)} = (w_0^{(i)}, ..., w_{s_i-1}^{(i)})$, $i = 1, 2, ..., u$, we denote by $\dot{w}^{(1)}w^{(2)} \cdots \dot{w}^{(u)}$ the periodic sequence ξ with $\xi_{-1} = w_{s_u-1}^{(u)}, \xi_0 = w_0^{(0)}, \xi_1 = w_1^{(0)}, ..., \xi_{s_1+s_2+\cdots+s_u-1} = w_{s_u-1}^{(u)}$. For $q_1, q_2 \in \partial Q$, $q_1 \rightarrow q_2$ means that q_1 and q_2 are connected by a line segment drawn from q_1 to q_2. Let $\xi^{(0)} = \dot\alpha\dot\alpha$ and $\xi^{(k)} = \underbrace{\dot\alpha \cdots \alpha}_{2k \text{ times}} \dot\beta$

for $k \geq 0$. We write as $\Phi(\sigma^i\xi_{(k)}) = x_i^{(k)} = (q_i^{(k)}, v_i^{(k)})$. We note that we have

$$|q_{ks+i}^{(k+1)} - q_i^{(0)}| \leq C\rho^{ks} \tag{9}$$

from Lemma 5 in the next section, where $C > 0$ and $0 < \rho < 1$ are independent of k.

Let τ_k be the corresponding closed orbit of S_t starting from $x_0^{(k)}$ and ℓ_k be the length of τ_k for $k \geq 0$. From [St], ℓ_k can be characterized as a strict local minimum of the function

$$F_k(q_0, q_1, ..., q_{2ks+t-1}) = |q_1 - q_0| + \cdots + |q_{2ks+t-1} - q_{2ks+t-2}| + |q_0 - q_{2ks+t-1}| \tag{10}$$

with $q_i \in K_i$, where K_i denotes a suitably chosen compact convex domain so that a subarc of $\Gamma_{j(x_i^{(k)})}$ containng $q_i^{(k)}$ is a subset of its boundary.

Next we make a fictitious closed orbit τ'_{k+1} inserting a part of τ_{k+1} into τ_k between the $ks - 1$-st collision and the ks-th collision so that the collisions occur as

$$
\begin{aligned}
q_0^{(k)} \to q_1^{(k)} \to \cdots \to q_{ks-1}^{(k)} &\to q_{ks}^{(k+1)} \to q_{ks+1}^{(k+1)} \to \\
\cdots \to q_{(k+2)s-2}^{(k+1)} \to q_{(k+2)s-1}^{(k+1)} &\to q_{ks}^{(k)} \to q_{ks+1}^{(k)} \to \\
\cdots &\to q_{2ks+t-1}^{(k)} \to q_0^{(k)}.
\end{aligned} \tag{11}
$$

From the estimate (9), we have

$$
\ell'_{k+1} \leq \ell_k + 2\ell_0 + C's\rho^{ks}, \tag{12}
$$

for some C' independent of k, where ℓ'_{k+1} is the length of τ'_{k+1}. If k is large, we can show that $\ell_{k+1} < \ell'_{k+1}$ as follows. We write $\bar{q}$ for he vector $(q_0^{(k+1)}, ..., q_{2(k+1)s+t-1}^{(k+1)})$ for simplicity. The corresponding vector to the fictitious orbit τ'_{k+1} will be written as $\bar{p}$. Clearly $\ell_{k+1} = F_{k+1}(\bar{q})$ and $\ell'_{k+1} = F_{k+1}(\bar{p})$. We notice that the distance between the consecutive collisions are uniformly bounded from below since we are working on a finite number of rectangles. Consider the vector $q(t) = t\bar{q} + (1-t)\bar{p} = (q_0(t), ..., q_{2(k+1)s+t-1}(t))$ for $0 \leq t \leq 1$. From the above notice and Lemma 5, we can connect $q_i(t)$ and $q_{i+1}(t)$ by a line segment which does not intersect ∂Q outside of K_i and K_{i+1} for each i (mod $2ks + t$). If $\ell'_{k+1} \leq \ell_{k+1}$ holds, we have

$$
F_{k+1}(q(t)) \leq t F_{k+1}(\bar{q}) + (1-t) F_{k+1}(\bar{p}) \leq F_{k+1}(\bar{q})
$$

since F_{k+1} is a convex function. Letting $t \to 1$ we can see that this contradicts the fact that $F_{k+1}(\bar{q})$ is a strict local minimum. From this fact we have seen that (8) and (12) implies

$$
\ell_{k+1} \leq \ell_k + 2\ell_0 + a. \tag{13}
$$

On the other hand we construct a fictitious closed orbit τ''_k by ignoring the collisions from the ksth one to the $(k+2)s - 1$st one so that the collisions occur as

$$
q_0^{(k+1)} \to q_1^{(k+1)} \to \cdots \to q_{ks-1}^{(k+1)} \to q_{(k+2)s}^{(k+1)} \to \cdots \to q_{2(k+1)s+t-1}^{(k+1)} \to q_0^{(k+1)}. \tag{14}
$$

In the same way as in the above, we can show that $\ell''_k < \ell_k$ holds for any k sufficiently large, where ℓ''_k is the length of τ''_k. Combining this with the estimate (9), we obtain

$$
\ell_{k+1} \geq \ell_k + 2\ell_0 - C''s\rho^{ks}, \tag{15}
$$

where C'' is a positive constant independent of k. Clearly, (15) contradicts (13). Now the proof of Theorem 1 is complete if we assume Lemma 5 and Proposition 2. $\qquad \square$

PROOF OF PROPOSITION 2

The last section is devoted to the proof of Proposition 2. We use the (r, φ)-coordinates in the sequel. A curve $\gamma \in M_-$ is said to be an *increasing* (resp. *decreasing*) curve if it is expressed as $\varphi = \varphi(r)$ with an increasing (resp. decreasing) function φ in r. For a positive integer k, an increasing (resp. decreasing) curve γ is said to be $k-increasing$ (resp. $k-decreasing$) if T^{-k} (resp. T^k) is continuous on γ and $T^{-k}\gamma$ (resp. $T^k\gamma$) is an increasing (resp. a decreasing) curve. We consider two kinds of length functions for a C^1-curve. One is the Euclidean length $s(\gamma)$ with respect to the (r,φ)-coordinates and the other is the so-called p-length defined by

$$p(\gamma) = \int_a^b -\cos\varphi\, dr, \tag{16}$$

when γ have the expression $\varphi = \varphi(r)$, $a \leq r \leq b$.

For a point $x = (j, r, \varphi) \in M_-$, we put $j_n = j(T^n x)$, $r_n = r(T^n x)$, $\varphi_n = \varphi(T^n x)$, $c_n = \cos\varphi_n(x)$, $t_{+,n} = t_+(T^n x)$, $t_{-,n} = t_-(T^n x)$ and so on if these are defined. K_n denotes the curvature of Γ_{j_n} at r_n.

The proof of Proposition 2 is divided into a series of lemmas. To begin with, elementary calculations give us the following.

Lemma 2. *Let γ be a C^1-curve expressed as $\varphi = \varphi(r)$, $a \leq r \leq b$. Assume that T (resp. T^{-1}) is continuous on γ and the image $T\gamma$ (resp. $T^{-1}\gamma$) can be expressed as $\varphi_1 = \varphi_1(r_1)$, $a_1 \leq r_1 \leq b_1$ (resp. $\varphi_{-1} = \varphi_{-1}(r_{-1})$, $a_{-1} \leq r_{-1} \leq b_{-1}$). Then we have*

$$\frac{d\varphi_1}{dr_1} = K_1 + c^{-1}c_1\left(-t_+c^{-1} + (\tfrac{d\varphi}{dr} + K)^{-1}\right)^{-1}$$
$$\left(resp.\ \ \frac{d\varphi_{-1}}{dr_{-1}} = -K_{-1} + c^{-1}c_{-1}\left(-t_-c^{-1} + (\tfrac{d\varphi}{dr} - K)^{-1}\right)^{-1}\right) \tag{17}$$

$$\frac{dr_1}{dr} = cc_1^{-1}\left(1 - c^{-1}t_+(\tfrac{d\varphi}{dr} + K)\right)$$
$$\left(resp.\ \ \frac{dr_1}{dr} = cc_{-1}^{-1}\left(1 - c^{-1}t_-(\tfrac{d\varphi}{dr} - K)\right)\right). \tag{18}$$

Here we give an easy but important corollary to Lemma 2.

Corollary. *Let γ be a 1-invreasing (resp. 1-decreasing) curve expressed as in Lemma 2. Then we have*

$$K_{\min} \leq \frac{d\varphi}{dr} \leq K_{\max} + t_+^{-1}$$
$$\left(resp.\ \ -K_{\max} + t_-^{-1} \leq \frac{d\varphi}{dr} \leq -K_{\min},\right) \tag{19}$$

where $K_{\min}$ and $K_{\max}$ denote the infimum and the supremum of the curvature of ∂Q, respectively.

The condition (A.3) guarantees us to find a positive integer n_0 and a positive number t_0 such that

$$t_+(x) + t_+(Tx) + \cdots + t_+(T^{n_0-1}x) \geq t_0 \tag{20}$$

holds for any $x \in M_-$ for which T^{n_0} is defined.

The assertion of the following lemma is slightly weaker than the corresponding one in [B.S.C]. But it is enough to get our results.

Lemma 3. *Let u be any number greater than 3. There are positive numbers C_1 and C_2 depending only on the billiards table Q and the number u such that*

$$C_1 p(\gamma) \le s(\gamma) \le C_2 p(\gamma)^{\frac{1}{u}}$$

$$(21)$$

holds for any n_0-increasing curve and any n_0-decreasing curve.

Proof. Let γ be an n_0-increasing curve expressed as in Lemma 2. Using the definition (16) the first inequality holds with $C_1 = 1$. Note that C_1 can be chosen to be $\sqrt{1 + K_{\min}^2}$ if we use the inequality (19).

Lemma 2.7 in [B.S.C] asserts that for any δ with $0 < \delta < 2$, there exists a positive number $C_{(1)}$ depending only on Q and δ such that

$$\int_a^b \left(\frac{d\varphi}{dr} \right)^{\delta} dr < C_{(1)}$$

$$(22)$$

holds uniformly in n_0-increasing curves. For $u > 3$ we can choose $p > 1$ and $q > 1$ such that $\frac{1}{p} + \frac{1}{q} = 1$ and $u > \max(p, 2q)$. From (19) we have

$$s(\gamma) = \int_a^b \sqrt{1 + \left(\frac{d\varphi}{dr} \right)^2} \, dr \le \sqrt{1 + K_{\min}^{-2}} \int_a^b \frac{d\varphi}{dr} dr.$$

Applying the Hölder's inequality twice to the last term we have

$$\int_a^b \frac{d\varphi}{dr} dr \le \left(\int_a^b \frac{d\varphi}{dr} |\cos \varphi|^{-\frac{p}{u}} dr \right)^{\frac{1}{p}} \left(\int_a^b \frac{d\varphi}{dr} |\cos \varphi|^{\frac{q}{u}} dr \right)^{\frac{1}{q}}$$

$$\le \left(\int_{-\frac{\pi}{2}}^{\frac{\pi}{2}} (\cos \varphi)^{-\frac{p}{u}} d\varphi \right)^{\frac{1}{p}} \left(\int_a^b \left(\frac{d\varphi}{dr} \right)^{\frac{u}{u-q}} dr \right)^{\frac{u-q}{u}} p(\gamma)^{\frac{1}{u}}.$$

Therefore we obtain the second inequality in (21) from the inequality (22). $\qquad \square$.

The next lemma is an easy consequence of the inequality (18) in Lemma 2 and the existence the numbers n_0 and t_0 in (20).

Lemma 4. *Let γ be an increasing (resp. a decreasing) curve on which T^{n_0} (resp. T^{-n_0}) is continuous. Then we have*

$$p(T^{n_0}\gamma) \ge (1 + t_0 K_{\min}) p(\gamma)$$
$$(resp. \quad p(T^{-n_0}\gamma) \ge (1 + t_0 K_{\min}) p(\gamma)).$$

$$(23)$$

For $u > 3$, we put

$$\theta = (1 + t_0 K_{\min})^{-\frac{1}{u n_0}}.$$

$$(24)$$

The we. have the lemma assumed in the proof of Theorem 1.

Lemma 5. *There exists a positive constant C_3 depending only on Q and θ in (24) such tat*

$$\sqrt{(r(x) - r(y))^2 + (\varphi(x) - \varphi(y))^2} \le C_3 \theta^n$$

$$(25)$$

holds whenever x and y are in the same increasing (resp. decreasing) curve on which $T, T^2, ..., T^n$ (resp. $T^{-n}, T^{-(n-1)}, ..., T^{-1}$) are all continuous.

Proof. Since ∂Q has only a finite number of smooth components (see (A.1)), $p(\gamma)$ is bounded uniformly in γ. Therefore, in virtue of Lemma 4, it is easy to see that

$$p(\gamma) \leq C_{(2)}(1 + t_0 K_{\min})^{\frac{n}{un_0}}$$

for some constant $C_{(2)}$ depending only on Q. Combining this with Lemma 3, we obtain the lemma. $\square$

Now we are ready to prove Proposition 2.

Proof of Proposition 2. Assume that ξ and ζ are periodic points in Σ with $\xi_i = \zeta_i$ for $|i| \leq n$. Put $x = \Phi(\xi)$ and $y = \Phi(\zeta)$. For a curve $\gamma \in M_-$ we write $\gamma[x_1, x_2]$ for a segment in γ joining x_1 and x_2. Recall that each $R_j \in \eta$ is a rectangle and is contained in a connected domain where T and T^{-1} are continuous. Therefore we can see that T^i, $i = 1, 2, ..., n$ are all continuous on $\gamma^u(x)[x, [x, y]]$. Since $\gamma^u(x)$ is a increasing curve, we have $|r(x) - r([x, y])| \leq C_3 \theta^n$ in virtue of Lemma 5. In the same way we can show that $|r(y) - r([x, y])| \leq C_3 \theta^n$ holds by using $\gamma^s(y)$. Thus we have $|r(x) - r(y)| \leq 2C_3 \theta^n$. Since parameter r is defined by the arc length, it is easy to see that $|t_+(x) - t_+(y)| \leq |r(x) - r(y)| + |r(Tx) - r(Ty)|$. Hence we obtain the desired inequality. $\square$

References

[B.S.C]   L. A. Bunimovich, Ya. G. Sinai, and N. J. Chernov, *Markov partitions for two-dimensional hyperbolic billiards*, Russian Math. Surveys **45**, (1990), 105-152.

[C.F.S]   I. P. Cornfeld, S. V. Formin, and Ya. G. Sinai, *Ergodic theory*, Springer, New York. (1982)

[E]   B. Eckhardt, *Periodic orbit theory*, preprint

[G]   M. C. Gutzwiller, *Chaos in classical and quantum mechanics*, Springer, New York. (1990).

[I]   M. Ikawa, *Singular perturbation of symbolic flows and poles of the zeta functions*, Osaka J. Math., **27**, (1990), 281-300.

[M]   T. Morita, *The symbolic representation of billiards without boundary condition*, Trans. Amer. Math. Soc. **325**, (1989), 819-828.

[P.P]   W. Parry and M. Pollicott, *An analogue of the prime number theorem for closed orbits of Axiom A flows*, Ann. of Math. **118**, (1983), 573-591.

[S]   Ya. G. Sinai, *Dynamical systems with elastic reflections*, Russian Math. Surveys, **25**, (1970), 137-189.

[St]   L. Stojanov, *An estimate from above of the number of periodic orbits for semi-newline dispersed billiards*, Commun. Math. Phys. **124**, (1989), 217-227.

CONTINUED FRACTIONS, GEODESIC FLOWS AND FORD CIRCLES

Hitoshi Nakada

Department of Mathematics
Keio University
3-14-1, Hiyoshi, Kohoku
Yokohama 223, Japan

§1 INTRODUCTION

A purpose of this paper is to give a short sketch of a relation between continued fractions and the hyperbolic geometry on the upper half plane, (the simple continued fractions case and a generalized case). Relations between continued fractions and the geodesic flows on the modular surface are well-known. For example, Adler and Flatto [1] showed that the continued fraction transformation is obtained as a cross-section map of the geodesic flow. Another interesting one is due to Moeckel [8], who proved a metrical property of continued fractions concerning to a distribution of digits by using the Farey tessellation and the ergodicity of geodesic flows.

In this paper, we start with some well-known facts of continued fractions and a classical notion of Ford circles. We will give a new view of these facts. Our fundamental idea is very simple. We only look at Ford circles as horocycles and their connection with geodesics. Then we have one method which is applicable to other problems. In §2-4, we give a sketch of a method stated above in the case of the classical continued fractions. Moreover, in §5-7, we apply this method to a problem of Rosen's continued fractions arising from a theory of a diophantine approximation associated with Hecke groups. A problem of the diophantine approximation associated with a zonal Fuchsian group was first proposed by Lehner [6]. Let G be a zonal Fuchsian group of the first kind, here "zonal" means that G has a cusp at infinity. If the group is of the first kind, then the set of limit points is $\mathbf{R} \cup \{+\infty\}$ and the set $G(\infty)$ is dense in $\mathbf{R}$. So it is natural to regard $G(\infty)$ as the set of "rational numbers" and we consider the following diophantine inequality for a real number x:

$$|x - g(\infty)| < h(c) \cdot 1/c^2 \tag{1}$$

Algorithms, Fractals, and Dynamics
Edited by Y. Takahashi, Plenum Press, New York, 1995

where $g \in G$ is expressed by a matrix

$$\begin{bmatrix} a & b \\ c & d \end{bmatrix}$$

and h is a positive valued symmetric function on $\mathbf{R}$. Here we only consider the case of a constant function h. Let $1/\kappa$ be the infimum of constants such that (1) has infinitely many solutions of $g(\infty)$'s for all x. We call such κ the Hurwitz constant for G. To discuss this constant for the Hecke group G_k, Rosen [12] introduced a notion of "the nearest integer type" continued fraction expansion of real numbers. By using this continued fractions, Lehner [7] determined the Hurwitz constant for the Hecke groups of even degree and gave a lower and an upper estimate of the constants for those of odd degrees. We will show some results for these continued fractions, in particular, in the case of Hecke groups of the even degree.

A similar application is also possible to the theory of complex continued fractions. We refer [9],[10],[11] and [13] for these.

§2 CONTINUED FRACTIONS

The simple continued fraction expansion of a real number $x, 0 < x < 1$, and its n-th convergent are the following:

$$x = \frac{1|}{|a_1} + \frac{1|}{|a_2} + \frac{1|}{|a_3} + \cdots , \tag{2}$$

$$\frac{p_n}{q_n} = \frac{1|}{|a_1} + \frac{1|}{|a_2} + \frac{1|}{|a_3} + \cdots + \frac{1|}{|a_n} \tag{3}$$

where $a_1, a_2, a_3, \ldots$ are positive integers and p_n and q_n are relatively prime positive integers. The following are well-known.

(i) The expansion (2) terminates at finitely many terms if and only if x is rational.

(ii) There exist positive integers n and m such that $a_{n+l+m} = a_{n+l}$ for any non-negative integer l if and only if x is a quadratic surd. In particular, $n = 1$ if and only if the conjugate root of x is less than -1.

(iii) For any $x, 0 < x < 1$, and any $n, n \geq 1$,

$$\left| x - \frac{p_n}{q_n} \right| < \frac{1}{q_n{}^2}. \tag{4}$$

(iv) If a rational number $p/q, 0 < p/q < 1$, satisfies

$$\left| x - \frac{p}{q} \right| < \frac{1}{q^2}. \tag{5}$$

then p/q is the n-th convergent of x for some positive integer n.

(v) For any irrational number $x, 0 < x < 1$, there exist infinitely many positive integers n such that

$$\left| x - \frac{p_n}{q_n} \right| < \frac{1}{\sqrt{5}q_n{}^2}. \tag{6}$$

If $c > \sqrt{5}$, then there exists an irrational number x such that

$$\left| x - \frac{p_n}{q_n} \right| < \frac{1}{cq_n{}^2}.$$

holds for at most finitely many positive integers n.

Because of (iv), (v) is equivalent to the following:

(v)' For any irrational number $x, 0 < x < 1$, there exist infinitely many positive rational numbers p/q such that

$$\left| x - \frac{p}{q} \right| < \frac{1}{\sqrt{5}q^2}. \tag{6'}$$

If $c > \sqrt{5}$, then there exists an irrational number x such that

$$\left| x - \frac{p}{q} \right| < \frac{1}{cq^2}.$$

holds for at most finitely many positive integers n.

To determine the coefficients of the continued fraction expansion of x, we define the following map of $I = [0,1)$ onto itself:

$$f(x) = \begin{cases} \frac{1}{x} - [\frac{1}{x}] & \text{if } 0 < x < 1 \\ 0 & \text{if } x = 0, \end{cases}$$

where $[s]$ denotes the integral part of a real number s. We put $a = a(x) = [1/x]$ for $x \neq 0$. Then we have

$$f(x) = \frac{-ax + 1}{x} \quad \text{if } x \in \left(\frac{1}{a+1}, \frac{1}{a} \right]$$

and

$$a_n = a_n(x) = a(f^{n-1}(x)).$$

If $f^n(x) = 0$ for some n, then the continued fraction expansion of x terminates at the n-th coefficient. In the sequel, we always assume $f^n(x) \neq 0$ for any n under consideration. We also note that

$$f^n(x) = \frac{1}{|a_{n+1}|} + \frac{1}{|a_{n+2}|} + \frac{1}{|a_{n+3}|} + \cdots.$$

We identify a matrix $\begin{bmatrix} p & r \\ q & s \end{bmatrix}$ with the linear fractional transformation $x \to \frac{px+r}{qx+s}$. Then we have

$$f^n(x) = \begin{bmatrix} -a_n & 1 \\ 1 & 0 \end{bmatrix} \begin{bmatrix} -a_{n-1} & 1 \\ 1 & 0 \end{bmatrix} \cdots \begin{bmatrix} -a_1 & 1 \\ 1 & 0 \end{bmatrix} (x) \tag{7}$$

$$\frac{p_n}{q_n} = \begin{bmatrix} 0 & 1 \\ 1 & a_1 \end{bmatrix} \begin{bmatrix} 0 & 1 \\ 1 & a_2 \end{bmatrix} \cdots \begin{bmatrix} 0 & 1 \\ 1 & a_n \end{bmatrix} (0) \tag{8}$$

$$= \begin{bmatrix} 0 & 1 \\ 1 & a_1 \end{bmatrix} \begin{bmatrix} 0 & 1 \\ 1 & a_2 \end{bmatrix} \cdots \begin{bmatrix} 0 & 1 \\ 1 & a_n \end{bmatrix} \begin{bmatrix} 0 & 1 \\ 1 & a_{n+1} \end{bmatrix} (\infty).$$

Note that $\begin{bmatrix} -a_n & 1 \\ 1 & 0 \end{bmatrix} = \begin{bmatrix} 0 & 1 \\ 1 & a_n \end{bmatrix}^{-1}$. In this sense, we rewrite (2) to a sequence of matrices

$$x = \begin{bmatrix} 0 & 1 \\ 1 & a_1 \end{bmatrix} \begin{bmatrix} 0 & 1 \\ 1 & a_2 \end{bmatrix} \cdots \begin{bmatrix} 0 & 1 \\ 1 & a_n \end{bmatrix} \cdots .$$

§3 GEODESICS AND FORD CIRCLES

We extend the map f on I to a map on a subset of geodesics over the upper-half plane. For this, we identify a matrix $\begin{bmatrix} p & r \\ q & s \end{bmatrix}$ with the map $z \to \frac{pz+r}{qz+s}$ if $ps - qr > 0$ or $z \to \frac{\overline{pz+r}}{qz+s}$ if $ps - qr < 0$. We define a set X of geodesics (with Poincare metric) as follows:

$$\begin{cases} \text{the terminal point } \alpha \text{ is in } I = [0,1), \\ \text{the initial point } \beta \text{ is less than } -1 \text{ or equal to the point } \infty. \end{cases}$$

We write such a geodesic (α, β). Note that (α, ∞) is the half straight line perpendicular to the real axis at α and $(\alpha, \beta), \beta \neq \infty$, is the half circle perpendicular to the real axis at α and β.

For $(\alpha, \beta) \in X$, we consider its image by $\begin{bmatrix} -a & 1 \\ 1 & 0 \end{bmatrix}$ with $a = a(\alpha)$. It is easy to see that the image is also in X. Now we define a map $\overline{f}$ of X into itself by

$$\overline{f}(\alpha, \beta) \begin{cases} = \begin{bmatrix} -a & 1 \\ 1 & 0 \end{bmatrix} (\alpha, \beta) = \left(\begin{bmatrix} -a & 1 \\ 1 & 0 \end{bmatrix} (\alpha), \begin{bmatrix} -a & 1 \\ 1 & 0 \end{bmatrix} (\beta) \right), & \text{if } \alpha \neq 0, \\[3mm] = \begin{bmatrix} 1 & 0 \\ 0 & 1 \end{bmatrix} (\alpha, \beta) = (\alpha, \beta), & \text{if } \alpha = 0. \end{cases}$$

It turns out that $\overline{f}$ is one-to-one onto X except for rational α's. Moreover, we have the following two lemmas.

Lemma 1. *It follows that*

$$\overline{f}^n(x, \infty) = (f^n(x), -\frac{q_n}{q_{n-1}})$$

for any $x \in I$.

Lemma 2. *For any $x \in I$ and $n \geq 1$,*

$$q_n{}^2 \cdot \left| x - \frac{p_n}{q_n} \right| = \left(f^{n+1}(x) + \frac{q_{n+1}}{q_n} \right)^{-1} .$$

From these two lemmas, it turns out that the quantity $q_n{}^2 \cdot |x - p_n/q_n|$ is determined by the difference of two components of $\overline{f}^{n+1}(x, \infty)$. Thus the inequality (3) is equivalent to the statement "$|\alpha - \beta| > 1$ if $(\alpha, \beta) \in X$", which is trivial.

We define $d(\alpha, \beta) = |\alpha - \beta|$ for $(\alpha, \beta) \in X$. Now we rewrite Lemma 2 in a different way.

Definition. For a rational number p/q, we denote by $F_{1/2}(p/q)$ the circle with the center $p/q + i \cdot 1/2q^2$ and the radius $1/2q^2$, which is tangent to the real axis at p/q. We call this *the Ford circle associated to p/q*, (we always assume that p and q are relatively prime and $0=0/1$).

It is easy to see that $F_{1/2}(p/q)$ is the image of $\{z : \mathbf{Im}z = 1\}$, the Ford circle associated to the point ∞, by a linear fractional transformation $\begin{bmatrix} p & r \\ q & s \end{bmatrix}$, $|ps - rq| = 1$ and $p, q, r, s \in \mathbf{Z}$.

Proposition 3. $F_{1/2}(p/q)$ and $F_{1/2}(p'/q')$ are tangent to each other if and only if $|pq' - p'q| = 1$. Otherwise, they are disjoint.

Next, we define a rank of Ford circles. For any rational number $p/q \in (0, 1)$, there exists a natural number k such that

$$f^{k-1}\left(\frac{p}{q}\right) = \frac{1}{u}, \quad (u \text{ is a positive integer}).$$

We call this k the rank of $F_{1/2}(p/q)$. Here, f^{k-1} corresponds to Euclidean algorithm for p and q.

Lemma 4. A geodesic $(x, \infty)(\in X)$ intersects with $F_{1/2}(p/q)$ if and only if $q^2|x-p/q| < 1/2$.

Suppose that the rank of $F_{1/2}(p/q)$ is equal to k and the inequality (4) holds for an irrational x. Then we see that (x, ∞) intersects to $F_{1/2}(p/q)$ and so $(f^{k-1}(x), -\frac{q_{k-1}}{q_{k-2}})$ to $F_{1/2}(1/u)$. In this case, we have $a_k = u$ or $u - 1$. If $a_k = u$, then we have $p/q = p_k/q_k$. On the other hand, if $a_k = u - 1$, then we have $a_{k+1} = 1$ and $p/q = p_{k+1}/q_{k+1}$ (see Fig.1). This shows the property (iv). Moreover, two Ford circles associated to p/q and p'/q' are of the same rank k with $p/q = p_k/q_k$ and $p'/q' = p_{k+1}/q_{k+1}$, if and only if $q < q'$ and $|pq' - p'q| = 1$. In this case, the Ford circle associated to p_{k+2}/q_{k+2} is of rank $k + 1$.

Now we generalize the notion of Ford circle. For $c > 0$, we denote by $F_c(p/q)$ the image of $\{z : \mathbf{Im}z = c/2\}$ by $\begin{bmatrix} p & r \\ q & s \end{bmatrix}$, $|ps - rq| = 1$ and $p, q, r, s \in \mathbf{Z}$. This image is the circle with the center $p/q + i \cdot 1/cq^2$ and the radius $1/cq^2$, which is also tangent to the real axis at p/q. It is clear that $F_c(p/q)$ and $F_c(p'/q'), p/q \neq p'/q'$, are disjoint to each other whenever $c > 2$.

Lemma 4'. A geodesic $(x, \infty)(\in X)$ intersects with $F_c(p/q)$ if and only if $q^2|x-p/q| < 1/c$, (see Fig.2).

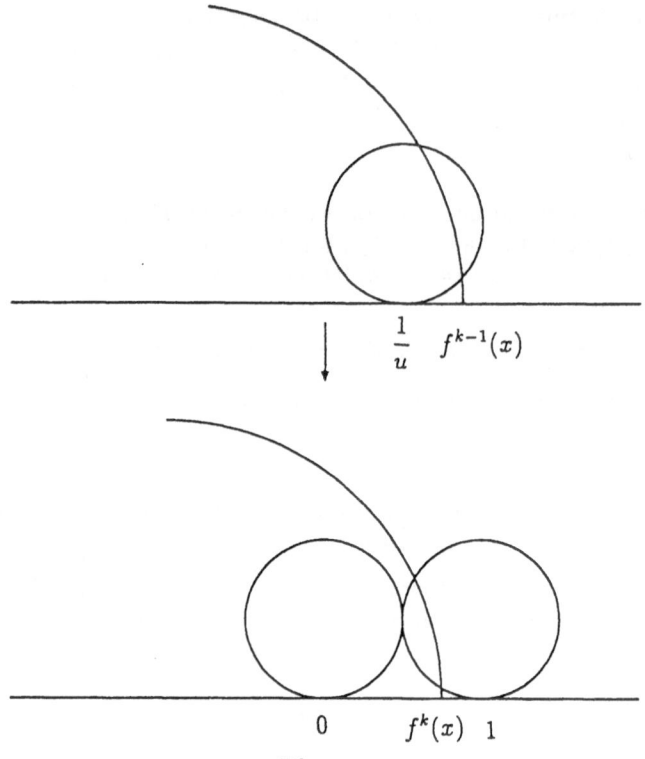

$$\downarrow \quad \frac{1}{u} \quad f^{k-1}(x)$$

$$0 \qquad f^k(x) \quad 1$$

Fig. 1

From (7) and (8), it turns out that

$$\infty = \begin{bmatrix} -a_{n+1} & 1 \\ 1 & 0 \end{bmatrix} \begin{bmatrix} -a_n & 1 \\ 1 & 0 \end{bmatrix} \cdots \begin{bmatrix} -a_1 & 1 \\ 1 & 0 \end{bmatrix} \begin{pmatrix} p_n \\ q_n \end{pmatrix}.$$

Then we see that $F_c(p_n/q_n)$ is mapped to $\{z : \text{Im} z = c/2\}$ and the geodesic (x, ∞) to $(f^{n+1}(x), -q_{n+1}/q_n)$ by the same linear fractional transformation. Consequently, (x, ∞) and $F_c(p_n/q_n)$ are tangent to each other if and only if the radius of the half-circle $(f^{n+1}(x), -q_{n+1}/q_n)$ is $c/2$, that is,

$$\frac{f^{n+1}(x) - (-q_{n+1}/q_n)}{2} = \frac{c}{2}.$$

Thus, we get Lemma 2 again. On the other hand, it is easy to see that

$$\lim_{n \to \infty} \{d(\overline{f}^n(x, w)) - d(\overline{f}^n(x, \infty))\} = 0 \tag{9}$$

In particular, if we take

$$x = \frac{\sqrt{5} - 1}{2} = \frac{1|}{|1} + \frac{1|}{|1} + \frac{1|}{|1} + \cdots$$

and

$$w = -\frac{\sqrt{5} + 1}{2},$$

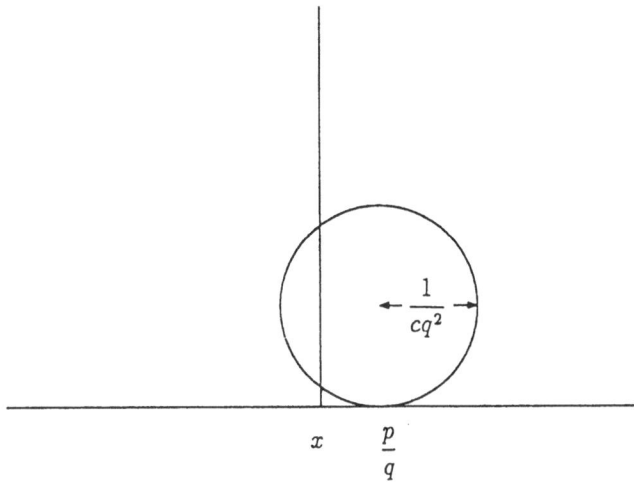

Fig. 2

then $\overline{f}(x,w) = (x,w)$ and $d(x,w) = \sqrt{5}$. This shows the later half of (v). Suppose there exists positive integer k such that $a_{k+l} = 1$ for $l \geq 1$ with

$$x = \frac{1|}{|a_1} + \frac{1|}{|a_2} + \frac{1|}{|a_3} + \cdots,$$

then the situation is the same with above. Now we suppose that there are infinitely many n such that $a_n \geq 2$. For such n's, we have

$$d(\overline{f}^n(x,\infty)) \geq \sqrt{5} \text{ or } d(\overline{f}^{n+1}(x,\infty)) \geq \sqrt{5}$$

and get the assertion (v).

Finally, we note that the expansion of x is periodic if and only if $(x,\overline{x}) \in X$, where $\overline{x}$ denotes the conjugate root of x. This also follows form (9).

§4 METRICAL PROPERTY OF CONTINUED FRACTIONS

It is easy to see that $\overline{f}$ (on X) preseves the hyperbolic measure, that is,

$$d\mu(\alpha,\beta) = \frac{1}{\log 2} \frac{d\alpha d\beta}{|\alpha - \beta|^2}, \quad (\alpha,\beta) \in X,$$

is an absolutely continuous invariant probability measure for $\overline{f}$. It is clear that the marginal distribution to the first coordinate is an invariant measure for f, which is called "Gauss measure". The transformation $\overline{f}$ on X is one-to-one and onto (μ-a.e.) and, in this sense, is a natural extension of f.

Proposition 5. $(\overline{f}, \mu)$ *is ergodic.*

Remark. It is well-known that f (with Gauss measure) is weak Bernoulli and so $(\overline{f}, \mu)$ is, too. But, here, we only need the ergodicity of $\overline{f}$.

From Lemma 1,2, the individual ergodic theorem and (9), we have the following.

Proposition 6([4]). *For almost all x,*

$$\lim_{N \to \infty} \frac{1}{N} \, \# \left\{ n : 1 \le n \le N, q_n{}^2 \left| x - \frac{p_n}{q_n} \right| < \frac{1}{c} \right\} = \frac{1}{\log 2} \frac{1}{c}$$

for any c ≥ 2.

On the other hand, from Lemma 4' and the ergodicity of the geodesic flow over the modular surface, we have the following.

Proposition 7. *For almost all x,*

$$\lim_{N \to \infty} \frac{1}{\log N} \, \# \left\{ \frac{p}{q} : q^2 \left| x - \frac{p}{q} \right| < \frac{1}{c}, 1 \le q \le N, (p, q) = 1 \right\} = \frac{12}{\pi^2} \frac{1}{c} \quad \cdot$$

for any c > 0.

From Proposition 7, with (iv), we have

$$\lim_{N \to \infty} \frac{1}{\log q_n} \, \# \left\{ n : 1 \le n \le N, q_n{}^2 \left| x - \frac{p_n}{q_n} \right| < \frac{1}{c} \right\} = \frac{12}{\pi^2} \frac{1}{c}$$

for any c ≥ 2 (a.e.). Consequently, from Proposition 6, we get the following.

Proposition 8. *For almost all x,*

$$\lim_{n \to \infty} \frac{1}{n} \log q_n = \frac{1}{\log 2} \frac{\pi^2}{12}.$$

Remark. Usually, we calculate this value by the ergodicity of f and the integration

$$\int_0^1 \frac{\log x}{1 + x} dx,$$

(see [3]).

§5 ROSEN'S CONTINUED FRACTIONS

In the sequel, we discuss some property of the nearest integer type continued fractions associated to Hecke groups.

Let $G_k, k \ge 3$, be the group generated by

$$S = \begin{bmatrix} 1 & \lambda_k \\ 0 & 1 \end{bmatrix} \quad \text{and} \quad T = \begin{bmatrix} 0 & -1 \\ 1 & 0 \end{bmatrix}, \quad \text{where } \lambda_k = 2 \cos \pi/k.$$

It is clear that G_3 is the modular group $SL(2, \mathbf{Z})$. G_k is a zonal Fuchsian group of the first kind with the width λ_k. The nearest integer type continued fraction expansion was introduced by Rosen [12]. Here, we define this expansion by a map defined on $[-\lambda_k/2, \lambda_k/2]$. As before, we regard S and T also linear fractional transformations. The continued fractions, defined here, is not exactly the same with Rosen's original. An advantage of our definition is that the equivalence relation arising from our continued fraction coincides with the G_k- equivalence relation, see Proposition 9 (iii). For a real number x, we define

$$[x]_k = m \quad (m \text{ is an integer}),$$

if $m \cdot \lambda_k - \lambda_k/2 < x \leq m \cdot \lambda_k + \lambda_k/2$. We define a map f_k on $I_k = [-\lambda_k/2, \lambda_k/2]$ by

$$f_k(x) = \begin{cases} -\frac{1}{x} - [\frac{1}{x}]_k & \text{if } x \in I_k, x \neq 0, \\ 0 & \text{if } x = 0. \end{cases}$$

We put

$$a_n(x) = a_{k,n}(x) = \left[\frac{-1}{f_k^{n-1}(x)} \right]_k$$

if $f_k^{n-1}(x) \neq 0$. It is easy to see that

$$f_k(x) = S^{-a_1(x)} T(x).$$

Similar to the case of simple continued fractions, we have

$$\begin{aligned} x &= -\frac{1\,|}{|a_1\lambda_k} - \frac{1\,|}{|a_2\lambda_k} - \frac{1\,|}{|a_3\lambda_k} - \cdots \\ &= T S^{a_1} T S^{a_2} T S^{a_3} \cdots. \end{aligned}$$

We also define the n-th convergent of x by

$$\begin{aligned} \frac{p_n}{q_n} &= T S^{a_1(x)} T S^{a_2(x)} \cdots T S^{a_n(x)}(0) \\ &= T S^{a_1(x)} T S^{a_2(x)} \cdots T S^{a_n(x)} T S^{a_{n+1}(x)}(\infty) \end{aligned}$$

If we regard T and S as matrices, then we can define p_n and q_n by

$$\begin{bmatrix} p_{n-1} & p_n \\ q_{n-1} & q_n \end{bmatrix} = T S^{a_1(x)} T S^{a_2(x)} \cdots T S^{a_n(x)}$$

up to $\pm$. We always assume that $q_n > 0$.
  We have the following.

Proposition 9
(i) The expansion of x terminates at finitely many terms if and only if x is a parabolic point of G_k.
(ii) The expansion of x is periodic after a certain number of steps if and only if x is a hyperbolic point.
(iii) For two numbers x and $y \in I_k$, there exists $V \in G_k$ with $y = Vx$ if and only if there exist natural numbers n and m such that $a_{n+l}(x) = a_{m+l}(y)$ for all $l \geq 1$.

  To find a condition on a hyperbolic point x having a periodic expansion, we need more detail discussion. From now on, we assume that k is even, $(k = 2l = 4u$ or $= 2(2u+1))$. In this case, we can apply the same argument stated above in §2-4.

§6 GENERALIZED FORD CIRCLES

For a parabolic point $p/q(= g(\infty)$ for some $g = \begin{bmatrix} p & * \\ q & * \end{bmatrix} \in G_k)$, we denote by $F_{k,c}(p/q)$ the image of $\{z : \mathrm{Im}\, z = c/2\}, c > 0$, by g. In particular, we call it the (generalized) Ford circle associated to p/q, when $c = 2$. The next lemma corresponds to Lemma 4'.

Lemma 10. *A geodesic (x, ∞) intersects with $F_{k,c}(p/q)$ if and only if $q^2|x - p/q| < 1/c$.*

Thus, if we can choose a good set of geodesics, similar to §3, then we get the similar results with §3 and 4.

For $\lambda_k = 2 \cos \pi/2l, (k = 2l)$, we put

$$\alpha_n = \frac{\cos \frac{n+1}{2l}\pi}{\cos \frac{n}{2l}\pi}$$

$$\beta_n = \frac{\cos \frac{n+1}{2l}\pi - \cos \frac{n+2}{2l}\pi}{\cos \frac{n}{2l}\pi - \cos \frac{n+1}{2l}\pi}$$

for $0 \leq n \leq l - 1$. Now we define X_k by the following. We denote by α the terminal point of a geodesic and by β its initial point: X_k is the set of geodesics with $\alpha \in I_k$ and

$$\begin{cases} \alpha \in [\alpha_{n+1}, \alpha_n) \Longrightarrow \beta \in [\beta_n, +\infty) \text{ or } \in [-\infty, -1) \\ \alpha \in [-\alpha_n, -\alpha_{n+1}) \Longrightarrow \beta \in [1, +\infty) \text{ or } \in [-\infty, -\beta_n) \end{cases} \tag{10}$$

We identify X_k with the set of pairs of real numbers (and ∞) (α, β) satisfying (10), (see Fig.3). Next we define a map $\overline{f}_k$ on X_k by

$$\overline{f}_k(\alpha, \beta) = (S^{-a}T(x), S^{-a}T(y))$$

when $[1/x]_k = a$.

We note that the expansion of a hyperbolic fixed point $x \in I_k$ is periodic if and only if $(x, \overline{x}) \in X_k$, where $\overline{x}$ is the conjugate hyperbolic point of x.

Proposition 11. *$\overline{f}_k$ is one-to-one and onto (μ-a.e.). Thus we see that $\overline{f}_k$ preserves the hyperbolic measure*

$$d\mu(\alpha, \beta) = \frac{d\alpha d\beta}{|\alpha - \beta|^2}, \quad \text{for } (\alpha, \beta) \in X_k,$$

and the marginal distribution of μ is an invariant measure for f_k.

By Lemma 10, it turn out that $\min\{|\alpha - \beta| : (\alpha, \beta) \in X_k\}$ bounds the values $q_n^2 \cdot |x - p_n/q_n|$. The following hold.

Proposition 12. *For any $n \geq 1$,*

$$\left| x - \frac{p_n}{q_n} \right| < \frac{1}{\kappa_0 \cdot q_n^2} \quad \text{with } \kappa_0 = \beta_u - \alpha_u.$$

Here κ_0 is the best possible constant.

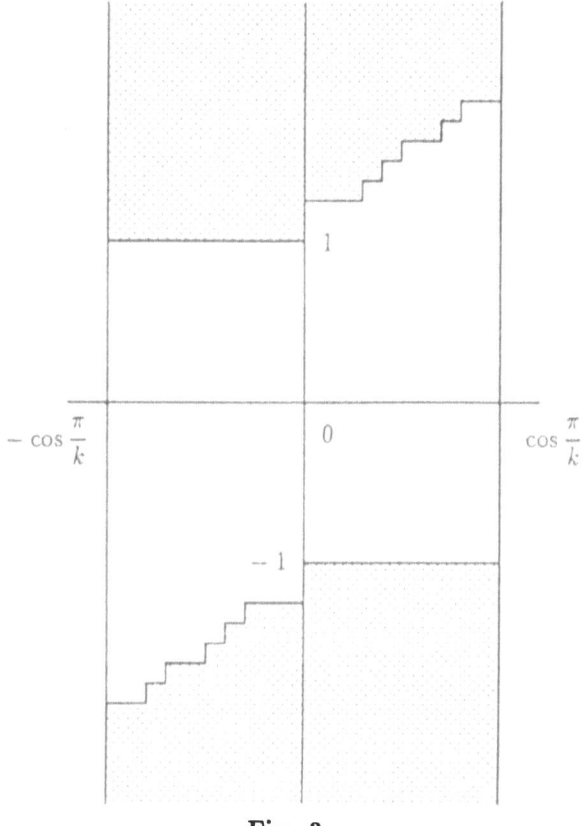

$$-\cos\frac{\pi}{k} \qquad 0 \qquad \cos\frac{\pi}{k}$$

Fig. 3

Proposition 13. *Suppose*

$$|x - g(\infty)| < \frac{1}{\kappa_1 q^2}$$

with

$$g = \begin{bmatrix} p & * \\ q & * \end{bmatrix} \in G_k,$$

then $g(\infty) = \frac{p_n}{q_n}$ for some positive integer n. Here,

$$\kappa_1 = 1 + \frac{1}{\cos\frac{\pi}{2l}}$$

and this is the best possible constant.

§7 METRICAL THEORY OF ROSEN'S CONTINUED FRACTIONS

It is easy to see that f_k is a Markov map (see Bowen and Series [5]) and is ergodic with respect to the invariant measure induced from the hyperbolic measure. Thus we can apply the same method stated in §4.

Proposition 14. *For almost all $x \in I_k$,*

$$\lim_{N \to \infty} \frac{1}{N} \, \# \left\{ n : 1 \leq n \leq N, q_n{}^2 \left| x - \frac{p_n}{q_n} \right| < \frac{1}{c} \right\} = \frac{1}{L_k} \frac{1}{c}$$

for any $c \geq \kappa_1$, where

$$L_k = 2 \left[\log(1 + \cos \frac{\pi}{2l}) - \log \sin \frac{\pi}{2l} \right].$$

On the other hand, we have the following by the ergodicity of the geodesic flow over H/G_k.

Proposition 15. *For almost all $x \in I_k$,*

$$\lim_{N \to \infty} \frac{1}{\log N} \, \# \left\{ g(\infty) : \left| x - \frac{p}{q} \right| < \frac{1}{cq^2}, 1 \leq |q| \leq N, g = \begin{bmatrix} p & * \\ q & * \end{bmatrix} \right\} = \frac{8l \cos(\pi/2l)}{c(l-1)\pi^2}.$$

Finally, from Proposition 14 and 15, we have the following.

Proposition 16. *For almost all $x \in I_k$,*

$$\lim_{N \to \infty} \frac{1}{N} \log q_N = \gamma,$$

where

$$\gamma = \frac{(l-1)\pi^2}{4l[\log(1 + \cos \pi/2l) - \log \sin \pi/2l]}.$$

Remark. Recently, T. Schmidt[14] got the similar result in the case of $l = 2$. His result is based on Rosen's original continued fractions. It is easy to see that our continued fractions are essentially the same with Rosen's original.

References

1. R. L. Adler and L. Flatto, *Cross section maps for geodesic flows I (the modular surface)*, "Ergodic Theory and Dynamical Systems, Vol. 2", Proceed. Special Year Md. (1979-1980), Progress in math., Birkhauser, Boston, Basel and Stuttgart, pp. 103-161.
2. R. L. Adler and L. Flatto, *Geodesic flows, interval maps and symbolic dynamics*, Bull. Amer. Math. Soc., **25**, (1991), 229-334.
3. P. Billingsley, *Ergodic Theory and Information*, John Wiley & Sons, Inc., NY, (1965).
4. W. Bosma, H. Jager and F. Wiedijk, *Some metrical observations on the approximation by continued fractions*, Indag. Math., **45**, (1983), 281-299.
5. R. Bowen and C. Series, *Markov maps associated with Fuchsian groups*, IHES, Publ. Math., **50**, (1979), 153-170.
6. J. Lehner, *A Diophantine property of the Fuchsian groups*, Pacific J. Math., **2**, (1952), 327-333.

7. J. Lehner, *Diophantine approximation on Hecke groups*, Glasgow Math. J., **27**, (1985), 117-127.

8. R. Moeckel, *Geodesics on modular surfaces and continued fractions*, Ergod. Th. and Dyn. Sys., **2**, (1982), 69-83.

9. H. Nakada, *On ergodic theory of A. Schmidt's complex continued fractions over Gaussian field*, Monatsh. Math., **105**, (1988), 131-150.

10. H. Nakada, *On metrical theory of diophantine approximation over imaginary quadratic field*, Acta Arithmetica, **51**, (1988), 392-403.

11. H. Nakada, *The metrical theory of complex continued fractions*, Acta Arithmetica, **56**, (1991), 279-289.

12. D. Rosen, *A class of continued fractions associated with certain properly discontinuous groups*, Duke Math. J., **21**, (1954), 549-563.

13. A. L. Schmidt, *Diophantine approximation of complex numbers*, Acta Math., **134**, (1975), 1-85.

14. T. A. Schmidt, *Remarks on the Rosen λ-continued fractions*, "Number theory with an emphasis on the Markoff spectrum" (A. D. Pollington and W. Moran, eds), Lecture Note in pure and applied Mathematics vol.147, Dekker, 1993, pp.227-238.

A SHORT PROOF OF EVEN α-EQUIVALENCE

Kyewon Koh Park

Department of Mathematics
Ajou University
Suwon, KOREA 441-749

Abstract. α-equivalence of two discrete dynamical systems is defined via continuous actions and vice versa. We present a short proof of the equivalence, bypassing the whole machinery of the restricted orbit equivalence.

§1 Introduction

Ambrose has shown that any measure preserving ergodic R-action $(\Omega, \mathcal{L}, \lambda, S^t)$ can be represented under a function with a measure preserving discrete Z-action on a cross section $(X, \mathcal{F}, \mu, T)$. The function in a representation is called a ceiling function and the cross section is called a base [Am]. Since then, relations between discrete actions and continuous action have been explored in several directions. (See [ORW],[Pa1] and [Sh.]) Discrete actions and continuous actions each have properties which have been studied via the other. One of these properties is Kakutani equivalence: We say two discrete actions $(X_1, \mathcal{F}_1, \mu_1, T_1)$ and $(X_2, \mathcal{F}_2, \mu_2, T_2)$ are equivalent if and only if a flow built with $(X_1, \mathcal{F}_1, \mu_1, T_1)$ as a base can be rescaled so that it can be built with $(X_2, \mathcal{F}_2, \mu_2, T_2)$ as its base [Ka]. This is to say that each of $(X_1, \mathcal{F}_1, \mu_1, T_1)$ and $(X_2, \mathcal{F}_2, \mu_2, T_2)$ is isomorphic to a cross section of the same flow. Likewise two flows $(\Omega_1, \mathcal{L}_1, \lambda_1, S_1^t)$ and $(\Omega_2, \mathcal{L}_2, \lambda_2, S_2^t)$ are equivalent if and only if there exists a discrete action $(X, \mathcal{F}, \mu, T)$ such that each of the flows can be represented with a base isomorphic to $(X, \mathcal{F}, \mu, T)$.

Ornstein and Weiss defined the even Kakutani equivalence. We say two descrete actions are even Kakutani equivalent if and only if any flow built with one discrete action can be built with the other discrete action. This is clearly a more restrictive equivalence relation because time rescaling is not allowed. Even Kakutani equivalence preserve the entropy, unlike Kakutani equivalence. A.delJunco [dJR] characterized the even Kakutani equivalence in terms of orbit equivalence. D.Rudolph has investigated these equivalence relations further and established a theory of restricted orbit equivalence [Ru2]. It covers a general class of equivalence relation from Dye's orbit equivalence to isomorphism. The notion of α-equivalence relation to be discussed shortly is a restricted orbit equivalence in the sense of [Ru2].

We denote the flow built under a function f with a base $(X, \mathcal{F}, \mu, T)$ by $[X, \mathcal{F}, \mu, T, f]$. We say two discrete actions of probability spaces are even α-related if and only if there exists a flow $(\Omega, \mathcal{L}, \lambda, S^t)$ which can be represented under a function of values 1 and $1 + \alpha$ with each of these as a base. (We do not put a restriction on the measure of Ω.) This can be shown to be an equivalence relation via the whole machinery of [Ru2]. It is called an even α-equivalence relation. We present here an elementary proof of the even α-equivalence. It is also known that this is a stronger equivalence relation than even Kakutani equivalence. That is, two actions can be even Kakutani equivalent, but not even α-equivalent (See [dJFR] for detail). We can also define even α-equivalence of continuous actions. We say two flows are α-equivalent if and only if both of them can be represented under 1 and $1 + \alpha$ with a same base $(X, \mathcal{F}, \mu, T)$. If we denote the ceiling functions by f and g respectively, then we require $\int f d\mu = \int g d\mu$. It is also shown that even α-equivalence of flow is stronger than even Kakutani equivalence [Pa3].

We say that the n-long P-name of x satisfies the ergodic theorem within ϵ with respect to a partition $P = \{P_1, \cdots, P_k\}$ if

$$\left| \frac{1}{n} \sum_{i=0}^{n-1} \chi_{P_j}(T^i x) - \mu(P_j) \right| < \epsilon$$

for every atom P_j of the partition. If the partition P and n are clear in the context, then we use the term "name of x" instead "n-long P-name of x". An analogous definition holds for a continuous action. We denote a point in $[X, \mathcal{F}, \mu, T, f]$ by (x, t) where $x \in X$ and $0 \le t < f(x)$. If a ceiling function takes finitely many values, it gives rise to a natural partition on the base, according to the values of the function. For the simplicity of our argument, we assume $\alpha < 1$.

§2 Even α-equivalence

Throughout the section, we assume that ceiling function takes values 1 and $1 + \alpha$.

Our proof is based simply on the skyscrapers of bases and flows. One advantage of the proof, in addition to bypassing the theory of restricted orbit equivalence, is that the same idea works for the proofs of both discrete actions and continuous actions.

Theorem 2.1. *Even α-relation is an equivalence relation.*

Proof. Let $(X_1, \mathcal{F}_1, \mu_1, T_1)$ and $(X_2, \mathcal{F}_2, \mu_2, T_2)$ be α-related and $(X_2, \mathcal{F}_2, \mu_2, T_2)$ and $(X_3, \mathcal{F}_3, \mu_3, T_3)$ be α-related. We denote $(\Omega_1, \mathcal{L}_1, \lambda_1, S_1^t)$ the flow which is built with $(X_1, \mathcal{F}_1, \mu_1, T_1)$ and $(X_2, \mathcal{F}_2, \mu_2, T_2)$ as its bases. There exist f and f' of values 1 and $1 + \alpha$ such that $[X_1, \mathcal{F}_1, \mu_1, T_1, f]$ and $[X_2, \mathcal{F}_2, \mu_2, T_2, f']$ are isomorphic to $(\Omega_1, \mathcal{L}_1, \lambda_1, S_1^t)$. Let $(\Omega_2, \mathcal{L}_2, \lambda_2, S_2^t)$ denote a flow which has a representation with $(X_2, \mathcal{F}_2, \mu_2, T_2)$ and $(X_3, \mathcal{F}_3, \mu_3, T_3)$ as its bases. We denote by g and g' the respective ceiling functions in the representations of $(\Omega_2, \mathcal{L}_2, \lambda_2, S_2^t)$.

It is sufficient to show that the flow $(\Omega_2, \mathcal{L}_2, \lambda_2, S_2^t)$ can be built with $(X_1, \mathcal{F}_1, \mu_1, T_1)$ as its base. In fact, what we show is that any flow built under 1 and $1 + \alpha$ with the base $(X_2, \mathcal{F}_2, \mu_2, T_2)$ can be built under a function of 1 and $1 + \alpha$ with the base $(X_1, \mathcal{F}_1, \mu_1, T_1)$.

We denote by $P^1 = \{P_0^1, P_1^1\}$ the partition of X_1 according to the values of f. That is,

$$P_0^1 = \{x \in X_1 : f(x) = 1 + \alpha\}$$

and

$$P_1^1 = \{x \in X_1 : f(x) = 1\}.$$

We denote by $Q^1 = \{Q_0^1, Q_1^1\}$ and $Q^2 = \{Q_0^2, Q_1^2\}$ the partitions of $(X_2, \mathcal{F}_2, \mu_2, T_2)$ according to the values f' and g respectively. We note that $\lambda_1(P_1^1) = \lambda_2(Q_1^1)$. We let $P = P^1 \vee T_1^{-1} P^1$ and $Q' = Q^1 \vee Q^2 = \{Q_1, Q_2, Q_3, Q_4\}$. Let φ denote an isomorphism between $[X_1, \mathcal{F}_1, \mu_1, T_1, f]$ and $[X_2, \mathcal{F}_2, \mu_2, T_2, f']$. Without confusion, we denote the set $\varphi(X_1)$ sitting in the flow $[X_2, \mathcal{F}_2, \mu_2, T_2, f']$ by X_1. By our assumption that α is less than 1, there are at most two t's, say t_1 and t_2, such that $(x, t_1) \in X_1$ and $(x, t_2) \in X_1$ for each $x \in X_2$. We refine the partition Q' to $Q = \{Q_1, \cdots, Q_k\}$ such that each atom of the partition satisfies one of the following:

(i.1) None of the points in Q_j has a point above such that $(x, t) \in X_1$,

(i.2) If there exists t_1 such that $(x, t_1) \in X_1$, then $\{(x, t_1) : x \in Q_j\}$ is contained in one of the atoms of the partition P.

We choose $\epsilon \ll \min\{\mu_2(Q_1), \cdots, \mu_2(Q_k)\}$. Construct a skyscraper with base $(X_2, \mathcal{F}_2, \mu_2, T_2)$. Let B denote the bottom level set. For each $x \in B$, we denote by $q(x)$ the height of the skyscraper. We divide the skyscraper into columns so that each column has a unique Q-name. We construct a skyscraper high enough so that $q(x)$-long names of the points in B satisfy the ergodic theorem within ϵ with respect to the partition Q. Let A_i denote the ith column and $B_i \subset B$ denote the bottom level set of A_i,

We also build the flow skyscraper with $[X_2, \mathcal{F}_2, \mu_2, T_2, f']$, using the base skyscraper. Each column of the base gives to a flow column. Denote the flow column by C_i corresponding to the base column A_i. We call X_1 which is embedded in each flow column by the level sets of X_1. We define $t_x = \min\{t : (x, t) \in X_1\}$ for each $x \in B$. By the definition of the partition Q, we note that for each column C_i the set $\{(x, t_x) \in X_1 : x \in B_i\}$ has unique P-name along the level sets of X_1 up to the top of the column. For $x \in B_i$, let $r(x)$ denote the number of the level sets of X_1 in the column. That is, $r(x)$ is the largest integer such that $\{T_1^j(x, t_x)\}_{j=0}^{r(x)-1}$ is contained in the flow column C_i. We may assume that B is chosen so that $r(x)$-long P-names of the set $\{(x, t_x) : x \in B\}$ satisfy the ergodic theorem within ϵ with respect to the partition P. For each column C_i, let

$$u_i = \sum_{i=0}^{q(x)-1} f'(T_2^i(x)).$$

and

$$v_i = \sum_{i=0}^{r(x)-1} f(T_1^i(x, t_x)).$$

(u_i is the height of the flow column C_i.)

Nest to this skyscraper, we build the skyscraper with $[X_2, \mathcal{F}_2, \mu_2, T_2, g]$ using the base skyscraper with $(X_2, \mathcal{F}_2, \mu_2, T_2)$. For each flow column D_i with the base B_i let

$$w_i = \sum_{i=0}^{q(x)-1} g(T_2^i(x)).$$

We make the following observations:

(a) u_i, v_i and w_i are linear sums of 1 and $1 + \alpha$.

(b) $|u_i - w_i|$ is a multiple of α.

(c) $|u_i - w_i|$ is a within $2\epsilon q_i \alpha \mu_2(Q_1^1 \triangle Q_1^2) < 2\epsilon q_i \alpha$, where q_i denotes the height of the ith column B_i.

(d) $|q_i - r_i| < 2\epsilon q_i$, where r_i is the number of level sets of X_1 contained in C_i.

(e) If $u_i = s_i$ for all i, then the two flows are clearly isomorphic.

Suppose $u_i > w_i$. Let $u_i - w_i = m_i < 2\epsilon q_i \alpha$. We call length from a next level set of X_1 (or X_2) to the next level set of X_i (or X_2) directly above the height of the level set. The height of a level set except for the top level set of X_1 in the first skyscraper is either 1 or $1 + \alpha$. We note that we can easily change the column C_i to the column D_i by cutting off the top flow block of length $u_i - w_i$. The issue here is that we should change the flow column so that we do not cut off the level sets X_1 and keep the heights of level sets to remain 1 or $1 + \alpha$.

For each column C_i let L_1^i denote the first level set of X_1 contained in P_0^1. We push down all the level sets of X_1 above L_1^i by α. The height of L_1^i has been changed from $1 + \alpha$ to 1. We also shift down the level sets of X_2 by α. That is, let M_1^i denote the first level set of X_2 contained in Q_0^1. We push down all the level sets of X_2 above M_1^i by α. We successively push down the level sets of X_1 so that the first m_i level sets of height $1 + \alpha$ now have height 1. When we push down level sets of X_1, we do the same for the level sets of X_2 as above. After we shift down all the level sets above L_j^i successively for $j = 1, 2, \cdots, m_i$, we have a flow block of height $m_i \alpha$ from the top which does not contain any level sets of X_1 or X_2. We cut out this flow block of length $m_i \alpha$ from the top.

The ith column of the base skyscraper of X_2 has at least $(1 - \epsilon)q_i \mu_2(Q_0^1)$ many sets of height $1 + \alpha$. Hence by our choice of ϵ, in each column we have enough level sets whose heights are $1 + \alpha$ so that their heights are changed to 1. Also we note that in the ith flow column, there are at least

$$(1 - \epsilon)\frac{q_i}{1 + \alpha}\mu_1(P_0^1) > \frac{1}{2}(1 - \epsilon)q_i \mu_2(Q_0^1)$$

many level sets of X_1 whose heights are $1 + \alpha$.

If $u_i < w_i$, then instead of cutting out a flow block from the top, we lengthen the flow column from u_i to w_i as follows. Let $w_i - u_i = m_i \alpha$. We put a flow block of length $m_i \alpha$ on top of the flow column C_i. Now the top level set has height bigger then $1 + \alpha$. Let L_1^i denote the first level set of X_1 of height 1 and shift up all the level sets above L_1^i by α. Now L_1^i has the height $1 + \alpha$. We repeat this for the first m_i level sets of X_1 of height 1. These level sets now have heights $1 + \alpha$. Let M_1^i denote the first level set of X_2 of height 1. Shift up all the level sets of X_2 above M_1^i by α so that M_1^i has height $1 + \alpha$. Each time we shift up the level sets of X_1, we do the same for the level sets of X_2 as above.

The flow column C_i of height u_i has been changed to a new flow column C_i' of height w_i. We repeat this for each column of the flow $[X_2, \mathcal{F}_2, \mu_2, T_2, f]$, comparing it with the corresponding flow column of $[X_2, \mathcal{F}_2, \mu_2, T_2, g]$. Let X_1' be the union of the new level sets shifted up or down from the level sets of X_1, depending on $u_i < w_i$ or $u_i > w_i$. We define X_2' analogously. We note that the number of level sets of X_1' in C_i' is r_i. On these new level sets except for the top one, we define $T_1'(x')$ to be the point directly above on the next new level set. On the new top level set, we define $T_1'(x') = (y, t_y)$ if $T_2(x) = y$ where x is the point directly above or below x' on the top level set of X_1. We note that the bottom level sets of X_1 and X_2 have not been changed. Define T_2' on X_2' in a similar way as T_1' on X_1'. The continuous action on the new flow skyscraper is well defined to be consistent with these new discrete actions.

It is clear that all the level sets have heights 1 or $1 + \alpha$. It is also clear that the level sets except the top of X_1' have height 1 or $1 + \alpha$. Since the length from the top level set of X_1' to the top of the flow column C_i' remains the same as the length from the top level set of X_1 to the top of the flow column C_i and we have not changed the first level set, the return time from the top level set of X_1' to X_1' remains 1 or $1 + \alpha$, depending on the return time from the top level set of X_1 to X_1.

Now the following observations are clear:

(a) The flow on the new flow skyscraper is isomorphic to
$$[X_2, \mathcal{F}_2, \mu_2, T_2, g] = [X_2, \mathcal{F}_2, \mu_2, T_2, g'].$$

(b) The new flow is built under 1 and $1 + \alpha$ with $(X_1', \mathcal{F}_1', \mu_1', T_1')$ where $\mathcal{F}_1'$ and μ_1' denote the obvious σ-algebra and the obvious measure on X_1'.

(c) $(X_1', \mathcal{F}_1', \mu_1', T_1')$ is isomorphic to $(X_1, \mathcal{F}_1, \mu_1, T_1)$

Remark. We can build the flow skyscraper long enough so that $u_i > w_i(u_i < w_i)$ for all i if $\mu_2(Q_0^1) > \mu_2(Q_0^2)$ $(\mu_2(Q_0^1) > \mu_2(Q_0^2))$.

(II) Continuous case.

Since the idea of the proof is same as for a discrete case, we will be brief. Let $(\Omega_1, \mathcal{L}_1, \lambda_1, S_1^t)$ and $(\Omega_2, \mathcal{L}_2, \lambda_2, S_2^t)$ be α-related and $(\Omega_2, \mathcal{L}_2, \lambda_2, S_2^t)$ and $(\Omega_3, \mathcal{L}_3, \lambda_3, S_3^t)$ be α-related. We have discrete actions $(X_1, \mathcal{F}_1, \mu_1, T_1)$ and $(X_2, \mathcal{F}_2, \mu_2, T_2)$ satisfying the following:

(1) There exists a ceiling funciton f such that
$(\Omega_1, \mathcal{L}_1, \lambda_1, S_1^t)$ is isomorphic to $[X_1, \mathcal{F}_1, \mu_1, T_1, f]$.

(2) There exists f' and g such that $(\Omega_2, \mathcal{L}_2, \lambda_2, S_2^t)$ is isomorphic to $[X_2, \mathcal{F}_2, \mu_2, T_2, f']$ and $[X_2, \mathcal{F}_2, \mu_2, T_2, g]$.

(3) There exists g' such that $(\Omega_3, \mathcal{L}_3, \lambda_3, S_3^t)$ is isomorphic to $[X_2, \mathcal{F}_2, \mu_2, T_2, g]$

(Although we do not put any restriction on the measure of X_1 or X_2, $\mu_2(X_1)$ and $\mu_2(X_2)$ are between 1 and $\frac{1}{1+\alpha}$ because $\lambda_1(\Omega_1) = \lambda_2(\Omega_2) = \lambda_3(\Omega_3) = 1$.)

What we need to show is that $(\Omega_3, \mathcal{L}_3, \lambda_3, S_3^t)$ can be build under 1 and $1 + \alpha$ with $(X_1, \mathcal{F}_1, \mu_1, T_1)$ as its base. As in a discrete case, we denote by $Q' = Q^1 \vee Q^2$ the partition of X_2 according to the values of g and g'. Let P' denote the partition of X_1 according to the values of f and f'. Let $P = P' \vee T_1^{-1} P'$. Without confusion we write the isomorphic image of the set $(X_1, \mathcal{F}_1, \mu_1, T_1)$ in $[X_2, \mathcal{F}_2, \mu_2, T_2, g]$ by $(X_1, \mathcal{F}_1, \mu_1, T_1)$.

We refine the partition Q' into $Q = \{Q_1, \cdots, Q_k\}$ such that each atom of the partition satisfies (i.1) and (i.2).

We choose $\epsilon \ll \min\{\mu_2(Q_1), \mu_2(Q_2), \cdots, \mu_2(Q_k)\}$. We build a skyscraper of $(X_2, \mathcal{F}_2, \mu_2, T_2)$ and let q_i denote the height of the ith column of the skyscraper. We build the skyscraper long enough so that the bottom level set of each column satisfies the ergodic theorem within ϵ with respect to the partition Q. Construct two flow skyscraper of $(\Omega_2, S_2^t, \mathcal{L}_2, \mu_2)$ and $(\Omega_3, \mathcal{L}_3, \lambda_3, S_3^t)$ next each other using the base skyscraper. Let r_i denote the height of the skyscraper of $(X_1, \mathcal{F}_1, \mu_1, T_1)$ contained in the flow column C_i of $(\Omega_2, S_2^t, \mathcal{L}_2, \mu_2)$. The bottom level set of $(X_1, \mathcal{F}_1, \mu_1, T_1)$ in C_i has a unique r_i-long P-name. We may assume that P-name of each column also satisfies the ergodic theorem within ϵ. We note that $\frac{1}{2} < \frac{1}{1+\alpha} < \frac{r_i}{q_i} < 1 + \alpha < 2$. Let u_i and w_i denote the height of the ith flow column of $(\Omega_2, \mathcal{L}_2, \lambda_2, S_2^t)$ and $(\Omega_3, \mathcal{L}_3, \lambda_3, S_3^t)$ respectively.

Suppose $u_i > w_i$. Let $u_i - w_i = m_i \alpha$. We shift down all the level sets above the first level set L_1^i of X_1 of height $1 + \alpha$ by α. The level set L_1^i has a flow height 1. We repeat this for the first m_i level sets of X_1 of height $1 + \alpha$. We cut the flow column of length $m_i \alpha$ from the top. this new column now has height w_i. Since we have at least

$$(1 - \epsilon) r_i \frac{\mu_1(P_0^1)}{\mu_1(X_1)} > \frac{(1 - \epsilon)}{2} \frac{\mu_1(P_0^1)}{\mu_1(X_1)} q_i > 2\epsilon q_i$$

-many level sets of X_1 of height $1 + \alpha$, we have enough level sets whose heights are to be reduced from $1 + \alpha$ to 1.

If $u_i < w_i$, then instead of shortening the flow column we lengthen the column by $m_i \alpha$ as in the proof of a discrete case. We also shift up the level set of X_1 successively so that the first m_i level sets of height 1 now have height $1 + \alpha$. After we repeat this for all columns, it is obvious how to define T_1' on the new set X_1' which is a union of new shifted level sets. The flow of the new flow skyscraper is isomorphic to $(\Omega_3, \mathcal{L}_3, \lambda_3, S_3^t)$ and it is built under 1 and $1 + \alpha$ with $(X_1', \mathcal{F}_1', \mu_1', T_1')$. Since the number of level sets of X_1' in each new flow column remains r_i, $(X_1', \mathcal{F}_1', \mu_1', T_1')$ is isomorphic to $(X_1, \mathcal{F}_1, \mu_1, T_1)$. This completes the proof.

There is another notion of equivalence which is stronger than Kakutani equivalence but weaker than even α-equivalence. We say a flow S_1^t is an α-change of S_2^t if S_1^t can be obtained from S_2^t by changing the orbits through the measurable operation of adding or removing the length of multiples of α from the orbits of S_2^t. The measure of the set removed is not necessarily the same as the measure of the set added. If S_1^t is an α-change of S_2^t and if S_2^t is an α-change of S_3^t, then it is clear that S_1^t is an α-change of S_3^t. Hence this defines an equivalence relation. We note that the entropies of two equivalent flows may differ, unlike the even α-equivalence of the flow described above.

We say $(X_1, \mathcal{F}_1, \mu_1, T_1)$ and $(X_2, \mathcal{F}_2, \mu_2, T_2)$ are related, denote by $T_1 \sim T_2$, if and only if a flow over X_1 can be α-changed so that the new flow can be built over X_2. The measures of X_1 and X_2 may be different. Since we only consider the flows built under the values of 1 and $1 + \alpha$, any two flows over the same base are α-change from each other. Hence it is easy to see that if there exists a flow over X_1 which can be α-changed to be built over X_2, then any flow over $X_1(X_2)$ can be α-changed to be built over $X_2(X_1)$. Hence $T_1 \sim T_2$ is an equivalence relation. This equivalence needs to be studied further. [Pa4]

We remark that this equivalence, called α-equivalence, can be also studied, using the idea of the proofs of even α-equivalence.

Acknowledgement. This research is supported in part by NSF DMS 8902080 and GARC-KOSEF

References

[Am]   W. Ambrose, *Representation of ergodic flows,* Ann. of Math., **42**, (1941), 723-739.

[dJFR]   A. delJunco, A. Fieldsteel, D. Rudolph, *α-equivalence : Refinement of Kakutani equivalence,* preprint.

[dJR]   A. delJunco and D.Rudolph, *Kakutani equivalence of ergodic Z^n-actions,* J. of Ergodic Theory and Dyn. Sys., **4**, (1984), 89-104.

[Ka]   Y. Kakutani, *Induced measure preserving transformations,* Imp. Acad. Tokyo, **19**, (1943), 635-641

[ORW]   D. S. Ornstein, D. Rudolph and B. Weiss, *Equivalence of measure preserving transformations,* Memoirs of the AMS, **262**, (1982).

[Pa1]   K. K. Park, *Three Bernoulli factors that generate an ergodic flow,* Lecture notes in Mathematics, Springer-Verlag (1988), 608-616.

[Pa2]   K. K. Park, *Two nonisomorphic flows with same Very Weak Bernoulli partition on a base,* J. of Math. Analy. and Appl., **113**, (1986), 255-265.

[Pa3]   K. K. Park, *An induced mixing flow under 1 and α,* preprint.

[Pa4]   K. K. Park, *α-equivalence,* in preparation.

[Ru1]   D. Rudolph, *A two valued step coding for ergodic flows,* Math. Zeit. (1976), 201-220.

[Ru2]   D. Rudolph, *A restricted orbit equivalence,* Memoirs of the AMS, **323**, (1985).

[Sh]   P. Shields, *The Theory of Bernoulli Shifts,* U. of Chicago Press.

ing a finite volume. We assume that Q is complete, which means that each geodesic can be infinitely continued. The Riemannian metric on Q is denoted by dq^2, the Riemannian volume on Q is $d\sigma(q)$. The essential assumption is $\sigma(Q) < \infty$.

The phase space of the geodesic flow is the unit cotangent bundle M. The points of M are pares $x = (q, v)$ where $q \in Q$ and $v \in T_q^*, T_q^*$ is the linear space of cotangent vectors at q, $\|v\| = 1$. The Riemannian structure makes T_q^* an Euclidean space and this gives a possibility to identify T_q^* and T_q where T_q is the tangent space. M is also a Riemannian manifold with the metric $ds^2 = dq^2 + dv^2$.

Denote by $S_q \subset T_q$ the unit sphere, i.e. $\|v\| = 1$ for $v \in Q_1$. Introduce the Liouville measure μ on M where for any bounded $f \in C(M)$

$$\int_M f(x)d\mu(x) = \int_Q d\sigma(q) \int_{S_q} f((q, v))d\omega_q(v)$$

and ω_q is the Lebesgue measure on S_q. We may assume that μ is normed.

Each point $x = x_0$ defines uniquely the oriented geodesic $g(x)$ (see Fig. 1) which can be considered as a curve in M. Denote by x_t the tangent vector $g(x)$ at the point q_t whose distance from q_0 to q_t along $g(x)$ is t. Then $S^t : x_0 \to x_t$ is a smooth transformation of M and $S^{t_1} \circ S^{t_2} = S^{t_1+t_2}$.

Definition 1 *The one-parameter group $\{X^t\}$ is called the geodesic flow.*

One can give also the definition of geodesic flows using some notions of symplectic geometry. The cotangent bundle $M = \{(q, p), q \in Q, p \in T_q^*\}$ of any Riemannian manifold Q carries a natural symplectic structure. Consider the Hamilton function $H(q, p) = \|p\|^2$. Then for the Hamiltonian flow corresponding to H each manifold of constant energy $H(q, p) = E$ is invariant under the flow. The geodesic flow is the restriction of this flow to the submanifold corresponding to $E = 1$. Another way of expressing this property is to say that the geodesic flow is the motion by inertia on the configuration space Q. We shall explain this in more detail. In the Lagrangian approach to classical mechanics people consider Riemannian manifolds Q as configuration space and describe the dynamics with the help of Lagrangean functions $L(q, \dot{q})$, where $\dot{q}$ is a tangent vector (see [Ar]). A system is called natural if in some local coordinates $L(q, \dot{q}) = \sum q_{ij}(q)\dot{q}_i\dot{q}_j + V(q)$. The first term is a kinetic energy which $V(q)$ is the potential of external forces. The Jacobi variational principle says that the trajectories of our Lagrangian system corresponding to the value of energy $H = E$ are extremals of the functional

$$\int \sqrt{(E - V(q))}(\sum a_{ij}(q)dq_i dq_j)^{1/2}$$

i.e. they are geodesics with respect to the metric $dq^2 = (E - V(q)) \sum a_{ij}(q)dq_i dq_j$. The motion takes place on the subset $V(q) \leq E$ which can be a submanifold with boundary. However the velocity of the motion is not equal to 1. This means that natural systems arise from geodesic flows with the help of "change of velocity." Some discussion of these questions can be also found in [AS].

The Liouville measure μ is invariant under the geodesic flow, i.e. for any bounded measurable function f

$$\int_M f(S^t x)d\mu(x) = \int_M f(x)d\mu(x).$$

This is a particular case of the well-known Liouville theorem. In statistical mechanics the measure μ is called microcanonical distribution.

GEODESIC FLOWS ON MANIFOLDS OF NEGATIVE CURVATURE

Ya. G. Sinai

Landau Institute of Theoretical Physics
Russian Academy of Science, Moscow, Russia

This text is based on the lectures given in the Summer school on Dynamical Systems in Trieste, June, 1992. The main motivation was to expose one of the most beautiful and classical chapters of ergodic theory using some basic achievements in the entropy theory of dynamical systems. Another reason was more pragmatic. The interest to geodesic flows on manifolds of negative curvature grew enormously during the last years due to the development of quantum class. A lot of numerical and qualitative facts discovered have mainly by physicists suggest difficult and important problems concerning the connection of eigen-values of Laplacians on compact manifolds of negative curvature and geodesics, especially closed geodesics on such manifolds. We believe that the theory which is explained below can be useful for attacking these problems.

The plan of this paper is the following. In §1 we introduce the general notion of geodesic flows and explain their main properties. In §2 we present a short information on manifolds of negative curvature and discuss properties and the geometrical meaning of solution of general needed fasts related to stable or unstable horospheres, and in particular their mutual non-integrability. In the next section we describe the construction of the invariant family of σ-finite measures on horospheres and their connection with the Liouville measure. In §4 we recall some general facts from ergodic theory. In §5 we use them for proving ergodicity and K-property of geodesic flows. §6 contains apparently a new result. Namely, we show the strict ergodicity or horospheres equipped with the above-mentioned σ-finite measures. The problem was motivated by a remark by D. Zagier [Z] concerning the connections between the uniform distribution of closed horocycles on the modular surface and Riemann hypothesis. We believe also that our result can be useful for the estimation of decay of correlation functions for the geodesic flows on manifolds of negative curvature which is still an open question.

§1. Geodesic Flows : definition and general properties

Let Q be a n-dimensional C^∞-Riemannian manifold, not necessary compact, but hav-

During these notes we shall use the so-called Fermi coordinates which have a simple geometrical meaning. Take $x_0 = (q_0, v_0) \in M$ and consider the directed geodesic $\{q_t\}$ (see Fig. 2). Choose $(n-1)$ unit vectors $e_1(0), e_2(0), \cdots, e_{n-1}(0)$ which together with v_0 give the orthonormal basis of the vector space T_q. Let us displace in parallel all vectors $v_0, e_j(0)$ along the geodesic $\{q_t\}$. Since the parallel displacement which constitute the orthonormal basis $v_t, e_j(t)$ of the space T_{q_t}.

Using coordinates connected with these bases we can compare points in different T_{q_t}. Especially we shall need the coordinates $u = (u_1, \cdots, u_{n-1})$ in the $(n-1)$-dimensional subspaces $T_{q_t}^{(0)}$ orthogonal to v_t. These coordinates are called Fermi coordinates.

Let us consider the following geometrical question. Assume that we are given a local $(n-1)$-dimensional C^2-submanifold $\tilde{R}_0 \subset Q$ passing through q_0 and such that the tangent space to $\tilde{R}_0$ at q_0 is $T_{q_0}^{(0)}$. Denote by R_0 the bundle of unit normal vectors $n(q), q \in \tilde{R}_0$ and $v_0 \in R_0$. Take $dq \in T_{q_0}^{(0)}$, then $n(q + dq) = n(q) + B_0 dq + o(dq)$. Here B_0 is a self-adjoint operator in the space $T_{q_0}^{(0)}$ which is called the operator of the second fundamental form. Such operators will play an important role below. If $B_0 > 0$ then locally $\tilde{R}_0$ is convex and the bundle R_0 is diverging at least for small times. If $B_0 < 0$ then $\tilde{R}_0$ is concave and the bundle is converging (see Fig. 3). In the multi-dimensional case these can be mixtures of convex and concave subbundles. Since $R_0 \subset M$ we may consider $R_t = s^t r_0$. It is rather well-known that at least for small t R_t will be a unit normal bundle of a local submanifold $\tilde{R}_t \subset Q, q_t \in \tilde{R}_t, (q_t, v_t) = S^t(q_0, v_0) \subset R_t$. Then R_t can be characterized by the corresponding operator of the second fundamental form B_t. If we write Bt in Fermi coordinates then their dependence on t is described by the Ricatti-type equation

$$\dot{B}_t + B_t + K(q_t) = 0 \tag{1}$$

Here $K(q_t)$ is a self adjoint matrix. The value of the quadratic form $(K(q_r)e, e)$ for any $e \in T_{q_t}^{(0)}, \|e\| = 1$, is equal to the Gaussian curvature of a local two-dimensional surface formed by geodesics going out of q_t along the two-dimensional plane formed by e and v_t.

Solutions of (1) often display some singularities which correspond to the so-called local and conjugate points. Geometrically it means that during the dynamics $\tilde{R}_t$ becomes singular.

The content of this text is the analysis of ergodic properties of the geodesic flow $\{S^t\}$ with respect to measure μ in the case of manifolds of negative curvature. A priori it is clear that such properties should depend on the geometry of Q. In the case of negative curvature this connection is especially clear.

§2. Some Information about Manifolds of Negative Curvature

We recall basic facts concerning manifolds of constant negative curvature $K = -1$. In Poincare model of the n-dimensional Lobachevsky space one considers the unit disk $D^{(n)}$ equipped with the metric $dq^2 = \frac{\sum dq_i^2}{(1-\sum q_i^2)^2}$. The geodesics in this metric are circle arcs orthogonal to the absolute $\partial D^{(n)}$.

The manifolds we are interested in are factor-spaces $D^{(n)}/\Gamma$ where Γ is a discrete subgroup of motions of $D^{(n)}$. The condition $vol(D^{(n)}/\Gamma) < \infty$ is some restriction to Γ. In the case $n = 2$ the Teichmüller theory describes the space of compact surfaces of constant negative curvature and given genus up to conformal equivalence. It follows from this theory that $area(D^{(n)}/\Gamma)$ can be arbitrary small. For $n \geq 3$ a famous theorem by Margulis says that $vol(D^{(n)}/\Gamma) \geq const.$

One case deserves a special consideration. Let $n = 2$. Consider the Poincare model of Lobachevsky plane on the upper half-plane $H^{(2)}$ of the complex plane C. The group G of all motions of $H^{(2)}$ consists of Mobius transformations $g(q) = \frac{aq+b}{cq+d}$ where all a, b, c, d are real. Since (a, b, c, d) and $(-a, -b, -c, -d)$ generate the same transformation the group G is $SL(2, R)/(\pm e)$.

Recall the classification of Mobius transformations.

1. Elliptic transformations. The transformation $g = \begin{pmatrix} a & b \\ c & d \end{pmatrix}$ elliptic if it has a fixed point. In this case it is a non-euclidean rotation around this point to some angle.

2. Hyperbolic transformation. The transformation has an invariant geodesic. The restriction of g to this geodesic is a shift to some distance $\rho > 0$.

3. Parabolic transformation. The transformation $g = \begin{pmatrix} a & b \\ c & d \end{pmatrix}$ has a fixed point on the absolute $\partial H^{(2)}$. If this point is ∞ then $g = \begin{pmatrix} 1 & t \\ 0 & 1 \end{pmatrix}$ or $q \to q + t$. In the general case g is conjugate with such transformation with the help of the transformation $g(q) = \frac{1}{-q+r_0}$.

The discrete subgroups generating — surfaces consist only of hyperbolic transformations. The presence of elliptic transformations implies some singularities in the analytic structure of Q. If Γ contains parabolic transformations then Q is non-compact.

Consider the case $\Gamma = SL(2, Z)$. This group is called $g_1(z) = z + 1$ and $g_2(z) = -1/z$ satisfying the relation $g_1^2 = (g_2 g_1)^3 = identity$. The factor space $H^{(2)}/\Gamma$ is called modular surface. The corresponding fundamental domain is given on the Fig. 4. The modular surface has two singular points (marked on the Fig. 4).

The modular surface is connected with many problems of number theory. One reason for this can be explained as follows. Take any lattice on the usual Euclidean plane for which the fundamental parallelogram has area equal to one. The choice of any basis of this lattice can be described by a matrix $g = \begin{pmatrix} a & b \\ c & d \end{pmatrix} \in SL(2, R)$. Two matrices g and $\gamma g, \gamma \in SL(2, Z)$ generate the same lattice. Thus $Q = SL(2, R)/SL(2, Z)$ can be considered as the space of two-dimensional lattices. It is easy to see that $area(Q) < \infty$. The geodesic flow on the modular surface was studied already by ε. Artin (Ein mechanisches System mit quasiergodichen Bahnen, Ab. Math. Sem. Univ. Hamburg 3,1924, 170-175) where the connections of this geodesic flow with continued fractions were observed. There exist interesting multi-dimensional generalizations of the modular surface like $SL(n, R)/SL(n, Z)$.

We shall deal with general manifolds of negative curvature having a finite volume. It is assumed that the manifolds are C^∞ and the curvature along any two-dimensional plane lies between two negative boundaries $-K_1$ and $-K_2, K_1 > K_2$. In the two-dimensional situation of compact surfaces the metrics are conformally equivalent to metrics of constant curvature i.e. can be written in the form $dq^2 = a(q)d_0q^2$ where d_0 is the metric of constant curvature, $a(q) > 0$. In the multi-dimensional case it is not necessarily so. Our analysis will be purely geometrical and in a sense local. The connections with metrics of constant negative curvature will not be used.

Return now to the Jacobi-Ricatti equation (1). In the case of manifolds of negative curvature it has the following formal property:

i) If B_0 is self-adjoint then B_t is self-adjoint for all $t \geq 0$. Indeed, B_t^* satisfies the equation

$$\dot{B}_t^* + (B_t^*)^2 + K(q_t) = 0.$$

Since $B_0^* = B_0$ then $B_t^* = B_t$ for all $t > 0$ due to the uniqueness of solutions.

ii) If $B_0 \geq$ then $B_t > 0$ for all $t > 0$. If $B_0 \leq$ then $B_t < 0$ for all $t < 0$. We shall prove only the first statement. Also we may assume that $B_0 > 0$. The general case can be obtained by the limit transition.

Suppose that one can find $t_0 > 0$ and a vector $e_0, \|e_0\| = 1$ such that $(B_{t_0}e_0, e_0) = (B_{t_0}e_0, B_{t_0}e_0) = 0$. Let t_0 to be the minimal number with this property. Then

$$\frac{d}{dt}(B_{t_0}e_0, e_0) = -(B_{t_0}e_0, B_{t_0}e_0) - (K(q_{t_0})e_0, e_0)$$
$$> K_1,$$

i.e. $B_t e_0$ is a growing function in a neighbourhood of $t = t_0$. But this contradicts minimality of t_0.

We shall give another non-formal proof of 2) inspired by the theory of billiards. Take small $\delta > 0$ and write down the difference equation corresponding to (1):

$$
\begin{aligned}
B_{(m+1)\delta} &= B_{m\delta} - \delta B_{m\delta}^2 - \delta K_{m\delta} = B_{m\delta}(I - \delta B_{m\delta}) - \delta K_{m\delta} \\
&= -\delta K_{m\delta} + B_{m\delta}(I + \delta B_{m\delta})^{-1} + o(\delta) \\
&= -\delta K_{m\delta} + (\delta I + \delta B_{m\delta}^{-1}) + o(\delta).
\end{aligned}
$$

This gives a possibility to represent $B_{m\delta}$ as an operator-valued continued fraction

$$B_{(m+1)\delta} = -\delta K_{m\delta} + \cfrac{I}{\delta I + \cfrac{I}{B_{m\delta}}} + o(\delta)$$

$$= -\delta K_{m\delta} + \cfrac{I}{\delta I + \cfrac{I}{-\delta K_{(m-1)\delta} + \cfrac{I}{\delta I + \cfrac{I}{-\delta K_{(m-2)\delta} + \cdots}}}}$$

Now we see that in the main order of magnitude $B_{(m+1)\delta}$ is represented as an infinite continued fraction. This is ———— to the negative curvature all terms in this fraction are strictly positive. The theory of such fractions is similar in many respects to the theory of the usual continued fraction.

We shall now construct special solutions of the equation (1) which are called limiting solutions. Fix $x_0 \in M$ and consider the initial datum $B|_{t=t_0} = B \geq 0$ where B is an arbitrary fixed matrix and $t_0 < 0$. Denote the corresponding solution of (1) for $t = 0$ by $B_0(t_0)$.

Theorem 1 *There exists the limit*

$$B^{(u)}(x_0) = \lim_{t_0 \to -\infty} B_0(t_0) \tag{2}$$

not depending on B.

In the same way consider $t_0 > 0$ and the initial datum $B|_{t=t_0} = B \leq 0$. Denote again the corresponding solution of (1) for $t = 0$ by $B_0(t_0)$.

Theorem 1' *There exists the limit*

$$B^{(s)}(x_0) = \lim_{t_0 \to \infty} B_0(t_0)$$

not depending on B'.

The statement of the theorems is a property of the ODE-system (1). In fact it is valid if the initial data depend on t_0 but remain within a bounded set of the space of non-negative (or non-positive) operators. Moreover the convergence in (2) is exponential.

It is easy to show that $B^{(u)}(x_0)$, $B^{(s)}(x_0)$ satisfy the Hölder condition and this smoothness cannot be better even in the case of analytical manifolds.

Each of $B^{(u)}(x_0)$, $B^{(s)}(x_0)$ determines a $(n-1)$-dimensional subspace in the tangent space T_{x_0} to M at $x_0 = (q_0, v_0)$. Namely, any pair $(dq, B^{(u)}(x_0)dq)$ or $(dq, B^{(s)}(x_0)dq)$ can be considered as a vector of T_{x_0} where the first component describes the variation of q_0 while the second component describes the variation of v_0. The whole family $\{B^{(u)}(x_0)\}$ or $\{B^{(s)}(x_0)\}$ is a distribution of $(n-1)$-dimensional subspace. A very important theorem says that both families are integrable.

Theorem 2 *For any point $x_0 = (q_0, v_0)$ one can find $\alpha(x_0) > 0$ and a $(n-1)$-dimensional local C^∞-submanifold $\tilde{R}^{(u)}(x_0) \subset Q$ and its framing $R^{(u)}(x_0) \subset M$ by unit normal vectors such that*

(i) $x_0 \in R^{(u)}(x_0)$.

(ii) $dist(q_0, \partial \tilde{R}^{(u)}(x_0)) \geq \alpha$;

(iii) *for any $x = (q, v) \in R^{(u)}$ the operator of the second fundamental form of $\tilde{R}^{(u)}$ at q is $B^{(u)}(x)$;*

(iiii) *for any $x \in R^{(u)}$ and some $\lambda > 0$*

$$dist(S^{-t}x, S^{-t}x_0) \leq conste^{-\lambda t}, t \geq 0$$

where const depends on x and x_0.

Similar statement is true for the distribution $\{B^{(s)}(x)\}$. It gives local submanifolds $R^{(s)}(x_0)$. In the property (iiii) should write $dist(S^t x, S^t y)$. In the case of compact manifolds $\alpha(x_0)$ can be chosen independently on x_0.

Theorem 2 is a particular case of the so-called Hadamard-Perron theorem which is valid for a wide class of hyperbolic dynamical system or hyperbolic trajectories (see [A].[HPS],[P]). From the point of view of non-euclidean geometry (iiii) points $x \in R^{(u)}(x_0)$ or $x \in R^{(s)}(x_0)$ determine the bundles of asymptotically converging geodesics

(as $t \to -\infty$ or $t \to \infty$) and $R^{(u)}(x_0), R^{(s)}(x_0)$ are orthogonal submanifolds to these bundles. They are called horospheres (local) or horocycles (local) in the two-dimensional case. $R^{(u)}(x_0)$ is convex while $R^{(u)}(x_0)$ is concave. Following the general terminology of hyperbolic dynamical system we shall call them local unstable and local stable manifolds or lum and lsm. Their form for $n = 2$ is given on the Fig. 5.

Unstable manifolds $\Gamma^{(u)}(x_0)$ of $x_0 \in M$ consists of all x satisfying (iiii). In the compact case when $\alpha(x) = \alpha$ it can be obtained with the help of the formula

$$\Gamma^{(u)}(x_0) = \bigcup_{t>0} S^t R^{(u)}(S^{-t}x_0).$$

In the case of manifolds of finite volume the construction is similar. In the same way one defines the stable manifolds $\Gamma^{(s)}(x_0)$. The construction of um $\Gamma^{(u)}$ and sm $\Gamma^{(s)}$ can be also done with the help of general hyperbolic methods.

The families $\Gamma^{(u)} = \{\Gamma^{(u)}(x_0)\}$, $\Gamma^{(s)} = \{\Gamma^{(s)}(x_0)\}$ constitute the unstable and stable foliations. They are invariant under the geodesic flow in the following sense

$$S^t \Gamma^{(u)}(x) = \Gamma^{(u)}(S^t x),$$
$$S^t \Gamma^{(s)}(x) = \Gamma^{(s)}(S^t x) \qquad -\infty < t < \infty.$$

Lum and lsm have a very important property which we shall call the mutual non-integrability. It will be used during the proofs later.

Take $x \in M$ and a small neighbourhood U of x.

Theorem 3 *For any $x', x'' \in U$ there exist $z', z'' \in U$ such that $z' \in R^{(s)}(x'), z'' \in R^{(u)}(z'), x'' \in R^{(u)}(z'')$.*

Fig. 6 explains the meaning of the statement of the theorem for $n = 2$. In the general case it follows easily from the inequalities $B^{(u)}(x) > 0, B^{(s)}(x) < 0$.

§3. Measure-Theoretic Properties of Stable and Unstable Foliations $\Gamma^{(s)}, \Gamma^{(u)}$

We shall deal with $\Gamma^{(u)}$, the construction for $\Gamma^{(s)}$ is similar. Our goal now to construct for each $\Gamma^{(u)}(x)$ a σ-finite measure $\nu_{\Gamma^{(u)}(x)}$ which is absolutely continuous with respect to the Lebesgue measure on $\Gamma^{(u)}(x)$ and which has the following property of invariance:

$$\frac{\nu_{\Gamma^{(u)}(x)}(C')}{\nu_{\Gamma^{(u)}(x)}(C'')} = \frac{\nu_{\Gamma^{(u)}(S^t x)}(S^t C')}{\nu_{\Gamma^{(u)}(S^t x)}(S^t C'')} \qquad (3)$$

for any compact $C', C'' \subset \Gamma^{(u)}(x)$ and any $t \in R^1$. It is clear that if $\nu_{\Gamma^{(u)}(x)}$ satisfies (3) then the measure $c \cdot \nu_{\Gamma^{(u)}(x)}$ for any constant c also satisfies (3). Thus it is better to say that we intend to construct for each $\Gamma^{(u)}(x)$ a class of proportional measures for which (3) holds.

Consider $\Gamma^{(u)}(S^{-t}x)$ for $t > 0$. Then S^t generates a smooth expanding mapping $\Gamma^{(u)}(S^{-t}x) \to \Gamma^{(u)}(x)$. Denote by $\lambda_t(y), y \in \Gamma^{(u)}(x)$ the Jacobian which corresponds to its image on $\Gamma^{(u)}(x)$. In other words we take an infinitesimally small volume $d\omega$ on $\Gamma^{(u)}(S^{-t}x)$, take its image $S^t d\omega$ and calculate its volume on $\Gamma^{(u)}(x)$ using the inner metric induced by the Riemannian metric in M. The ratio is exactly $\lambda_t(y)$.

Lemma 1 *For any $x', x'' \in \Gamma^{(u)}(x)$ there exists the limit $\lim_{t \to \infty} \frac{\lambda_t(x')}{\lambda_t(x'')} = \kappa(x', x'')$. For this limit $\kappa(x', x'') \cdot \kappa(x'', x''') = \kappa(x'', x''')$.*

The proof is easy it follows directly from the construction of semi-trajectories going out of $\Gamma^{(u)}(x)$ as $t \to -\infty$.

Now define the measure $\nu_{\Gamma^{(u)}(x)}$ on $\Gamma^{(u)}(x)$ as an absolutely continuous with respect to the Lebesgue measure whose density is $\kappa(x', x)$. It follows easily from the construction that (3) is true. The main property of measure $\nu_{\Gamma^{(u)}(x)}$ is described in the following theorem.

Theorem 4 *Let ξ be an arbitrary measurable partition whose element $C_\xi(x)$ are open subset of lum $R^{(u)}(x)$ such that the Lebesgue measure of the boundary $\partial C_\xi(x)$ is zero. Then the conditional measure on $C_\xi(x)$ induced by the Liouville measure μ has the form*

$$
\mu(A|C_\xi(x)) \quad = \quad \frac{\int_{A \cap C_\xi(x)} d\nu_{\Gamma^{(u)}(x)}(y)}{\int_{C_\xi(x)} d\nu_{\Gamma^{(u)}(x)}(y)}
$$
$$
= \quad \frac{\int_{A \cap C_\xi(x)} \kappa(y, x) d\sigma_{\Gamma^{(u)}(x)}(y)}{\int_{C_\xi(x)} \kappa(y, x) d\sigma_{\Gamma^{(u)}(x)}(y)} \tag{4}
$$

The last expression does not depend on the choice of $\nu_{\Gamma^{(u)}(x)}$. Here $d\sigma_{\Gamma^{(u)}(x)}$ is the volume generated by the Riemannian metric on $\Gamma^{(u)}(x)$

The proof consists of two steps. At the first step one proves a general property which says that the conditional measures on elements of the partition ξ are given by densities with respect to the Riemannian volume on um. It is based upon the absolute continuity of the so-called canonical isomorphism of lum and lsm (see the discussion in [AS] [KS] [S]). At the second step one shows that if the conditional measure has a density then the density has the form (4). This follows from the invariance of μ. Namely, take the partition $S^{-t}\xi$. Its elements are very small subsets of lum. ¿From the previous discussion it follows that conditional measures on elements $C_{S^{-t}\xi}$ induced by μ are "almost" uniform. The one takes into account that under dynamics conditional measures are transformed in conditional measures.

Similar measures $\nu_{\Gamma^{(s)}(x)}$ can be constructed for sm $\Gamma^{(s)}(x)$.

Another class of measures $\bar{\nu}_{\Gamma^{(u)}(x)}, \bar{\nu}_{\Gamma^{(s)}(x)}$ can be obtained in the following way. Take $C', C'' \subset R^{(u)}(x)$. Then $\bar{\nu}_{\Gamma^{(u)}(x)}$ is defined with the help of the limit

$$
\frac{\bar{\nu}_{\Gamma^{(u)}(C')}}{\bar{\nu}_{\Gamma^{(u)}(C'')}} = \lim_{t \to \infty} \frac{\int_{S^t C'} \kappa(y, S^t x) d\sigma_{\Gamma^{(u)}(S^t x)}(y)}{\int_{S^t C''} \kappa(y, S^t x) d\sigma_{\Gamma^{(u)}(S^t x)}(y)}
$$

I don't know an elementary proof of the existence of the last limit. Apparently $\bar{\nu}_{\Gamma^{(u)}(x)}, \bar{\nu}_{\Gamma^{(s)}(x)}$ are connected with the so-called Margulis measure and in general are singular with respect to the Lebesgue measures.

§4. Several Basic Facts from General Ergodic Theory

The geodesic flow $\{S^t\}$ preserves the measure μ and is a flow in the sense of general ergodic theory. The Birkhoff-Khinchin ergodic theorem says that for any function

$f \in L^1(M, \mu)$ there exist and coincide a.e the limits

$$\bar{f}^+(x) = \lim_{T \to \infty} \frac{1}{T} \int_0^T f(S^t x) dt,$$

$$\bar{f}^-(x) = \lim_{T \to \infty} \frac{1}{T} \int_0^T f(S^{-t} x) dt,$$

$$\bar{f}(x) = \lim_{T \to \infty} \frac{1}{2T} \int_{-T}^T f(S^t x) dt,$$

$$\bar{f}^+(x) = \bar{f}^-(x) = \bar{f}(x) \quad a.e..$$

We shall need also the notion of K-flow (or Kolmogorov flow, see, e.g.[CFS]).

Definition 2 *A measure-preserving flow $\{S^t\}$ is called K-flow if there exists a measurable partition ξ_0 such that*

1° $\xi_t = X^t \xi_0 > \xi_0$ *(mod 0) for all $t > 0$;*

2° $\lim_{t \to \infty} \xi_t = \varepsilon$ *(mod 0) where ε is the largest measurable partitions whose elements are points ;*

3° $\lim_{t \to -\infty} \xi_t = \nu$ *(mod 0) where ν is the smallest trivial partitions which has only one element M ;*

Any partition for which $1°, 2°, 3°$ hold is called K-partition.

Certainly the properties $1° - 3°$ can be formulated in terms of σ-subalgebras of the whole σ-algebra of measurable subsets of M. Also we assume that M is the Lebesgue space having no points of positive measure.

K-flows are ergodic, mixing, have countable Lebesgue spectrum in the subspace of functions whose expectation is zero. For K-flows it is reasonable to study other stochastic properties like the decay of time correlation functions, central limit theorem for fluetuations, Bernoulli property, etc.

We shall need some refinements of the properties described in the Definition 2. Assume that we have constructed a measurable partition ξ_0 such that

1° $\xi_t = X^t \xi_0 > \xi_0$ (mod 0) for all $t > 0$;

2° $\lim_{t \to \infty} \xi_t = \varepsilon$ (mod 0)

Denote by $\eta^{(inv)}$ the measurable partition of M onto ergodic components. Then it is easy to show that

$$\eta^{(inv)} \leq \lim_{t \to -\infty} \xi_t = \xi^-.$$

An additional statement says that if $H(\xi_t|\xi_0) = th(\{S^t\}), t > 0$ where H is the conditional entropy and $h(\{S^t\})$ is the measure-theoretic entropy of the flow $\{S^t\}$ the the intersection $\bigwedge_t \xi_t$ is the maximal partition of the zeroth-entropy or Pinsker partition $\pi(\{S^t\})$(see [RS] for automorphisms of Lebesgue spaces and [Ru],[Bl],[G] for the generalization to flows).

Pinsker partition is always invariant, i.e. $S^\tau(\pi(\{S^t\})) = \pi(\{S^t\}), -\infty < \tau < \infty$, $\eta^{(inv)} \leq \pi(\{S^t\})$ and a flow $\{S^t\}$ is a K-flow iff $\pi(\{S^t\}) = \nu$.

In the next section we show how all these notions can be used for the analysis of geodesic flows.

§5. Ergodicity and K-property of Geodesic Flows on Manifolds of Negative Curvature and Finite Volume

Denote by $\eta^{(u)}, \eta^{(s)}$ the measurable hulls of the partitions of M onto um $\Gamma^{(u)}$ and sm $\Gamma^{(s)}$. In other words consider σ-subalgebras $M^{(u)}$, $M^{(s)}$ such that if $A \in M^{(u)}$ $(A \in M^{(u)})$ then A consists mod 0 of um (sm). In Lebesgue spaces such σ-algebra is generated by a measurable partition, i.e. subsets of this subalgebra consist mod 0 of elements of this measurable partition. Therefore we take as $\eta^{(u)}, \eta^{(s)}$ the measurable partitions corresponding to $M^{(u)}$, $M^{(s)}$ respectively.

Take any continuous function $f \in C(M)$ with compact support. Then it is easy to show $\bar{f}^+$ is constant (mod 0) on each $\Gamma^{(s)}$ where it is defined and thus is measurable with respect to $\eta^{(s)}$. (see [H]). By the same reason $\bar{f}^+$ is measurable with respect to $\eta^{(u)}$. Since $\bar{f}^+ = \bar{f}^- = \bar{f}$ a.e. we conclude that $\bar{f}$ is measurable with respect to $\eta^{(u)}$ and $\eta^{(s)}$. This immediately gives $\eta^{(inv)} \le \eta^{(u)}$, $\eta^{(inv)} \le \eta^{(s)}$ i.e. $\eta^{(inv)} \le \eta^{(u)} \wedge \eta^{(s)}$. The mutual non-integrability of $\Gamma^{(u)}$ and $\Gamma^{(s)}$ (see §3) easily yields $\eta^{(u)} \wedge \eta^{(s)} = \nu$. Thus we come to the first important result.

Theorem 5 *Geodesic flows on manifolds of negative curvature and finite volume are ergodic.*

The next step is to construct measurable partitions ξ_0 satisfying $1°$ and $2°$ (see above).

Theorem 6 *There exists a measurable partition ξ_0 satisfying $1°$ and $2°$ and such that for a.e. $x \in M$ the element $C_{\xi_0}(x)$ of ξ_0 containing x is an open subset $C_{\xi_0} \subset R^{(s)}(x)$, $\sigma(\partial C_{\xi_0}(x)) = 0$. Moreover, for a.e. x there exists an interval $(\tau^-, \tau^+), \tau^- < 0 < \tau^+$ such that $S^t C_{\xi_0}(x) = C_{\xi_0}(S^t x)$ for all $t \in (\tau^-, \tau^+)$ (see Fig. 7).*

The construction of ξ_0 is done in a purely geometric way. We begin with a finite partition α of M which has the last property and then take $\alpha^- = \bigvee_{t>0} S^{-1}\alpha = \xi_0$. The main problem is to show that ξ_0 is non-trivial, i.e. that its elements are not points. But this follows easily from the exponential divergence of lsm as $t \to -\infty$.

Since ξ_0 is generated by a finite partition α it is not too difficult to show that the conditional entropy $H(\xi_t|\xi_0) = th(\{S^t\})$. The general theory (see above) now says that $\bigwedge_t \xi_t = \pi(\{S^t\})$. Also it is easy to see that $\bigwedge_t \xi_t = \eta^{(s)}$. Thus $\pi(\{S^t\}) = \eta^{(s)}$. Changing $tt_0 - t$ and replacing lsm by lum we immediately get $\pi(\{S^t\}) = \eta^{(u)}$, i.e. $\pi(\{S^t\}) = \eta^{(s)} = \eta^{(u)} = \eta^{(s)} \wedge \eta^{(u)}$. But we already saw that $\eta^{(s)} = \eta^{(u)} = \eta^{(inv)} = \nu$, i.e.

$$\eta^{(s)} = \eta^{(u)} = \pi(\{S^t\}) = \nu.$$

This gives the following theorem.

Theorem 7 *Geodesic flows on manifolds of negative curvature and finite volume are K-flows.*

§6. Some Refinements

Each surface of constant negative curvature can be represented as a factor-space of the upper half-plane H^2 with respect to some discrete subgroup if motion of H. Accordingly $M = \Gamma \setminus SL(2, R)$ where Γ consists of hyperbolic transformation and if $\gamma \in \Gamma$ then $-\gamma \in \Gamma$. We can also write $M = P\Gamma \setminus PSL(2, R)$ where $PSL(2, R)$ is the factorization of $SL(2, R)$ with respect to $\pm e$ and $P\Gamma = \Gamma / \pm e$. The geodesic flow can be represented as the right action of the subgroup $g_t = \begin{pmatrix} e^{t/2}, & 0 \\ 0, & e^{-t/2} \end{pmatrix}$. It is easy to check that stable manifolds which are stable horocycles are trajectories of the one-parameter subgroup $h_t^+ = \begin{pmatrix} 1, & t \\ 0, & 1 \end{pmatrix}$ while unstable horocycles are trajectories of the subgroup $h_t^- = \begin{pmatrix} 1, & 0 \\ t, & 1 \end{pmatrix}$. Both subgroups preserve the Liouville measure μ. There is a well-known theorem by Hedlund which says that in the case of compact surfaces of constant negative curvature each horocycle is everywhere dense and strictly ergodic. The last property follows from the fact that if f is a continuous function then

$$\left| \frac{1}{T} \int_0^T f(h_t^+ x)dt - \int f d\mu \right| \leq \varepsilon_T$$

where ε_T depends only on T, $\varepsilon_T \to 0$ as $T \to \infty$.

In the case of surfaces of finite area this statement is wrong. Here there exists a finite number of families of closed horocycles. All horocycles which do not belong to one of these families are everywhere dense and uniformly distributed. In this section we shall discuss some multi-dimensional generalization also to the case of variable curvature of some of quoted results. Another motivation for our analysis was a remark by D Zagier [Z] according to which the Riemann hypothesis for zeros of ζ-function can be derived from the properties of the uniform distribution of closed horocycles on the modular surface.

Let us formulate the general question which we are going to discuss. Take a continuous function f with compact support on M and for any $x \in M$ consider the sm $\Gamma^{(s)}(x_f)$. According to the above-described construction each $\Gamma^{(s)}$ is equipped with the natural σ-finite measure $\nu_{\Gamma^{(s)}}$. This measure is not originated from any dynamics for which $\Gamma^{(s)}$ are orbits but comes from the connection with the geodesic flow. Consider now a sequence of compact open subsets $O_j \subset \Gamma^{(s)}(x)$ such that

$a_1)$ $O_1 \subset O_2 \subset \cdots, \bigcup_{j \geq 1} O_j = \Gamma^{(s)}$;

$a_2)$ for any $R > 0$ let $O_j(R) \subset O_j$ be a subset of points whose distance to ∂O_j (in the induced metric on $\Gamma^{(s)}$) is greater than R ; then

$$\frac{\nu_{\Gamma^{(s)}(x)}(O_j(R))}{\nu_{\Gamma^{(s)}(x)}(O_j)} \to 1$$

as $j \to \infty$.

Definition 3 *The um $\Gamma^{(s)}(x)$ is uniformly distributed if*

$$\lim_{j \to \infty} \frac{\int_{O_j} f(y) d\nu_{\Gamma^{(s)}(x)}(y)}{\int_{O_j} d\nu_{\Gamma^{(s)}(x)}(y)} = \int f(z) d\mu(z). \qquad (5)$$

The main result of this section is the following theorem.

Theorem 8 *If Q is compact then each $\Gamma^{(s)}(x)$ is uniformly distributed.*

Corollary *Each $\Gamma^{(s)}$ is everywhere dense.*

This is also a general property of hyperbolic systems, see [A].

During the proof of theorem 8 we introduce some notions which might be useful in other situations. Take an open lum $R^{(u)}(x)$.

Definition 4 *An s-cell is a set*

$$R^{(s)} = \bigcup_{y \in R^{(u)}(x)} R^{(s)}(y)$$

where $R^{(s)}$ is a family of open lsm.

In the same way one defines u-cells $R^{(u)}$ (see Fig.7) for $n = 2$.

Each $R^{(s)}$ carries a natural partition $\xi^{(s)}(R^{(s)})$ onto $R^{(s)}(y)$ while each $R^{(u)}$ carries a natural partition $\xi^{(u)}(R^{(s)})$ onto $R^{(u)}(y)$.

Take a finite collection of s-cells $\{R_1^{(s)}, R_2^{(s)}, \cdots, R_m^{(s)}\}$. For any $z_0 \in R_j^{(s)}$ consider the smallest $\tau(x_0) > 0$ such that $S^{\tau(z_0)} z_0 \in \bigcup_{j=1}^{m} R_j^{(s)}$.

Definition 5 *The collection $\{R_1^{(s)}, R_2^{(s)}, \cdots, R_m^{(s)}\}$ is called Markov if for any $z_0 \in R_j^{(s)}$ for which τ is continue at z_0 one can find a neighbourhood $Y(z_0) \in R_j^{(s)}$ consisting of elements of $z_{\xi^{(s)}}(R^{(s)})$ and such that*

(i) *$\tau(z)$ is constant on each $R^{(s)}(y) \subset U(z_0)$ and $S^{\tau(z)}(R^{(s)}(z)) \subset R(S^{\tau(z)}(z)) \in R_k^{(s)}$ for some $k, 1 \le k \le m$;*

(ii) *for any $R^{(s)}(\bar{z}) \subset R_k^{(s)}$ one can find $\bar{z}_0 \in U(z_0)$ such that $S^{\tau(\bar{z}_0)} R^{(s)}(\bar{z}_0)) \subset R^{(s)}(\bar{z})$.*

(iii) *the partition of $R^{(u)}(x)$ produced by different $U(z_0)$ is finite.*

The meaning of this definition can be seen from the Fig. 8. Our definition is a version of the definition of Markov partitions in the general theory of hyperbolic dynamical system (see the discussion in the book by R. Bowen [B]). It is worthwhile to stress that our definition uses only the dynamics in the positive direction of time.

Theorem 9 *If Q is compact then the finite Markov collections $\{R_1^{(s)}, R_2^{(s)}, \cdots, R_m^{(s)}\}$ exist.*

This theorem is a particular case of a general theorem almost the existence of Markov partitions (see [B] and [S]). For two-dimensional geodesic flows Markov collections were constructed by M. Ratner [R1]. Apparently for non-compact manifolds there exist countable Markov collections but I don't know whether such a construction was done explicitly. Any Markov collection induces a special representation of the geodesic flow (concerning special representations see e.g. [CFS]). Namely, take any s-cell $R_j^{(s)}, 1 \le$

$j \leq m$ and its finite partition onto subsets $U_{jr}, 1 \leq r \leq r_j$ where $U_{jr} = U(z_{jr})$. Denote by τ_{jr} the function τ defined on U_{jr} (see definition 5). An additional statement to the theorem 9 says that all τ_{jr} are bounded, $\tau_{jr} \leq const$.

Now the phase space M can be reprented as a union of towers built under each U_{jr} (see Fig. 9). In other words for a.e. $z \in M$ are can bind $z_0 \in R_j^{(s)}$ and $t, 0 \leq t \leq \tau(z_0)$, such that $z = S^t z_0$. One can use Markov collections for constructing symbolic representations of geodesic flows.

For us it is important that any Markov collection generates a K-partition (see definition 2).

Namely consider a tower $T_{jr} = \bigcup_{z \in U_{jr}} \bigcup_{0 \leq t \leq \tau(z)} S^t z$. Each U_{jr} is decomposed onto lsm $R^{(s)}(z) \subset U_{jr}$. If $x = S^t z$ then we put $C_{\zeta_0}(x) = S^t R^{(s)}(z)$. It is clear that so defined partition ζ_0 is measurable and is a K-partition. The property 1° follows from Markov, 2° is obvious, 3° follows from $\eta^{(s)} = \nu$.

Now we shall prove (1). We may assume that $\int f d\mu = 0$. Doob's theorem about the convergence of conditional probabilities gives for μ-a.e. x

$$\lim_{t \to \infty} \frac{\int_{C_{\zeta^{(-t)}}(x)} f(y) d\nu_{\Gamma^{(s)}(x)}(y)}{\int_{C_{\zeta^{(-t)}}(x)} d\nu_{\Gamma^{(s)}(x)}(y)} = 0.$$

Here $C_{\zeta^{(-t)}}(x)$ is the element of the partition $\zeta^{(-t)}$ containing x. Therefore for each $\alpha_1, \alpha_2 > 0$ one can find $t_0 = t_0(\alpha_1, \alpha_2)$ and a measurable subset A_{t_0} consisting mod 0 of elements of $\zeta^{(-t_0)}$ such that $\mu(A_{t_0}) \geq 1 - \alpha_1$, and for any $x \in A_{t_0}$

$$\left| \frac{\int_{C_{\zeta^{(-t)}}(x)} f(y) d\nu_{\Gamma^{(s)}(x)}(y)}{\int_{C_{\zeta^{(-t)}}(x)} d\nu_{\Gamma^{(s)}(x)}(y)} \right| \leq \alpha_2 \tag{6}$$

for all $t \geq t_0$.

Take now $\varepsilon > 0$. We shall show that for every x

$$\left| \frac{\int_{C_{\zeta^{(-t)}}(x)} f(y) d\nu_{\Gamma^{(s)}(x)}(y)}{\int_{C_{\zeta^{(-t)}}(x)} d\nu_{\Gamma^{(s)}(x)}(y)} \right| \leq \varepsilon$$

if $t \geq t_0$ which is equivalent t_0 the statement of theorem 8.

For any N construct subtowers $T_{jrq}, 0 \leq q < N$ where $T_{jrq} = \bigcup_{z \in R_{jr}^{(s)}} \bigcup_{\frac{q}{N}\tau(z) \leq \tau \leq \frac{q+1}{N}\tau(z)} S^t z$. The crucial role is due to the following lemma.

Lemma *For any $\delta > 0$ one can find $N = N(\delta, f)$ and take so small $R_{jr}^{(s)}$ that for all $t > 0$ and any $x', x'' \in T_{jrq}$*

$$\left| \frac{\int_{S^{-t}C_{\zeta^{(0)}}(x')} f(y) d\nu_{\Gamma^{(s)}}(S^{-t}x')(y)}{\int_{S^{-t}C_{\zeta^{(0)}}(x')} d\nu_{\Gamma^{(s)}}(S^{-t}x')(y)} \right.$$
$$\left. - \frac{\int_{S^{-t}C_{\zeta^{(0)}}(x'')} f(y) d\nu_{\Gamma^{(s)}}(S^{-t}x'')(y)}{\int_{S^{-t}C_{\zeta^{(0)}}(x'')} d\nu_{\Gamma^{(s)}}(S^{-t}x'')(y)} \right|$$
$$\leq \delta$$

213

Proof of the lemma follows from the non-integrability of stable and unstable foliations and from the possibility to establish a one-to-one correspondence between large subsets of $C_{\zeta^{(0)}}(x')$ and $C_{\zeta^{(0)}}(x'')$ in such a way that the distances between the shifts long positive semi-trajectories of corresponding points will be arbitrarily small provided that N is sufficiently large.

Take $\delta = \frac{\varepsilon}{2}$, $\alpha_2 = \frac{\varepsilon}{2}$, $\alpha_1 < \frac{1}{2}\min \mu(t_{jrs})$. Consider $S^t x \in T_{jrq}$. One can find $\bar{x}_{jrq} \in T_{jrq}$ such that $S^{-t}\bar{x}_{jrq} \in A_{t_0}$. In view of the lemma

$$
\left| \frac{\int_{C_{\zeta(-t)}(x)} f(y) d\nu_{\Gamma^{(s)}(x)}(y)}{\int_{C_{\zeta(-t)}(x)} d\nu_{\Gamma^{(s)}(x)}(y)} \right|
$$
$$
\leq \left| \frac{\int_{C_{\zeta(-t)}(x)} f(y) d\nu_{\Gamma^{(s)}(x)}(y)}{\int_{C_{\zeta(-t)}(x)} d\nu_{\Gamma^{(s)}(x)}(y)} - \frac{\int_{C_{\zeta(-t)}(S^t\bar{x}_{jrq})} f(y) d\nu_{\Gamma^{(s)}(S^t\bar{x}_{jrq})}(y)}{\int_{C_{\zeta(-t)}(S^t\bar{x}_{jrq})} d\nu_{\Gamma^{(s)}(S^t\bar{x}_{jrq})}(y)} \right|
$$
$$
+ \left| \frac{\int_{C_{\zeta(-t)}(S^t\bar{x}_{jrq})} f(y) d\nu_{\Gamma^{(s)}(S^t\bar{x}_{jrq})}(y)}{\int_{C_{\zeta(-t)}(S^t\bar{x}_{jrq})} d\nu_{\Gamma^{(s)}(S^t\bar{x}_{jrq})}(y)} \right|
$$
$$
\leq \varepsilon,
$$

Q.E.D.

* References

[A] Anosov D.V., *Geodesic Flows on Closed Riemannian Manifolds of Negative Curvature*, Proc. of Steklov Institute **90** (1967).

[AS] Anosov D.V., Sinai Ya.G., *Some Smooth Dynamical Systems*, Russian Math. Surveys (Vspehi) **22,5** 107-172. (1986).

[Ar] Arnold V.I., *Mathematical Methods of Classical Mechanics*, Springer-Verlag New-York,Heidelberg,Berlin 462. (1978).

[Bl] Blanchard, *Partition Extremales de Flots d'entropie Infinie*, Aeitschrift fur Wahrscheinlichkeits theorie **36,2** 129-136, (1976).

[B] Bowen R., *Equilibrium States and the Ergodic Theory of Anosov diffeomorphisms*. Lecture Notes in Mathematics **470** (1975), Springer-Verlag

[CFS] Cornfeld I.P., Formin S.V., Sinai Ya.G., *Ergodic Theory* Springer-Verlag 486, (1982).

[G] Gurevich B.M., *Perfect Partitions for Ergodic Flows*. Funct. Anal. Appl.**11,3** 20-23, (1977).

[HPS] Hircsh M., Pugh C., Shub M., *Invariant Manifolds* Lecture Notes in Math. **583** Springer-Verlag (1977).

[KS] Katok A.B., Streleyn I-M., *Invariant Manifolds Entropy and Billiards, Smooth Maps with Singularities*. Lecture Notes in Math. **1222** Berlin Heidelberg New-York Springer-Verlag (1986).

[P] Pesin Ya.B., *Characteristic Lyapunov Exponents and Smooth Ergodic Theory*. Russian Math. Surveys (Vspehi) **32,4** (1977), 55-112.

[R] Ratner M.E., *Invariant Measures for U-flows on Three-dimensional Manifolds*. Doklady **186,3** 519-521, (1969).

[RS] Rokhlin V.A., Sinai Ya.G., *Construction and Properties of Invariant Measurable partition* Doklady **141,5** 1038-1041, (1961).

[Ru] Rudolph D., *A two-valued step-coding for ergodic flows*. Proc. of the International Conference on Dynamical Systems in Math. Physics. Rennes, Sept. 14-21, (1975).

[S1]   Sinai Ya.G., *Markov Partitions and U-diffeomorphisms*. Funct. Anal.
         Appl. **2,1** 64-89, (1968).

[S2]   Sinai Ya.G., *Gibbs Measures in Ergodic Theory*. Russian Math. Surveys
         (Vspehi) **27,4** 21-64, (1972).

[Z]   Aagier D., *Eisenstein Series and the Riemann Zeta-FUnction Automor-
         phic forms*. Representation Theory and Arithmetic. Tata Institute of Fun-
         damental Research, Bombay, Springer-Verlag, Berlin Heidelberg New-
         York, 275-301, (1981).

GEOMETRIC REALIZATIONS OF HYPERELLIPTIC CURVES

William A. Veech *

Rice University
Mathematics Department
Houston, Texas 77251

INTRODUCTION

Every elliptic curve $w^2 - z(z - 1)(z - y) = 0, y \neq 0, 1$ is a torus and, in particular, can be represented as an identification space of a parallelogram. The gluing maps are translations. The present paper is concerned with the question of a corresponding realization of hyperelliptic curves

$$w^2 - \prod_{j=0}^{n}(z - y_j) = 0 \qquad (0.1)$$

where $y \in \mathbb{C}^{n+1}, y_a \neq y_b, a \neq b$. The curve (0.1) has genus $[\frac{n}{2}]$, where $[\cdot]$ is the greatest integer function. We shall prove

Theorem 0.1. *Each curve (0.1) can be realized as the identification space of a centrally symmetric simple planar $2n$-gon P_y with opposite sides glued by translation. For an open set of y, of full measure in the parameter space, P_y can be taken to be convex.*

In genus one the exceptional set of Theorem 0.1 is empty. We shall prove

Theorem 0.2. *If $g > 0$, the curve $w^2 - (1 - z^{2g+1}) = 0$ cannot be realized as the identification space of a centrally symmetric convex $4g$-gon.*

The first statement in theorem 0.1 is a consequence of the analysis of a natural map from a certain space of polygons to the moduli space of punctured spheres. The second statement is shown in Section 4 to be a consequence of known facts about an action of $G = SL(2, \mathbb{R})$ on a circle bundle over the moduli space. Finally, Theorem 0.2 will be seen to be a consequence of a study of "periodic points" for this G action, points whose isotropy groups are lattices in G.

---
*Research supported by NSF.

The author wishes to dedicate this paper to the Hayashibara Company in recognition of its generous support for science in general and mathematics in particular, through the Hayashibara Forums.

1. SYMMETRIC POLYGONS

Fix $n > 1$, and define $\mathcal{P}(n)$ to be the set of pairs $p = (P, v)$ such that $P \subseteq \mathbb{C}$ is a simple, symmetric $2n$-gon and v is a vertex of P.

Given $p \in \mathcal{P}(n)$, set $v_0(p) = v$, and let $v_j(P)$, $0 \le j < 2n$ be the remaining vertices of P, arranged in counterclockwise order. The map $H : \mathcal{P}(n) \to \mathbb{C}^n$, define by

$$H(p) = (v_0(p), \cdots, v_{n-1}(p)) \tag{1.1}$$

is a one-to-one map of $\mathcal{P}(n)$ onto an open subset of $\mathbb{C}^n$. In particular, $\mathcal{P}(n)$ carries the natural structure of a complex manifold of dimension n.

Continuing with $p \in \mathcal{P}(n)$, denote the edges of P by $e_j(p) = [v_{j-1}(p), v_j(p)]$, $1 \le j \le 2n$. Glue e_j to e_{j+n} by parallel translation. The identification space is a Riemann surface with ideal points corresponding to the vertices of P. As the gluing defined above sends $v_j(p)$ to $v_{j+n-1}(p)$, the equivalence class $[v_j]$ is identified as the set of v_k such that k is in the orbit of j under the map $i \to i + n - 1 \pmod{2n}$. When n is even, there is one vertex class, and when n is odd, there are two, $[v_0]$ and $[v_n]$.

$X(p)$ denote the Riemann surface defined above. The total angle of P at the vertex class is $(n-1)2\pi$ when n is even. When n is odd the total angle at each of the two vertex classes is $(\frac{n-1}{2})(2\pi)$ (because the isometry $z \to -z$ of P interchanges these classes). If w_p is the holomorphic 1-form on $X(p)$ determined by dz, then w_p has, by the total angle count just made, one or two zeros whose total order is $n - 2$ (n even) or $n - 3$ (n odd). As this total order must also be $2g - 2$, $g = \text{genus}(X(p))$, we have

$$g = g(X_p) = \left[\frac{n}{2} \right]. \tag{1.2}$$

The involution $\tau(z) = -z$ induces a holomorphic involution, also denoted τ, of $X(p)$. This involution has fixed points the (equivalence class of) points 0, $u_j(p) = \frac{1}{2}(v_j(p) + v_{j-1}(p))$, $1 \le j \le n$. When n is even, the vertex class is fixed. As we have already mentioned, the vertex classes are interchanged by τ when n is odd.

Proposition 1.3. *If $p \in \mathcal{P}(n)$, then $X(p)$ is hyperelliptic.*

Proof. The involution τ has $n + 1$ fixed points if n is odd and $n + 2$ fixed points if n is even. In either case this number is $2 \left[\frac{n}{2} \right] + 2 = 2g(X(p)) + 2$, and the proposition obtains.

In what follows we shall use $u_0(p) = 0$, $u_j(p)$, $1 \le j \le n$, to denote the points defined above as points in $X(p)$. The proposition implies there exists a unique holomorphic map

$$F : X(p) \to \mathbb{C} \cup \{\infty\} \tag{1.4}$$

such that (a) $F \circ \tau = F$, (b) $F(0) = 0$, $F(u_1) = 1$, and $F([v_0]) = \infty$ and (c) F is a biholomorphism modulo τ. Of course, $F([v_n]) = \infty$ when n is odd.

With notations as above, define $y \in \mathbb{C}^{n+1}$ by $y_0 = F(0)$ and $y_j = F(u_j), 1 \leq j \leq n$. Set up the quadratic differential

$$q_y(z) = \frac{dz^2}{\prod\limits_{j=0}^{n}(z - y_j)}. \tag{1.5}$$

There exists $\pm \alpha \in \mathbb{C}^*/\pm 1$, $\mathbb{C}^* = \mathbb{C} \setminus 0$, such that

$$F^*(\alpha^2 q_y) = w_p^2. \tag{1.6}$$

Accordingly, we define

$$\Phi(p) = (y, (\pm\alpha)). \tag{1.7}$$

Define $\Omega_n \subseteq \mathbb{C}^{n+1}$ to be.

$$\Omega_n = \{y \in \mathbb{C}^{n+1} | y_0 = 0, y_1 = 1, y_a \neq y_b, a \neq b\}.$$

The map

$$\Phi : \mathcal{P}(n) \rightarrow \Omega_n \times \mathbb{C}^*/\pm 1 \tag{1.8}$$

is now well-defined.

Remark 1.9. *Let* $p = (P, v) \in \mathcal{P}(n)$ *and* $\beta \in \mathbb{C}^*$. *If* $\beta p = (\beta P, \beta v)$, *and if* $\Phi(p) = (y, \pm\alpha)$, *then clearly*

$$\Phi(\beta p) = (y, \pm\alpha\beta). \tag{1.10}$$

Theorem 1.11. *The map* Φ *in (1.8) is holomorphic, locally one-to-one and surjective.*

Theorem 1.11 well be proved in Sections 2 and 3.

2. SURJECTIVETY OF Φ

Fix $n > 1$ and $y \in \Omega_n$. Denote by X the Riemann surface of the curve $w^2 - \prod_{j=0}^{n}(z - y_j) = 0$, and let τ be the hyperelliptic involution. The 1-form $\frac{dz}{w}$ determines, up to a factor ± 1, a holomorphic 1-form w on X. The zero set E of w lies above ∞ and has one element (of order $n - 2$) when n is even and two elements (of order $\frac{n-3}{2}$) when n is odd.

Local solutions of $df = w$ determine an atlas $\mathcal{U}$ on $X \setminus E$ with transitions which are local translations. If $\theta \in \mathbb{R}$ let $\mathcal{F}(\theta)$ denote the oriented foliation of $X \setminus E$ which is the lift by $\mathcal{U}$ charts of the foliation of $\mathbb{C}$ by lines which make an angle θ with the horizontal. Leaves of $\mathcal{F}(\theta)$ are geodesic for the metric $|w|^2$. The fact $\tau^* w = -w$ implies $\mathcal{F}(\theta + \pi) = \tau \mathcal{F}(\theta)$.

We shall now make a construction. The first part of the construction will use only the properties (a) $\tau^* w = -w$, and (b) there exists $u_0 \in X \setminus E$ such that $\tau(u_0) = u_0$. (Of course, (a) and (b) imply $\tau^2 = Id$.)

Let γ_0 be an $|w|^2$-geodesic in $X \setminus E$ joining u_0 to a point $b \in E$. For example, a geodesic length minimizing path from u_0 to E will do. Replace w by $\zeta w, |\zeta| = 1$, if necessary and r elabel so that γ_0 is a segment of an incoming separatrix of $\mathcal{F}(0)$ at b. Define $a = \tau(b)$, and parametrize the union $\gamma_0 \cup \tau(\gamma_0)$ as a geodesic path γ from a to b, i.e., an $\mathcal{F}(0)$ saddle connection.

For all but a countable set of θ the foliation $\mathcal{F}(\theta)$ admits no saddle connection. Fix such a θ with $0 < \theta < \pi$. The path γ above is transverse to $\mathcal{F}(\theta)$, and Poincaré map of $\mathcal{F}(\theta)$ on γ decomposes $X \setminus E$ into a set $R_1, \cdots, R_k$ of maximal flowboxes with bases $I_1, \cdots, I_k$ aligned along γ from left to right. (cf. [3]) The facts $a, b \in E$ and $\mathcal{F}(\theta)$ admits no saddle connection imply that k exceeds by one the number of incoming separatrices of $\mathcal{F}(\theta)$. Indeed, the boxes R_i and R_{i+1} are joined along a common segment of incoming separatrix at $x_i \in E$ (the zero set of ω).

Remark 2.1. *When n, X and ω are as in the first paragraph of this section, one finds readily that there are $n-1$ incoming separatrices, whether n is even or odd, and therefore there are n flowboxes, i.e., $k = n$ above.*

$^.$ For each j the flowbox R_j is a parallelogram in $\mathcal{U}$-coordinates. That is, there exists a $\mathcal{U}$-chart function which maps R_j to a parallelogram based upon the real axis and having one angle θ. Let $x_0 = a$, and construct a path δ from a to b by connecting x_{i-1} to x_i by a geodesics in $R_i, 1 \leq i \leq k$. The portion of $\bigcup_{j=1}^{k} R_j$ which lies above γ and below δ is denoted P_0. It is evident there exists a $\mathcal{U}$-chart (P_0, f) such that $f(P_0)$ is a polygon in the upper half plane with base on the real axis centered at 0.

As $\mathcal{F}(\theta)$ and $\mathcal{F}(\theta+\pi)$ coincide but for orientation of their leaves, the same parallelograms are flowboxes for $\mathcal{F}(\theta+\pi)$. Now the top or R_j relative to $\mathcal{F}(\theta)$ is the base of R_j relative to $\mathcal{F}(\theta + \pi)$. The construction which led to δ above yields a path, denoted ϵ, from b to a. ϵ has the same segments as δ, but the orders of appearance and orientations are not the same. The region between γ and ϵ (below γ), denoted Q_0, is the domain of $\mathcal{U}$-chart (Q_0, h) such that $h(Q_0)$ is a polygon in the lower half plane with base on the real axis coinciding with the base of $f(P_0)$ above.

Let $\mathcal{O}$ be the region consisting of P_0, Q_0 and γ. The chart functions f and h coalesce on $\mathcal{O}$ to give $\mathcal{U}$-chart $(\mathcal{O}, F)$ such that $F(\mathcal{O})$ is a polygon with a diameter on the real axis, centered at 0.

As $\tau \mathcal{F}(\theta) = \mathcal{F}(\theta + \pi)$, it must be that $\tau(P_0) = Q_0$, and this implies $F \circ \tau = -F$. Therefore, $F(\mathcal{O})$ is a simple symmetric $2k$-gon, k = number of flowboxes above.

Thus far we have used only (a) and (b) above. It has been noted in Remark 2.1 that in the case of interest $k = n$. We shall find a further restriction imposed by the existence of $2g(X) + 2$ fixed points for τ in the hyperelliptic case. When n is even, this number is $n + 2$; when n is odd, it is $n + 1$. The center of $\mathcal{O}$ is fixed by τ, and no other τ-fixed point lies in $\mathcal{O}$. τ fixes the single vertex class when n is even or odd, the set $\delta \setminus E$ will contain n fixed point of τ. If the component δ_i of δ which connects x_{i-1} to x_i inside R_i contains a τ fixed point, then $\tau(R_i) = R_i, \tau(\delta_i) = \delta_i$, and R_i, i.e., δ_i, contains only one fixed point. As there are exactly n parallelograms, it must be that $\tau(R_i) = R_i$ for $1 \leq i \leq n$.

The path ϵ from b to a is in all cases comprised of segments $\tau(\delta_1), \tau(\delta_2), \cdots, \tau(\delta_n)$ in that order. However, when τ has $2g + 2$ fixed points, we have proved that $\tau(\delta_i) = \delta_i$ but for parametrization. It follows that X is the identification space of $F(\mathcal{O})$ obtained by gluing opposite edges, i.e., edges i and $i + n, 1 \leq i \leq n$.

Recall that we are seeking a pair $p = (P, v) \in \mathcal{P}(n)$ such that $\Phi(p) = (y, \pm \alpha)$, where y is as given above and $\alpha \in \mathbb{C}^*$. In our construction we have associated to (X, ω) a centrally symmetric simple polygon $P = F(\mathcal{O})$ which realizes X. The construction made use of an arbitrary Weierstrass point $u_0 \in X \setminus E$. It is therefore no loss of generality to suppose u_0 sits above 0 on the Riemann surface of $w^2 - \prod_{j=0}^{n}(z - y_j) = 0$.

Now choose $v \in P$ a vertex so that in the canonical ordering $u_1 = u_1(P, v)$ sits above $1 = y_1$. By construction of the map Φ there exists $\alpha \in \mathbb{C}^*$ such that $\Phi((P, v)) = (y, \pm\alpha)$. As $y \in \Omega_n$ is arbitrary, surjectivity of Φ is now a consequence of Remark 1.9.

3. Φ IS A LOCAL BIHOLOMORPHISM

The purpose of this section is to prove that the map $\Phi : \mathcal{P}(n) \to \Omega_n \times \mathbb{C}^*/\pm 1$ is a local biholomorphism and thus to complete the proof of Theorem 1.11. We shall first prove Φ is holomorphic, but only to establish its continuity. Using this continuity and a determinant calculation from [4] we shall then prove Φ admits a right inverse on a neighborhood of each image point.

Fix $p = (P, v) \in \mathcal{P}(n)$, and let P be triangulated by a symmetric triangulation t whose vertex set is the set of vertices of P. If $q = (Q, u)$ is sufficiently close to p, then t determines a triangulation $t(q)$ of Q with similar properties. Let $F_{p,q} : X(p) \to X(q)$ be the canonical PL map determined by the PL-structures $t(p)$ and $t(q)$. F_{pq} preserves the ordering of Weierstrass points. A standard calculation (cf.[4]) show that the Beltrami differential $\mu_{p,q}$ associated to F_{pq} varies holomorphically in q for p fixed. Moreover, if τ_p, τ_q are the hyperelliptic involutions, symmetry of t implies $F_{pq} \circ \tau_p = \tau_q \circ F_{pq}$ and therefore $\mu_{p,q} \circ \tau_p = \mu_{p,q}$. It follows that F_{pq} induces a quasiconformal homeomorphism H_{pq} from $X(p)/\tau_p$ to $X(q)/\tau_q$ and that Beltrami differential varies holomorphically with q. If $\Phi(p) = (y(p), \pm\alpha(p))$ and $\Phi(q) = (y(q), \pm\alpha)$, the definitions imply $H_{pq}(y(p)) = y(q)$, and therefore $y(q)$ varies holomorphically with q. It follows easily from the definition of Φ that $\pm\alpha(q)$ varies holomorphically with q.

In order to construct local right inverses fix $p = (P, v)$ with $\Phi(p) = (y, \pm\alpha(p))$, as above. Let $F : X(p) \to \mathbb{C} \cup \{\infty\}$ be the map (1.4) which is used in the definition of Φ. For $1 \leq j \leq n$ let γ_j be the segment of ∂P from $u_j(p) = \frac{1}{2}(v_j(p), v_{j-1}(p))$ to $v_j(p)$. Also, let γ_0 be a smooth path in P from 0 to $v_0 = v$. Define $\delta_j = F(\gamma_j), 0 \leq j \leq n$. δ_j is a path from $y_j(p)$ to ∞. Choose a version of $\frac{dz}{w} = \eta$ such that $F^*(\alpha\eta) = \omega$, and declare η to have values on δ_j which are limits from the lefthand side of δ_j. We have

$$\begin{aligned} v_0(p) &= 2\alpha \int_{\delta_0} \eta \\ v_j(p) - v_{j-1}(p) &= \int_{\delta_j} \eta. \end{aligned} \qquad (3.1)$$

As y varies in a small neighborhood of $y(p)$ it is possible to vary paths $\delta_j(y)$ (from y_j to ∞) and the definition of $\frac{dz}{w} = \eta_y$ in such a way that

$$\Psi(y, \beta) = \left\{ 2\beta \int_{\delta_j(y)} \eta_y \right\}_{1 \leq j \leq n} \qquad (3.2)$$

is holomorphic in (y, β). We restrict the subscripts to $1 \leq j \leq n$ because the remaining integral is minus the sum of the other integrals. There are n integrals and n parameters (y, β). If (y, β) is sufficiently close to $(y(p), \alpha(p))$ then $\Psi(y, \beta)$ determines a symmetric polygon whose distinguished vertex is a function (negative sum) of the integrals (3.2). As Φ is continuous, the relation (3.1) implies Φ is locally biholomorphic as soon as Ψ has this property.

The Jacobian determinant of the map Ψ has been calculated in [4] in a more general setting. One finds

$$\left(\det \frac{\partial \Psi}{\partial(\beta, y)}\right)^2 = \left(\pi^{n+1}\Gamma^2\left(\frac{n-3}{2}\right)\beta^{n-1} \prod_{0 \leq k < l \leq n}(y_k - y_l)^{-1}\right)^2. \tag{3.3}$$

It follows that Ψ is nowhere singular. Theorem 1.11 is thereby proved.

4. DYNAMICS OVER MODULI SPACE

Theorem 1.11 asserts that the map $\Phi : \mathcal{P}(n) \to \Omega_n \times \mathbb{C}^*/\pm 1$ is a surjective local biholomorphism. Let λ be the euclidean volume element on $\mathcal{P}(n)$. The determinant formula (3.3) suggests a prescription for a volume element ν on $\Omega_n \times C^*/\pm 1$

$$\nu = \left(\frac{i}{2}\right)^n |\beta|^{2n-2} \prod_{0 \leq k < l \leq n} |y_k - y_l|(d\beta \wedge d\bar\beta) \bigwedge_{j=2}^n dy_j \wedge d\bar y_j. \tag{4.1}$$

There exists a constant $c(n) > 0$ such that

$$\lambda = \Phi^*(c(n)\nu). \tag{4.2}$$

It follows that λ projects to a volume element on the equivalence relation determined by Φ.

If $p = (P, v) \in \mathcal{P}(n)$, we denote the area of P by $N(p)$. If $\Phi(p) = (y, \pm\alpha)$, then but for a dimensional constant

$$N(p) = \frac{i}{2}\int_{\mathbb{C}} |\alpha|^2 \frac{dz \wedge d\bar z}{|w|^2} \tag{4.3}$$

where $w^2 = \prod_{j=0}^n(z - y_j)$. In what follows we take (4.3) for the definition of $N(P)$. Also, express the right-hand side of (4.3) as $|\alpha|^2 M(y)$ so that

$$N(p) = |\alpha|^2 M(y). \tag{4.3'}$$

Set up the $(2n-1)$-form (with $\alpha = |\alpha|e^{i\theta}$) $\tilde\mu$, where

$$\tilde\mu = \frac{|\alpha|^{2n-1}}{M(y)} \prod_{0 \leq k < l \leq n} |y_j - y_l|^{-2} \bigwedge_{j=2}^n \left(\frac{i}{2}dy_j \wedge d\bar y_j\right) \wedge d\theta$$

and observe that, up to a scale factor, $d(|\alpha|^2 M(y)) \wedge \tilde\mu = \nu$. The restriction of $\tilde\mu$ to the constant norm surface $|\alpha|^2 M(y) = 1$ is the form

$$\mu = \frac{d\theta \wedge \bigwedge_{j=2}^n \left(\frac{i}{2}dy_j \wedge d\bar y_j\right)}{M(y)^n \prod_{0 \leq k < l \leq n} |y_k - y_l|^2}. \tag{4.4}$$

Recall from [4]: if $\Lambda_n = \{(y, \pm\alpha) | |\alpha|^2 M(y) = 1\}$, then

$$\int_{\Lambda_n} \mu < \infty. \tag{4.5}$$

Identify μ with the measure it defines on Λ_n.

For a moment we shall use G to denote the group $SU(1,1) \cong SL(2,\mathbb{R})$ of 2×2 complex matrices $A = \begin{pmatrix} \xi & \bar{\eta} \\ \eta & \bar{\xi} \end{pmatrix}$ such that $\det A = 1$. G acts $\mathbb{R}$-linearly on $\mathbb{C}$ by $Az = \xi z + \bar{\eta}\bar{z}$, and this induces an action of G on $\mathcal{P}(n)$, e.g. coordinatewise in (1.1). Denote this latter action by $p \to T_A p$.

If $p \in \mathcal{P}(n)$ and $A \in G$, then T_A induces a quasiconformal homeomorphism from $X(p)$ to $X(T_A p)$. As $A^* dz = \xi dz + \bar{\eta} d\bar{z}$, $A = \begin{pmatrix} \xi & \bar{\eta} \\ \eta & \bar{\xi} \end{pmatrix}$, the homeomorphism h_A, which is real analytic away from the vertex class(es), satisfies $h_A^* \omega_{T_A p} = \xi \omega_p + \bar{\eta}\bar{\omega}_p$.

Let $p_1, p_2 \in \mathcal{P}(n)$ be such that $\Phi(p_1) = \Phi(p_2)$. The identity map on $\mathbb{C} \cup \{\infty\}$ lifts to a biholomorphism $\phi : X(p_1) \to X(p_2)$ such that $\phi^* \omega_{p_2} = \omega_{p_1}$ and ϕ preserves the ordering of the Weierstrass points. If $h_A^j : X(p_j) \to X(T_A p_j)$ are as above, the composition $\phi_A : X(T_A p_1) \to X(T_A p_2)$ where $\phi_A = h_A^2 \circ \phi \circ (h_A^1)^{-1}$ satisfies $\phi_A^* \omega_{T_A p_2} = \omega_{T_A p_1}$, and therefore $\Phi(T_A p_1) = \Phi(T_A p_2)$. The action of G on $\mathcal{P}(n)$ descends to an action on $\Omega_n \times \mathbb{C}^* / \pm 1$.

It is obvious from the definitions that the G-action preserves the volume element λ on $\mathcal{P}(n)$. As λ projects to the volume element ν ((4.1)), the action of G which is induced upon $\Omega_n \times \mathbb{C}^* / \pm 1$ by Φ must preserve ν. As $\det A = 1$, $A \in G$, the G action preserves the area function $N(p)$. It follows from (4.3') and the definition (4.4) of the restriction volume from μ on Λ_n that μ is also G-invariant.

Theorem 4.6. *The triple (Λ_n, μ, G) is a real analytic, ergodic, finite measure preserving action.*

Proof. Finiteness has been noted above. (Λ_n, G) is a component of a stratum, in the sense of [6], and ergodicity of topological components is established in [6].

5. CLOSED ORBIT AND CONVEXITY

Theorem 4.6 implies that almost every $u \in \Lambda_n$ has a dense G-orbit. We turn now to a discussion of behavior at the opposite extreme, points u such that Gu is closed.

Denote by $\Gamma(u)$ the isotropy group of $u \in \Lambda_n$. We recall that $\Gamma(u)$ enjoys three properties for all u: 1. $\Gamma(u)$ is discrete. 2. $\Gamma(u)$ is not cocompact. 3. If $A \in \Gamma(u)$, then $\text{tr}(A)$ is an algebraic integer. Generically, $\Gamma(u)$ is trivial. However, for a dense set of u $\Gamma(u)$ is a lattice. For example, by the Remark, p.579 in [5] $\Gamma(u)$ is commensurable with $SL(2,\mathbb{Z})$ when $u = \Phi(p)$ is such that $H(p) \in \mathbb{C}^* \mathbb{Q}^n$ ($H(\cdot)$ is defined in (1.1).) Consideration of the integrals (3.1)-(3.2) yields a corresponding criterion for $u = (y, \pm\alpha)$ which depends upon y and integrals of the 1-form $\frac{dz}{w}$, $w^2 = \prod_{j=0}^{n}(z - y_j)$.

In this section we shall make use of consequences of an isotropy group being a lattice to establish

Theorem 5.1. *Let $X(p)$ be the Riemann surface of the equation $w^2 = 1 - z^{2g+1}$, and let ω_g be a lift to $X(g)$ of the 1-form $\frac{dz}{w}$. The pair $(X(g), \omega_g)$ cannot be realized as $(X(p), \omega_p)$ for any $p = (P, v) \in \mathcal{P}(2g)$ such that P is a **convex** polygon.*

For each n let $\mathcal{P}_c(n)$ be the set of $p = (P, v) \in \mathcal{P}(n)$ such that P is convex. Clearly, $G\mathcal{P}_c(n) = \mathcal{P}_c(n)$ and $\mathcal{P}_c(n)$ contains a nonempty open set. If $\Lambda_{n,c} = \Phi(\mathcal{P}_c(n))$, then Theorem 4.6 implies $\Lambda_{n,c}$ is a dense set of full measure; indeed, its interior has this property.

Question 5.2. Let $\Lambda_{n,b} = \Lambda_n \setminus \Lambda_{n,c}$. If $n > 3$, does there exist $u \in \Lambda_{n,b}$ such that Gu is not closed? If the answer is 'yes', does there exist $u \in \Lambda_{n,b}$ such that Gu is dense in $\Lambda_{n,b}$?

We shall now give the proof of Theorem 5.1. To begin let (X, ω) be a pair consisting of a closed Riemann surface X and a nontrivial holomorphic 1-form ω. Let $\mathcal{U}$ be the atlas on $X \setminus E$, $E = \omega^{-1}0$, as in Section 2. Denote by $\mathrm{Aff}(\mathcal{U})$ the group of orientation preserving homeomorphism ϕ of X which are affine in $\mathcal{U}$-coordinates. $\mathcal{F}(\theta), \theta \in \mathbb{R}$, has the same meaning as in Section 2.

Let $\Gamma = \Gamma(\mathcal{U})$ be the image in G of $\mathrm{Aff}(\mathcal{U})$ under the map which assigns to $\phi \in \mathrm{Aff}(\mathcal{U})$ its derivative $D\phi$ in $\mathcal{U}$-coordinates. $\Gamma(\mathcal{U})$ enjoys the properties 1-3 which were listed above for $\Gamma(u)([5])$.

If $\theta \in \mathbb{R}$, define $A(\theta) \subseteq \mathrm{Aff}(\mathcal{U})$ to be the set of ϕ such that $D\phi(\cos\theta, \sin\theta) = (\cos\theta, \sin\theta)$. Notice that $A(\theta)$ is a subgroup and for each $\phi \in A(\theta)$ $D\phi \in \Gamma(\mathcal{U})$ is unipotent.

Lemma 5.3. *Assume $\Gamma(\mathcal{U})$ is a lattice. The following are equivalent:*

(A) $\mathcal{F}(\theta)$ admits a saddle connection.

(B) $DA(\theta)$ is nontrivial.

(C) $\mathcal{F}(\theta)$ partitions $X \setminus E$ into cylinders of closed leaves.

A proof of the lemma may be found in [5].

Lemma 5.4. *Let (X, ω) be such that $\Gamma(\mathcal{U})$ is a lattice with a single cusp. If $\theta_1, \theta_2 \in \mathbb{R}$ are such that $\mathcal{F}(\theta_j)$ admits a saddle connection for $j = 1, 2$, there exists $\phi \in \mathrm{Aff}(\mathcal{U})$ such that $\phi\mathcal{F}(\theta_1) = \mathcal{F}(\theta_2)$ or $\mathcal{F}(-\theta_2)$.*

Proof. The assumptions combine to imply $DA(\theta_1)$ and $DA(\theta_2)$ are conjugate in $\Gamma(\mathcal{U})$. Choose $\psi \in \mathrm{Aff}(\mathcal{U})$ such that $(D\psi)^{-1}DA(\theta_2)(D\psi) = DA(\theta_1)$. If $\psi\mathcal{F}(\theta_1) \overset{\mathrm{def}}{=} \mathcal{F}(\theta)$ then θ is such that $DA(\theta) = DA(\theta_2)$, and this implies $\theta = \theta_2$ or $-\theta_2$ modulo 2π. That is, $\psi\mathcal{F}(\theta_1) = \mathcal{F}(\theta_2)$ or $\mathcal{F}(-\theta_2)$. The lemma is proved.

To apply the lemma let (X, ω) be such that $\Gamma(\mathcal{U})$ is a lattice with one cusp, and let $\theta_0 \in \mathbb{R}$ be such that $\mathcal{F}(\theta_0)$ decomposes $X \setminus E$ into cylinders of closed leaves. Denote the maximal such cylinders by $C_1, \cdots, C_r$, and let their heights be denoted $h_1, \cdots, h_r$. Now suppose $\theta \in \mathbb{R}$ is such that $\mathcal{F}(\theta_0)$ admits a saddle connection. Lemma 5.4 implies that there exists $\phi \in \mathrm{Aff}(\mathcal{U})$ and a choice of $\pm$ such that $\phi\mathcal{F}(\theta_0) = \mathcal{F}(\pm\theta)$. As $D\phi$ is a linear transformation, there exists $t > 0$ such that the cylinder ϕC_j has height $th_j, 1 \leq j \leq r$, relative to $\mathcal{F}(\theta)$.

Proof of Theorem 5.1. We take $(X, \omega) = (X_g, \omega_g)$, $g > 1$. It is proved in [5] that $\Gamma(\mathcal{U})$ is a $(2, 2g + 1, \infty)$ triangle group and, in particular, $\Gamma(\mathcal{U})$ has only one cusp. In [5] it is shown that for one choice of θ $\mathcal{F}(\theta)$ has exactly g maximal cylinders of closed leaves which, up to a common constant factor have lengths $h_j = \sin\left(\frac{2j-1}{2g+1}\pi\right), 1 \leq j \leq g$. If the cylinders are denoted $C_1, \cdots, C_g$, there are two additional facts to record for later reference. A. $\partial C_1 \subseteq C_2, \partial C_g \subseteq C_{g-1}$ and $\partial C_j \subseteq C_{j-1} \cup C_{j+1}, 1 < j < g$. B. If $1 < j < g$,

224

then up to a common constant factor each side of C_j is comprised of a pair of saddle connections of lengths $\sin \frac{2\pi j}{2g+1}$ and $\sin \frac{2\pi(j-1)}{2g+1}$. We also record the elementary inequality

$$2 \sin \frac{2\pi(j-1)}{2g+1} > \sin \frac{2\pi j}{2g+1} \qquad (2 \le j \le g) \tag{5.5}$$

$(\sin \frac{2\pi j}{2g+1} < \sin \frac{2\pi(j-1)}{2g+1} + \sin \frac{2\pi}{2g+1} \le \sin \frac{2\pi(j-1)}{2g+1}, 2 \le j \le g).$

Now suppose $p = (P, v) \in \mathcal{P}(2g)$ is such that P is convex and $(X(p), \omega_p)$ is isomorphic to $(X(g), \omega_g)$. We shall prove this leads to a contradiction. The vertices of P are ordered as $v = v_0, v_1, \cdots$ in the usual way.

Let l_j denote the oriented segment $\overrightarrow{v_{4g-j}v_j}, 1 \le j \le 2g$. We observe first that these segments cannot be pairwise parallel. For if they are, the foliation $\mathcal{F}(\theta)$, where θ is the common direction, has cylinders of lengths $\|l_1\|, \|l_1\| + \|l_2\|, \cdots, \|l_{g-1}\| + \|l_g\|$. Moreover, central symmetry and convexity imply $\|l_1\| \le \|l_2\| \le \cdots \le \|l_g\|$. When $g = 2$, one concludes that $h_2 > 2h_1$, contradicting $h_j = \sin\left(\frac{2j-1}{5}\pi\right), j = 1, 2$. When $g > 2$, one concludes $h_1 < h_2 < \cdot < h_g$ contradicting $h_j = \sin\left(\frac{2j-1}{2g+1}\pi\right)$.

Let k be the first positive integer such that l_k is not parallel to l_1. The edges $e_1, \cdots, e_{k-1}$ are cross-sections of cylinders $D_1, \cdots, D_{k-1}$ of closed leaves, and these cylinders have on each side a pair of saddle connections of lengths l_j and $l_{j-1}, 1 \le j \le k-1 (l_0 = 0)$. Property B and the fact every leaf of $\mathcal{F}(\theta)$ has length at least $\|l_1\|$ imply $\|l_j\| = \sin \frac{2\pi j}{2g+1}, 1 \le j < k$.

The left side of l_{k-1} is one of the saddle connections on the right hand side of D_k. Parallel segments from the vertices v_{2g+k} and v_{2g-k} do not coincide because, by assumption, l_k and l_{k-1} are not paralell. It follows that the length of the second saddle connection on the right side of D_k has length at least $2l_{k-1}$. This implies $\sin \frac{2\pi(k-1)}{2g+1} \le \sin \frac{2\pi k}{2g+1}$ contradicting (5.5). We have reached a contradiction, and Theorem 5.1 is proved.

Remark 5.6. *It is not difficult to see that each $p \in \mathcal{P}(3)$ admits a convex equivalent. We believe that an argument similar to the one above will show that the curves $w^2 = 1 - z^{2g+2}, g > 1$ equipped with $\frac{dz}{w}$, do not admit convex representatives in $\mathcal{P}(2g+1)$.*

6. CHARACTERIZATION OF CLOSED ORBITS

For any $u \in \Lambda_n$ the canonical map $G/\Gamma(u) \to Gu$ is continuous. Consideration of codimension three transversals to the G action shows that this map is a homeonorphism when Gu is closed in Λ_n.

Lemma 6.1. *If Gu is closed, and if $\Gamma(u)$ is viewed as a Fuchsian group in the disc, the limit set of $\Gamma(u)$ is all of S^1.*

Proof.   Let K be the rotation subgroup of G, and let $g_t = \begin{pmatrix} e^t & 0 \\ 0 & e^{-t} \end{pmatrix}, t \in \mathbb{R}$. According to [1] it is true for almost all $k \in K$ that the ω-limit set of ku (in Λ_n) relative to $\{g_t | t \in \mathbb{R}\}$ is nonempty. For the disc picture this translates to the statement that for almost all k the geodesic from zero in direction k does not diverge to ∞ in $\Gamma \backslash \Delta, \Delta = $disc. It follows in particular that $\Gamma(u)$ has no domain of discontinuity on S^1. That is, $\Gamma(u)$ has S^1 for its limit set.

Proposition 6.2. *If Gu is closed, and if $\Gamma(u)$ is finitely generated, then $\Gamma(u)$ is a lattice.*

Proof. A finitely generated Fuchsian group with limit set S^1 must be a lattice.

Question 6.3. *If $u \in \Lambda_n$, is $\Gamma(u)$ finitely generated?*

J.Smillie has observed that Proposition 6.2 is true without the assumption that $\Gamma(u)$ is finitely generated. Question 6.3 appears to be open.

We shall give an outline of a proof of Smillie's theorem (for the setting of (Λ_n, G)):

1. If ν_n is the probability measure on Λ_n which is the image of normalized Haar measure on K under $k \to ku$, the orbit $G\nu_u$ is relatively compact in the weak-$*$ topology of *probability* measures on Λ_n. This is implicit in [1] and follows from its techniques.

2. If ν is a cluster point of $\{g_t\nu_u | t \to +\infty\}$, then ν is invariant under the group N of upper triangular unipotent matrices. This is also from [1].

3. Let $D = \{v \in \Lambda_n | \lim_{t \to \infty} g_t v = \infty\}$. The facts $G\nu_u$ is relatively compact in the space of *probability* measure and ν above is a cluster point imply $\nu(D) = 0$. In particular, a.e. ergodic component ν_e of ν satisfies $\nu_e(D) = 0$.

4. If ν_e is as in 3., then Ratner's Main Theorem [2] implies $G\nu_e = \nu_e$.

Step 4 establishes the fact that when Gu is closed, $G/\Gamma(u)$ supports a finite G-invariant measure. Therefore, $\Gamma(u)$ is a lattice.

References

1. Kerckhoff,S., Masur H., Smille J., *Ergodicity of billiard flows and quadratic differentials*, Ann. of Math. **124**(1986), 293-311.
2. Ratner, M., *On Roghunathan's measure conjecture*, Ann. of Math. **134** (1991), 545-607.
3. Strebel, K., *Quadratic Differentials*, Berlin-Heidelberg-New York, Springer 1984.
4. Veech, W.A., *Flat Surfaces*, Am. J. of Math. **115**(1993), in press.
5. ——, *Teichmüller curves in moduli space, Eisenstein series and an application to triangular billiards*, Inv. Math. **97**(1989), 553-583.
6. ——, *The Teichmüller geodesic flow*, Ann. of Math. **124**(1986), 441-530.

INDEX